MOS Device and Circuit Design

MOS Device and Circuit Design

Oliver J. McCarthy

Senior Lecturer, Electronic Engineering,
National Institute for Higher Education,
Limerick, Republic of Ireland.

A Wiley–Interscience Publication

175 YEARS OF PUBLISHING
1807 1982

JOHN WILEY & SONS
Chichester · New York · Brisbane · Toronto · Singapore

Library of Congress Cataloging in Publication Data:

McCarthy, Oliver J.
MOS device and circuit design
'A Wiley–Interscience publication.'
Includes index.
1. Metal oxide semiconductors. I. Title.
TK7871.99.M44M36 621.3815'2 81-14645
ISBN 0 471 10026 9 AACR2

British Library Cataloguing in Publication Data:

McCarthy, Oliver J.
MOS device and circuit design
1. Metallic oxide semiconductors
I. Title
621.3815'2 TK7871.85

ISBN 0 471 10026 9

Printed in Northern Ireland at The Universities Press (Belfast) Ltd.

To Mag

Contents

Preface

This book is emerging at a time when MOS/LSI has already brought about a revolution in the field of computing. It provides a comprehensive treatment of MOS at device level ranging from the underlying physical principles through device characterization to static and dynamic logic configurations and analog circuit roles. It continues with some discussion of integrated circuit layout and provides some insights to the design of integrated systems, mainly digital, where the topics of circuit configuration and layout can no longer be separated.

A good understanding of semi-conductor conduction mechanisms and p–n junction behaviour is important in MOS design. For this reason, the basic physical theory of Chapter 1 is from first principles and could serve as a brief but adequate treatment for students with no previous knowledge of semi-conductor physics. This discussion moves quickly to introduce a number of important relationships. These are accepted as plausible and well-tested engineering results, while the burden of rigorous developments is omitted as being more appropriate to a detailed physics course.

Throughout the book an effort is made to impart a sound intuitive grasp of each topic before undertaking a quantitative assessment. Chapter 2 describes the action of the MOS transistor in very simple terms and then proceeds to describe the action of a variety of digital circuit configurations; the objective is to provide a broad qualitative appreciation of MOS circuitry before any detailed modelling is undertaken.

The planar fabrication process is described in Chapter 3. Chapter 4 continues that work on a quantitative level and is principally concerned with the specification of diffusion and oxidation cycles and with the electrical properties of diffused layers.

Chapters 5 through 8 are concerned with MOSFET device characterization, starting with a detailed discussion of threshold voltages in Chapter 5 and of current control in Chapter 6. In these chapters again the approach has been to generate reasonably accurate models through arguments which are plausible rather than rigorous while maintaining a 'feel' for the subject at all times. Design methods are based on process and device parameters which are easily measured. Actual measurement data for p- and n-channel devices are included in the appendices as well as derived parameter values. This data is used in some of the worked examples to predict the performance of certain circuit configurations. Though the current-control model is somewhat simplified, comparison with a more precise model is used to illustrate how accurate results can be obtained.

Chapters 7 and 8 provide a comprehensive analysis of the fundamental digital building block, the inverter. Important properties such as trigger level, rise time, and fall time are readily determined for different load types and varying supply conditions. The work is easily extended to arbitrary logic configurations and to buffer and interface designs.

Chapter 9 looks at some of the different technologies in the light of actual performance requirements and provides clarification of the relative merits and trade-offs, and the narrow ranges of parameters essential to satisfactory operation, particularly with CMOS. It also gives brief treatment of some special topics which include V-MOS power transistors and SOS technology.

Chapters 10 to 12 can be viewed as the 'applied' section of the book. They are concerned with analog and digital circuitry utilizing MOS devices and systems. However, they are not intended to provide instruction in analog design techniques or digital system design in a broad sense; their purpose is to show how the particular properties of MOS are best exploited. Of special importance is the topic of multi-phase dynamic logic and memory design. The computational methods of earlier chapters are fully applicable here, too, in dealing with topics such as charge sharing, refresh time, and such like. Chapter 12 is concerned mainly with digital systems design and the question of circuit layout from which it is inseparable. The treatment is no more than an introduction to a very large subject and is intended mainly to illustrate the interaction between electrical and topographical features and to demonstrate some very general approaches to the design of digital circuitry.

This book is intended primarily as an undergraduate text for Degree-level studies in Electronic Engineering. The material can be satisfactorily covered in courses ranging from 20 to 40 lecture hours over a three-month term or longer. It is also appropriate for use in intensive short courses spanning only a few weeks and was first used, in note form, for this purpose as a familiarization program for engineering staff at Analog Devices BV, Limerick. Accordingly, it is also recommended for use by practicing engineers whether they are engaged in actual integrated circuit design and fabrication or in a related peripheral activity. Its suitability as a design reference manual for engineers is supported by the inclusion of extensive graphical data in the appendices.

As a teaching instrument, the book can be used in a few ways. The detailed quantitative work is contained in Chapters 4 to 8 and in Chapter 10. While these represent the core material for design studies, it is possible also to teach a simplified and reasonably self-contained course which takes Chapters 1, 2, 3, 9, 11, and 12 only, in that order. Alternatively, this body of material might be taken first, and followed at a later time by Chapters 4 to 8 and 10. The main part of this work has been taught, for a few years now, as part of their final year Degree-level programme, to students of Electronic Engineering at the National Institute for Higher Education in Limerick.

It was felt that the purpose of the book would best be served by brevity in discussion and by offering a large number of numeric examples which serve to consolidate new ideas and to introduce reasonable values of the various process parameters. In addition, a number of problems appear in the form of a set of Case Studies which are used to tie together the various elements of the design activity. Answers to case study problems are also provided.

The writing of this book is in a large measure a consequence of my association with Analog Devices over a number of years as a design and training consultant. I am indebted to their Vice-President, Heinrich Krabbe, who provided both opportunity and encouragement and to Dr Philip Burton, Development Manager at ADI, for his every assistance. My thanks also to John C. Irvin of Bell Laboratories for his very helpful cooperation in furnishing graphical data on diffused layers.

The preparation of the manuscript has been greatly assisted by the cooperation of my colleagues in the Department of Electronic Engineering at NIHE, Limerick. In particular, I wish to thank Mr Gerard Stockil who provided programming support for inverter characterization, Ms Aine O'Brien who typed the manuscript, and Dr Grant Anderson, Head of Electronic Engineering, who provided full support and facilities.

May 1981 OLIVER J. MCCARTHY

List of Symbols

A	Cross-section area
BV	Breakdown voltage
β	Transistor 'gain', A/V^2
β'	Transistor 'gain', A/V^2, for square geometry
β_D	Driver gain
β_L	Load device gain
β_R	Ratio of (β_D/β_L)
C	Capacitance per unit area
C_L	Load capacitance
D	Diffusion coefficient
d_l	Depletion layer width above the junction
E	Electric field strength
E_c	Conduction band energy level
E_v	Valance band energy level
E_F	Fermi level
E_i	Intrinsic level
E_{go}	Gap energy
e_0	Permittivity of free space
F	Particle flux
GND	Symbol for 'ground' or reference potential
g_m	Device transconductance
I_s	Saturation current for diode or bipolar transistor
I_{DS}	Drain to source current
i_L	Instantaneous load current
i_D	Instantaneous driver current
I_o	Transistor current at entry to saturation
J	Current density
K_D	Depletion factor, V_I/V_{DD}, for depletion load
k	Boltzmann's constant
k_{si}	Relative permittivity of silicon
k_{ox}	Relative permittivity of silicon dioxide
L	Device length (from source to drain diffusion)
L'	Distance from end of channel to drain
ΔL	Horizontal penetration distance of diffusion
M	Body-effect factor
m	Relative drive factor for a non-saturated load
μ	Mobility
$\bar{\mu}$	Average mobility
n	Electron concentration
n_p	Electron concentration in p-type material

N	Impurity ion concentration
N_A	Acceptor concentration
N_D	Donor concentration
N_B	Ion concentration from doping sources in bulk
n^+	Heavily doped n-type region
n^-	Lightly doped n-type region
n_i	Intrinsic electron or hole concentration
p	Hole concentration per unit volume
p_n	Hole concentration in n-type material
p^+	Heavily doped p-type region
p^-	Lightly doped p-type region
Q	Electric charge or charge per unit area
Q_B	Depletion layer charge per unit area under gate with zero bulk bias
Q_{SS}	Oxide trapped charge per unit area for gate regions
Q_{Na^+}	Charge per unit area for mobile sodium ions in gate dielectric
q	Charge on an electron
R	Resistance
R_{on}	The ON-resistance of a driver or switch
R_s	Sheet resistance, ohms 'per square'
r_o	Small-signal output resistance
r_L	Small-signal load resistance
ρ	Resistivity
S	Strobe symbol
σ	Conductivity
T	Temperature
t_{ox}	Thickness of SiO_2 in gate region
T_D	Time constant, $C_L/\beta V_I$ for depletion loads.
T_f	Time constant (fall-time) for driver discharge
T_r	Time constant (rise-time) for MOS loads
T_{sc}	Unified time constant for enhancement loads and resistors
V_d	Drift velocity
V_{DS}	Drain-to-source potential difference
V_{GS}	Gate-to-source potential difference
V_{SB}	Source-to-bulk potential difference
V_{DB}	Drain-to-bulk potential difference
V_{CHS}	Channel-to-source potential difference
V_{CS}	Channel support voltage
V_{CB}	Channel-to-bulk potential difference
V_{GB}	Gate-to-bulk potential difference
V_{DD}	Drain supply level
V_{CC}	Supply level (CMOS circuits)
V_{GG}	Gate supply level
V_T	Threshold voltage in gate regions
V_{TF}	Field threshold voltage

V_{TO}	Threshold at source with $V_{BS}=0$
V_{TS}	Threshold voltage at source
V_{TD}	Threshold voltage at drain
V_{TCE}	Threshold voltage at channel end
V_{TL}	Load device threshold voltage
V_{TCEL}	Channel-end threshold for load device
V_{TP}	Threshold voltage for p-channel transistor
V_{TM}	Threshold voltage, mean of source and channel-end values
V_{TN}	Threshold voltage for n-channel transistor
V_{tr}	Trigger level
V_I	Inversion voltage, negative of V_T for a depletion-load device
V_{CHE}	Channel-end voltage using source reference
V_{OL}	Logic output low level
V_{OH}	Logic output high level
v_i	Instantaneous input voltage
v_o	Instantaneous output voltage
V_x	Transition voltage for a CMOS inverter
v_{ox}	Output voltage at transition between saturated and non-saturated conditions
v_{o1}, v_{o2}	Initial and final values of output voltage
v_s	Instantaneous signal voltage at input
W	Device width
ΔW	Lateral penetration distance of diffused layer under window
x_j	Junction depth
X_{mL}	Depletion layer width
ψ_0	Contact potential
$(2\phi_F)$	Contact potential based on N_B
$\phi_1, \phi_2, \ldots$	Clock phase labels

CHAPTER 1

Conduction in Semi-conductors: An Overview

This introductory chapter does not deal directly with MOS devices or circuits. It is concerned with the basic 'raw materials' of integrated circuit fabrication—the semi-conductors. An understanding of the electrical properties of semi-conductors is a prerequisite to a study of electronic devices, both bipolar and MOS. The treatment here is brief and is principally directed towards providing insights into the electrical behaviour of a semi-conductor p–n junction. Those properties, which will be important in later work, are emphasized. Theoretical derivations are usually avoided; they are to be found in the appropriate physics texts. The objective here is rather to present results which have wide acceptance as useful engineering approximations and to impart an ability to use these results intelligently.

Silicon (Si) and, to lesser extent, germanium (Ge) are semi-conductors of interest. Both materials furnish four valence electrons per atom and all four are required in forming stable bonds with neighbouring atoms (Fig. 1.1).

This highly ordered arrangement is valid at absolute zero temperature (0 K or −273 °C). Increase in temperature introduces a random 'vibratory' motion to the crystal, increasing the possibility that an electron may break loose from its bond—in doing so, it becomes a 'free' electron and can participate in conduction. It leaves behind a *hole*, i.e. a location of nett positive charge. An electron in a neighbouring bond may 'jump' to fill this hole and so create another so that the hole appears to move in the opposite direction. (The electron in question is *not* a free electron but it has no statistical preference for one location rather than the other.) As this process repeats (with different electrons) *the hole displays true mobility*. The key point is that the hole is as acceptable at one location as any other and, given a highly dynamic system, will shift location on a purely random basis. The electron (negative charge $-q$) and hole (positive charge $+q$) can both participate in conduction. The value of electron charge is $q = 1.6 \times 10^{-19}$ Coulombs.

The thermal 'birth' of electron-hole pairs is balanced by a 'death' mechanism. An electron, on 'meeting' a hole, fills it; they *recombine* and disappear. Defining:

n = mobile electrons per cubic centimetre

p = (mobile) holes per cubic centimetre

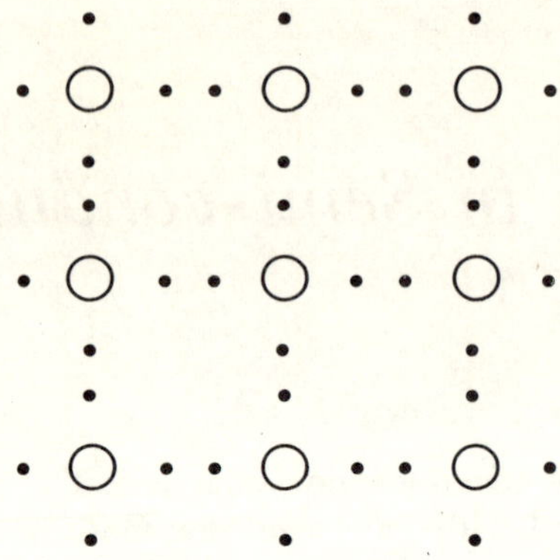

Fig. 1.1. Valence electrons in silicon and germanium.

a sample of pure (intrinsic) silicon or germanium has, by dynamic equilibrium, an intrinsic concentration $n_i = n = p$ which increases rapidly with temperature. A widely accepted expression for $n_i(T)$ is:

$$n_i(T) = 3.87 \times 10^{16} T^{3/2} e^{-E_{go}/(2kT/q)} \quad \text{per cm}^3 \tag{1.1}$$

with T in °kelvin, $k = 1.38 \times 10^{-23}$ Joules/°K is Boltzmann's constant, and E_{go} is a physical constant with a value of 1.21 for silicon. For example:
At 300 K (27 °C):

$$n_i = 1.41 \times 10^{10} \quad \text{per cm}^3$$

At 398 K (125 °C):

$$n_i = 6.80 \times 10^{12} \quad \text{per cm}^3$$

Conduction in silicon is, of course, in direct proportion to n_i and the figures represent poor conduction at room temperature—hence the term *semi-conductor.*

DONOR DOPING

Silicon or germanium may be doped with five-valence atoms (the most common is phosphorus). Doping is a process by which very low concentrations of a foreign material are added to the semi-conductor. It is usually achieved by diffusion of dopant atoms at very high temperatures and is described in detail later. A typical dose would be one phosphorus atom to 10^7 silicon atoms! Each phosphorus atom forms a bond with four silicon atoms in the usual way—which leaves a loosely held fifth electron, so loose that it is essentially a *conduction* electron. On entering conduction it leaves behind *an immobile positive ion.* The immobility stems from the fact that this atom is unlike any of its neighbours; it has little affinity for an electron in a neighbouring bond since it already has four bonding electrons.

A typical dose of phosphorus might be $N_D = 10^{16}$ atoms/cm^3 while silicon has about 10^{23} atoms/cm^3. Thus N_D augments n by 10^{16} atoms/cm^3 but since n_i is only 10^{10} approximately, this is a million-to-one increase! The new n value describes excellent conducting properties. The phosphorus atoms are referred to as 'donors' since each one 'donates' an electron to the conduction band. The large increase in n gives a boost to recombination so that p falls to quite a low level. Eventually, equilibrium is reached in a manner such that:

$$n_i^2(T) = np$$

is true for doped as well as intrinsic semi-conductors! (This is called the law of mass action). In this situation, electrons are *majority carriers* and holes are *minority carriers*, the latter contributing very little to current flow. To excellent approximation, $n = N_D$ and $p = n_i^2/N_D$. Since the majority carriers carry negative charge, the conduction mechanism, and the material so treated, are said to be 'n-type'.

ACCEPTOR DOPING

Doping with three-valence electron atoms (such as boron), in similar quantities to N_D, produces a crystal with occasional boron atoms bonding to silicon neighbours. In such a situation, one bond is incomplete (although neutral). It will readily take an electron from a normal bond nearby and thus become a stable immobile negative ion. Such three-valence atoms are called 'acceptors' since they accept an electron in order to form a stable bond. The electron capture leaves a hole which is essentially mobile. The concentration N_A of boron acceptor atoms contributes directly to p which increases vastly. The large hole population leads to a short-lived increase in recombination so that electron population falls and equilibrium is reestablished when $p = N_A$ (neglecting the small thermal contribution) and $n = n_i^2/N_A$.

If an n-type sample (i.e. one doped with donor atoms) is subsequently doped with an equal concentration of acceptor atoms, e.g. $N_A = N_D = 10^{16}$ atoms/cm^3, their effects cancel and the sample becomes intrinsic again. More generally, the effective doping level is $N_D - N_A$. Since electrical neutrality must be preserved, equating positive and negative total charge yields:

$$p + N_D = n + N_A \tag{1.2}$$

true for any doping conditions. Note that N_D, the concentration of donor atoms, is also the concentration of immobile positive ions and is approximately equal to the free electron concentration. Similar comments are valid for N_A.

The results, thus far, can be summarized as follows. Intrinsic semiconductors offer poor conduction with equal concentrations of holes and electrons which are thermally generated. Donor doping produces n-type material with large conductivity via electron flow. Acceptor doping produces p-type material with large conductivity via hole movement.

Our next task is to examine the two mechanisms which result in mass movement of conducting particles, positive or negative, and are thereby responsible for all external current flow.

CURRENT FLOW BY DRIFT

Drift is the familiar process whereby carriers (electrons or holes), in addition to their random thermal motion, show a nett movement in response to an applied electric field E volts per centemetre. Holes drift in the E direction, electrons in the opposite way (see Fig. 1.2). While E accelerates carriers

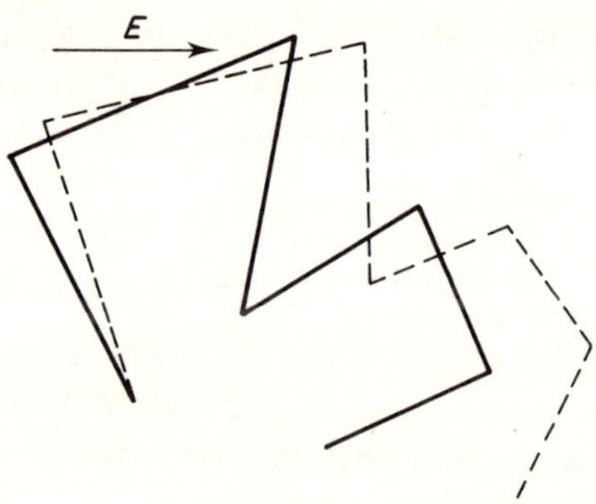

Fig. 1.2. Hole drift.
——— Random ($E=0$)
- - - - - - Forced ($E\neq 0$)

between collisions, the velocity component acquired is lost with each collision and carriers show an average *drift velocity* V_d due to E. To take hole conduction as an example, the resulting current density is:

$$J_p = qpV_d \qquad \text{amperes (A)/cm}^2 \tag{1.3}$$

This is better appreciated in dimensional form as:

$$\frac{\text{Charge}}{\text{Particle}} \times \frac{\text{particles}}{\text{cm}^3} \times \frac{\text{cm}}{\text{sec}} = \frac{\text{charge/sec}}{\text{cm}^2} = \frac{\text{A}}{\text{cm}^2}$$

The drift velocity V_d is proportional to E and to a *mobility* term:

$$V_d = \mu E \qquad \text{cm/sec} \tag{1.4}$$

The mobility (μ_p for holes, μ_n for electrons) depends on crystal structure, temperature, level of doping, and other factors. Roughly speaking, electrons are about twice as mobile as holes.

We also define *conductivity* (σ) by the simple statement $J=\sigma E$. Since $J=qp\mu E$, we then have:

$$\sigma = qp\mu \qquad \text{(for holes)} \tag{1.5}$$

or, more generally:

$$\sigma = q(p\mu_p + n\mu_n) \tag{1.6}$$

for conduction involving both holes and electrons. It is intuitively correct that conductivity, the ability to transport electric current, should be in proportion to carrier concentrations and to their freedom of movement, their mobility.

The reciprocal of conductivity is *resistivity* (ρ):

$$\rho = \frac{1}{\sigma} = \frac{1}{q(p\mu_p + n\mu_n)} \quad \text{ohm } (\Omega) \cdot \text{cm} \tag{1.7}$$

In turn, resistivity and physical dimensions give the resistance of a sample (Fig. 1.3):

$$R = \frac{\rho l}{A} \tag{1.8}$$

or

$$\rho = \frac{RA}{l} \tag{1.9}$$

A most important set of curves give the resistivity of silicon at 300 K as a function of impurity concentration, and doping type. These appear in Appendix D. They refer to uniformly doped semi-conductors and demonstrate the higher mobility of electrons for all but very high doping levels. (The appendix also gives similar data for germanium and for gallium arsenide).

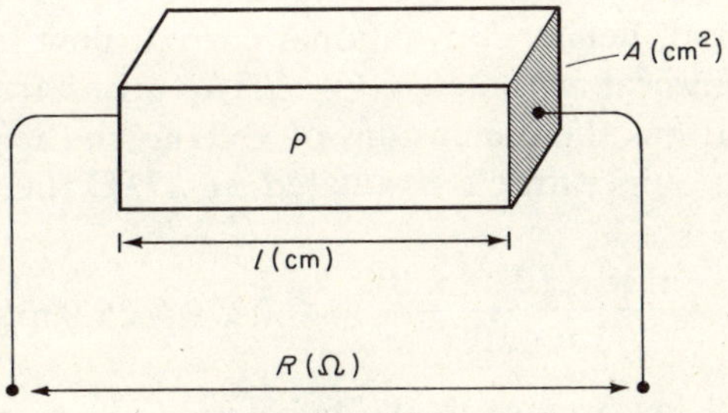

Fig. 1.3. Resistance of a rectangular section.

CURRENT FLOW BY DIFFUSION

In certain circumstances, concentration of a carrier type is non-uniform. Consider a case where hole density diminishes in the positive x direction (Fig. 1.4). Since these carriers are in random thermal motion, then, in a given time interval, more carriers will cross AA from left to right than in the other direction *simply because carriers are more numerous on the left.* This

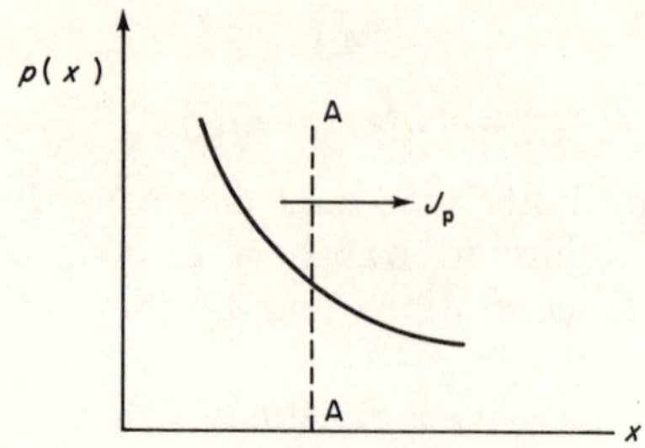

Fig. 1.4. Diffusion.

indicates a nett hole current density J_p amperes per square centimetre, as shown, which is quite plausibly proportional to the concentration gradient dp/dx (which determines particle flow rate) and the charge q per particle (since J is *charge* per square centimetre per second) with a proportionality factor known as the *diffusion coefficient* (D_p for holes, D_n for electrons). Thus, for hole diffusion:

$$J_p = -qD_p \frac{dp}{dx} \tag{1.10}$$

and, for electrons:

$$J_n = qD_n \frac{dn}{dx} \tag{1.11}$$

In checking the direction of J_n, remember that it represents *conventional* current flow. A positive electron gradient (dn/dx) causes electron diffusion from right to left and, hence, conventional current flow from left to right.

An important physical parameter $V_T = kT/q$ arises frequently in semiconductor theory. It has the dimensions of voltage and is referred to as the 'volt-equivalent of temperature'. Evaluated at 27 °C (i.e. 300 K), it has a value;

$$\frac{kT}{q} = \frac{1.38 \times 10^{-23} \times 300}{1.6 \times 10^{-19}} = 0.0259 = 25.9\ \text{mV}$$

It is of fundamental importance to the behaviour of a p–n junction. It also arises as the factor which links the two conduction mechanisms:

$$D_n = \frac{kT}{q} \mu_n \tag{1.12}$$

$$D_p = \frac{kT}{q} \mu_p \tag{1.13}$$

It indicates a direct proportionality between drift-induced flow and that resulting from diffusion. The diffusion effect is further promoted by higher temperature due to increased thermal motion.

AN IDEALIZED p–n JUNCTION

We consider a sample of silicon with an abrupt transition from p-type to n-type doping and the resulting electrical implications.

The example used, resembling an n+ diffusion into a p-type substrate, i.e. $N_D \gg N_A$, is quite typical. The figures in Fig. 1.5(b) are consistent with the law of mass action:

$$np = n_i^2(T)$$

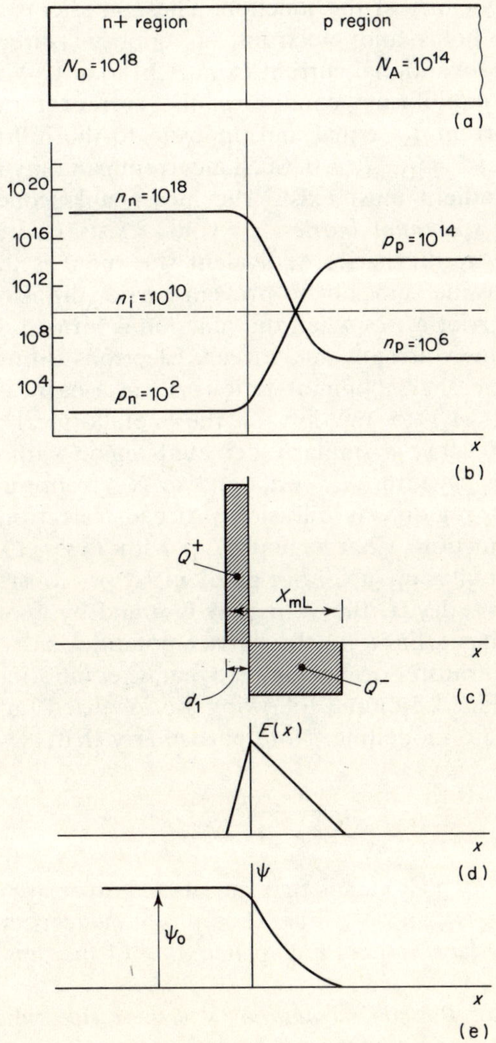

Fig. 1.5. A semi-conductor p–n junction.

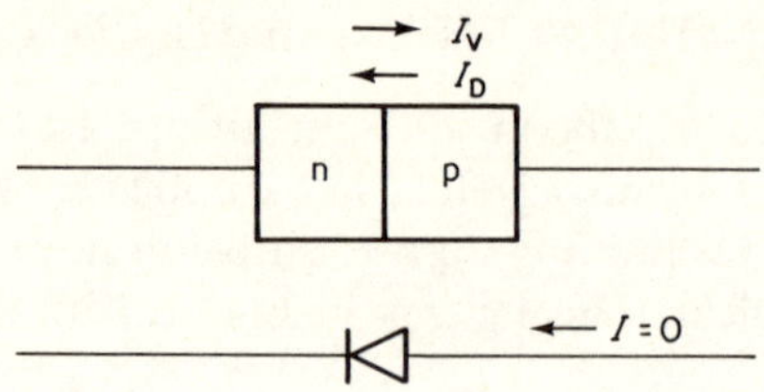

Fig. 1.6. An open-circuited diode.

at all points. The curves also make it clear that finite gradients of n and p much exist in the vicinity of the junction. These gradients are conducive to diffusion of both holes and electrons, in opposite directions, but both contributing to a conventional current from right to left. While the gradients *must* exist under equilibrium conditions, the current I cannot (Fig. 1.6). Hence another current I_V, equal and opposite to the diffusion current I_D, must exist such that $I = I_D - I_V = 0$. Such a current can only result from drift, i.e. a potential gradient must exist. The inescapable conclusion is that a voltage difference (*potential barrier*, ψ_0 volts) exists (see Fig. 1.5(e)) and accounts for I_V. (An alternative equivalent statement is that the potential barrier ψ_0 is of a value sufficient to prevent carrier diffusion.)

The potential barrier arises when the junction is formed. Initially, there is no barrier and some diffusion *does* occur. Electrons diffusing to the right leave a nett positive charge (immobile ions ⊞) in a narrow *depleted* region (of nominal width d_1) to the left of the metallurgical junction. Holes diffusing to the left leave a similarly depleted region with a negative ionic charge ⊟. This has a width $(X_{mL} - d_1)$ and so X_{mL} represents the width of the total depletion region. As diffusion proceeds, electrons and holes recombine at the junction. Charge builds up with $Q+ = Q-$, always (Fig. 1.5(c)). As charge builds up, a voltage gradient $E(x) = \text{const.} \int Q(x)\,\mathrm{d}x$ arises also and, since $E = -\mathrm{d}\psi/\mathrm{d}x$, the voltage ψ is found by another integration. Eventually ψ reaches a limit ψ_0, the barrier potential, at which $I_V = I_D$, no further nett charge transfer occurs, and a dynamic equilibrium is established.

The profiles of Fig. 1.5(c) and following are idealized but can be used to obtain fairly accurate modelling. Note particularly that:

$$\frac{d_1}{X_{mL} - d_1} = \frac{N_A}{N_D} \tag{1.14}$$

because $Q+ = Q-$ (which means that the shaded areas are equal) and ionic charge densities are N_D and N_A in n+ and p regions respectively. The point is that a *low doping level implies high penetration* of the depletion layer, and vice versa.

The application of *Poisson's equation* ($\nabla^2\psi = -q_v/ke_0$ where q_v is charge density) shows that ψ_0 is determined by end-region doping levels and does not depend on junction details. The law of mass action implies that $p_p n_p =$

$n_n p_n$ so that $(n_n/n_p)=(p_p/p_n)$. ψ_0 can then be shown theoretically to be:

$$\psi_0=\frac{kT}{q}\,\mathrm{Ln}\,\frac{n_n}{n_p}=\frac{kT}{q}\,\mathrm{Ln}\,\frac{p_p}{p_n} \tag{1.15}$$

Upon substituting $n_n=N_D$, $p_p=N_A$, $p_n=n_i^2/N_D$ and $n_p=n_i^2/N_A$, it follows that:

$$\psi_0=\frac{kT}{q}\,\mathrm{Ln}\,\frac{N_A N_D}{n_i^2}=\frac{kT}{q}\,\mathrm{Ln}\,\frac{N_A}{n_i}+\frac{kT}{q}\,\mathrm{Ln}\,\frac{N_D}{n_i} \tag{1.16}$$

$$\psi_0=\psi_p+\psi_n \tag{1.17}$$

Evaluation for $N_D=10^{18}$ atoms/cm^3, $N_A=10^{14}$ atoms/cm^3, and using $n_i=10^{10}$ and $kT/q=26$ mV gives:

$$\psi_0=\psi_p+\psi_n=0.239+0.479=0.718\text{ V}$$

Equations (1.1) and (1.16) jointly indicate the temperature dependence of ψ_0, where, in general, ψ_0 falls with increasing T.

Application of *forward bias* (Fig. 1.7) to a p–n junction simply reduces the potential barrier by an amount V. Forward current, I_f, results and a

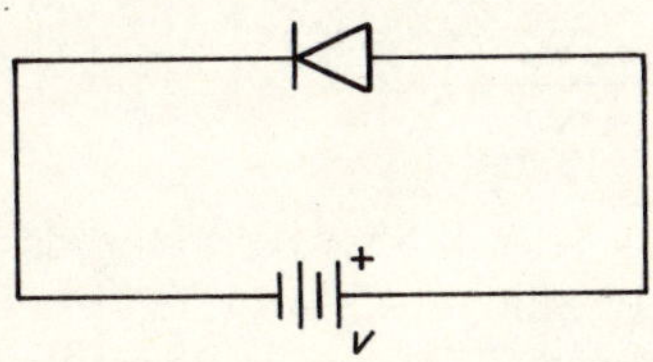

Fig. 1.7. Forward bias.

simplified analysis employing $N_D \gg N_A$ yields:

$$I_f=A\,\frac{qD_n n_p}{L_n}\,(e^{qV/kT}-1) \tag{1.18}$$

where $L_n=(D_n\tau)^{1/2}$, the diffusion length
$\tau=$ carrier lifetime
$A=$ junction area

The implication is that electrons injected into the p region will, on average, travel a distance L_n and exist for τ seconds before recombination occurs.

Equation (1.18) is surprisingly accurate at low and medium current levels. Exactly analogous results apply for a p+–n junction. This equation, *the rectifier equation*, is widely used in circuit design in the form:

$$I=I_s(e^{qV/kT}-1)=I_s(e^{V/V_T}-1) \tag{1.19}$$

I_s is typically in the region 10^{-14} to 10^{-15} A.

The equation also works in the case of *reverse bias* (V a negative number) which actually enlarges the potential barrier. This bias condition shows up a leakage effect: *minority* carriers on both sides of the junction see the 'barrier' as a *downhill* journey and fall across the junction to give a current I_s determined simply by the *availability* of such carriers. Thus it is not surprising that Eqn. (1.19) indicates a leakage which is essentially independent of reverse bias for voltages in excess of, say, 0.1 V (or $4kT/q$).

The leakage exists even for $V=0$ but, in that condition, ψ_0 is small enough to permit a compensating diffusion current. Reverse bias does not affect leakage but, by adding directly to ψ_0, it rapidly eliminates the small diffusion effect. Then, leakage *alone* produces a current which can be observed at the terminals.

The true situation is a little more complex than this, unfortunately. Actual leakage currents are in the nanoampere region for small silicon devices. Only a small part of observed leakage current levels can be attributed to I_s because other mechanisms, involving crystal defects, tend to dominate.

Since a potential barrier is made possible by provision of a *depletion region* with associated ionic charges $Q+$ and $Q-$, one can expect that depletion region width X_{mL} will increase with total barrier height $(\psi_0 - V)$, where *V is positive for forward bias,* and vice versa.

For the *one-sided step-junction* model of Fig. 1.5(c) which uses $N_D \gg N_A$, a theoretical development gives:

$$X_{mL} = \left[\frac{2k_{si}e_0(\psi_0 - V)}{qN_A}\right]^{1/2} \tag{1.20}$$

where e_0 is the free-space permittivity ($e_0 = 8.85 \times 10^{-14}$ farads (F)/cm) and k_{si} is the relative dielectric constant of silicon, $k_{si} = 11.7$.

A rather more realistic model for *some* situations replaces the discontinuous step from N_D to N_A with a *smooth* transition from the n-type to p-type having a finite slope dN/dx at the 'junction' (where $N_D = N_A$). In this case:

$$X_{mL} = \left[\frac{12k_{si}e_0(\psi_0 - V)}{q\,|dN/dx|}\right]^{1/3} \tag{1.21}$$

referred to as the *graded junction approximation.*

Since depletion region charges $Q+$ and $Q-$ are established by current flow (the removal of mobile carriers), this is essentially a capacitive effect. It is fortunate that the depletion region can be treated as a parallel-plate capacitor with plate separation X_{mL} and dielectric constant $k_{si} = 11.7$, as far as capacitive effects are concerned. By this analogy:

$$C = \frac{k_{si}e_0}{X_{mL}} \quad \text{F/cm}^2 \tag{1.22}$$

It will be noticed that since X_{mL} is controlled by applied bias V, Eqn. (1.22) describes a voltage-controlled capacitance. Reverse bias reduces C. This capacitance is immensely important in determining the switching times of MOS logic since it takes finite time to charge and discharge these junctions.

Substituting from Eqn. (1.20) into Eqn. (1.22) gives, for the one-sided step junction:

$$C=\left[\frac{qk_{si}e_0N_A}{2(\psi_0-V)}\right]^{1/2} \quad \text{F/cm}^2 \tag{1.23}$$

and, for the graded junction, using Eqn. (1.21):

$$C=(k_{si}e_0)^{2/3}\left[\frac{q\,|dN/dx|}{12(\psi_0-V)}\right]^{1/3} \quad \text{F/cm}^2 \tag{1.24}$$

The approximate models described here yield good results when applied, with care, to actual junctions.

JUNCTION BREAKDOWN

Junction breakdown voltage (BV) under reverse bias is determined by either the *Zener* or *avalanche* mechanism.

Zener breakdown occurs at 5 V or less and occurs with very heavily doped junctions. It is a quantum-mechanical tunnelling effect, not normally met in integrated structures.

If the impurity concentration on *either* side of the junction is under about 10^{18} atoms/cm^3, reverse breakdown is by avalanche multiplication. This occurs when free carriers acquire sufficient energy from the E field to have collisions which dislodge electrons from normal bonds; these electrons dislodge further electrons and carrier availability increases enormously yielding a conducting state at the breakdown voltage value. Avalanche breakdown voltage (BV) follows approximately the expression:

$$\text{BV}=2.7\times10^{12}N^{-2/3} \quad \text{V} \tag{1.25}$$

where N is the impurity concentration (atoms/cm^3) on the *lighter* doped side. Figure 1.5(d) and (e) show why this kind of relationship applies. Avalanche occurs at a certain critical field strength E_c. The associated BV is, in crude terms, $E_c\times$(depletion layer width). Because the width X_{mL} is largely on the lightly doped side, the doping level on that side has a dominant influence. Notice, for example, in Fig. 1.5(e) that most of the potential drop ψ_0 occurs in the lightly doped p region. It follows from these observations that low doping and consequent large X_{mL} are key factors in achieving high breakdown voltage.

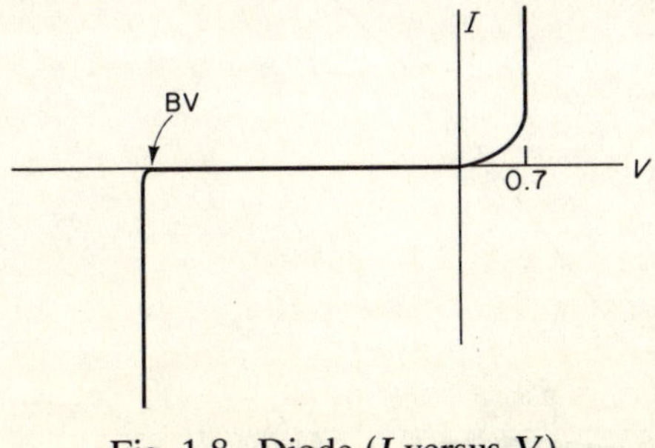

Fig. 1.8. Diode (I versus V).

Figure 1.8 shows a composite diode characteristic. Operation in the breakdown region offers voltage reference and stabilization capability. The breakdown is non-destructive provided current flow is suitably limited. Avalanche breakdown must, however, be avoided in most situations and tends to set an upper limit to the doping level (N_B) of the silicon wafers used in integrated circuit fabrication.

The chapter concludes with a few numerical examples, as will most other chapters throughout the book. Their purpose is to give the reader an indication of the actual parameter values which are typical of integrated circuits. A conscious effort should be made to remember typical values of parameters because only in this way can the trade-offs and the tolerance limits of integrated circuit design be appreciated.

EXAMPLE 1.1

Plot $n_i(T)$ in 25° steps from −50 to +125 °C.

Solution

$$n_i(T) = 3.87 \times 10^{16} T^{3/2} \exp\left[\frac{-1.21}{2 \times 1.38 \times 10^{-23} T/(1.6 \times 10^{-19})}\right] \quad \text{per cm}^3$$

−50 °C,	$T = 223$ K,	$n_i = 2.8 \times 10^6$	per cm³
−25 °C,	$T = 248$ K,	$n_i = 7.8 \times 10^7$	per cm³
0 °C,	$T = 273$ K,	$n_i = 1.2 \times 10^9$	per cm³
25 °C,	$T = 298$ K,	$n_i = 1.2 \times 10^{10}$	per cm³
50 °C,	$T = 323$ K,	$n_i = 8.3 \times 10^{10}$	per cm³
75 °C,	$T = 348$ K,	$n_i = 4.4 \times 10^{11}$	per cm³
100 °C,	$T = 373$ K,	$n_i = 1.9 \times 10^{12}$	per cm³
125 °C,	$T = 398$ K,	$n_i = 6.8 \times 10^{12}$	per cm³

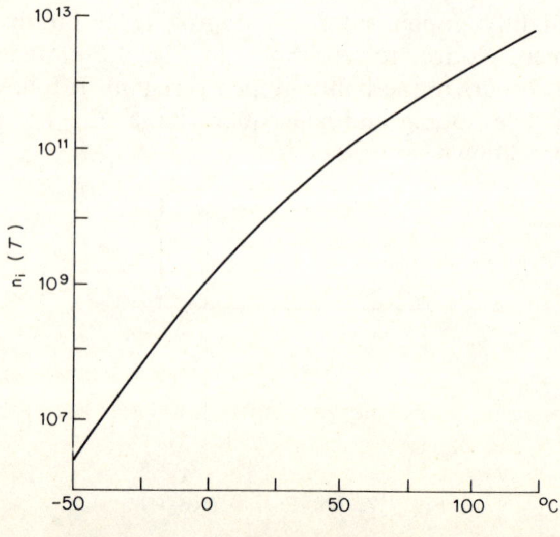

EXAMPLE 1.2

A p–n junction is described in idealized fashion by $N_D = 5 \times 10^{17}$, $N_A = 10^{15}$. Find minority carrier concentrations n_p and p_n. Find the contact potential ψ_0. Use the step-junction approximation to find X_{mL} and capacitance C per square centimetre for (a) zero bias and (b) 20 V reverse bias. Assume $T = 300$ K. Find the avalanche BV.

Solution

$$n_i = 1.41 \times 10^{10} \qquad \text{at 300 K}$$

$$n_p = \frac{n_i^2}{N_A} = 1.99 \times 10^5 \text{ per cm}^3$$

$$p_n = \frac{n_i^2}{N_D} = 397 \text{ per cm}^3$$

$$\psi_0 = \frac{kT}{q} \text{Ln} \frac{N_A N_D}{n_i^2} = 0.0259 \text{ Ln} \frac{10^{15} \times 5 \times 10^{17}}{1.41^2 \times 10^{20}} = 0.74 \text{ V}$$

(a) Zero bias;

$$X_{mL} = \left(\frac{2k_{si}e_0\psi_0}{qN_A}\right)^{1/2} = \left(\frac{2 \times 11.7 \times 8.85 \times 10^{-14} \times 0.74}{1.6 \times 10^{-19} \times 10^{15}}\right)^{1/2}$$

$$X_{mL} = 9.8 \times 10^{-5} \text{ cm} = 0.98 \text{ micrometres } (\mu\text{m})$$

$$C = \frac{k_{si}e_0}{X_{mL}} = \frac{11.7 \times 8.85 \times 10^{-14}}{9.8 \times 10^{-5}} = 1.05 \times 10^{-8} \text{ F/cm}^2$$

$$C = 1.05 \times 10^{-8} \times 6.45 \times 10^{-6} \text{ F/mil}^2 = 68.1 \times 10^{-15} \text{ F/mil}^2$$

$$C = 68.1 \text{ femtofarads (fF)/mil}^2$$

(b) 20 V bias:

$$X_{mL} = \left[\frac{2k_{si}e_0(20.74)}{qN_A}\right]^{1/2} = 5.18 \times 10^{-4} \text{ cm} = 5.18 \ \mu\text{m}$$

$$C = \frac{k_{si}e_0}{X_{mL}} = 12.9 \text{ fF/mil}^2$$

$$\text{Avalanche BV} = 2.7 \times 10^{12} N_A^{-2/3} = 270 \text{ V}$$

EXAMPLE 1.3

A p–n junction is formed by a rich n+diffusion ($N_D = 5 \times 10^{18}$) to a p-type substrate ($N_A = 2 \times 10^{14}$). Assume that $D_n = 20$ cm^2/sec and carrier lifetime $\tau = 0.1$ μsec. Using the step-junction approximation, calculate the forward current per square mil of junction area with 0.6 V applied. Determine also the theoretical leakage current per square mil for a reverse bias of (a) 0.1 V and (b) 10 V.

Actual leakage currents include a component due to crystalline defects. This can be shown to be responsible for a carrier generation rate of $g=(n_i/2\tau)$ particles per second per cubic centimetre of the depletion layer. Calculate the leakage current per square mil from this source with 10 V of reverse bias at 25 °C. What is the new leakage current at 125 °C?

Solution

The rectifier equation is applicable:

$$I_f = I_s(e^{V/(kT/q)} - 1) \qquad \text{with} \qquad I_s = \frac{AqD_n n_p}{L_n}$$

and

$$L_n = (D_n \tau)^{1/2}$$

Note that 1 $\text{mil}^2 = 6.45\times10^{-6}\ \text{cm}^2$. Also $n_i = 1.2\times10^{10}$ at 25 °C and then:

$$n_p = \frac{n_i^2}{N_A} = \frac{1.2^2\times10^{20}}{2\times10^{14}} = 7.2\times10^5$$

$$L_n = (D_n\tau)^{1/2} = (20\times10^{-7})^{1/2} = 1.41\times10^{-3}$$

Then:

$$I_s = \frac{6.45\times10^{-6}\times1.6\times10^{-19}\times20\times7.2\times10^5}{1.41\times10^{-3}}$$

$$I_s = 1.05\times10^{-14}\ \text{A/mil}^2$$

$$\frac{kT}{q} = \frac{1.38\times10^{-23}\times298}{1.6\times10^{-19}} = 0.0257\ \text{V} = 25.7\ \text{mV}$$

With 0.6 V applied:

$$I_f = 1.05\times10^{-14}(e^{0.6/0.0257} - 1)$$
$$= 1.05\times10^{-14}(e^{23.3}) = \underline{0.144\ \text{mA}}$$

(a) With −0.1 V applied:

$$I = 1.05\times10^{-14}(e^{-0.1/0.0257} - 1)$$
$$= 1.05\times10^{-14}(e^{-3.9} - 1)$$
$$= 1.05\times10^{-14}(-0.98) = -\underline{1.02\times10^{-14}\ \text{A}}$$

(b) With −10 V applied:

$$e^{-10/0.0257} \ll 1$$

and

$$I = -I_s = \underline{-1.05\times10^{-14}\ \text{A}}$$

Leakage current due to crystalline defects is:

$$I = gq(\text{volume of depletion layer})$$
$$= gqX_{\text{mL}} \times 6.45 \times 10^{-6}$$
$$X_{\text{mL}} = \left[\frac{2k_{\text{si}}e_0(\psi_0 - V)}{qN_{\text{A}}}\right]^{1/2}$$

and the contact potential is:

$$\psi_0 = \frac{kT}{q}\,\text{Ln}\,\frac{N_{\text{A}}N_{\text{D}}}{n_{\text{i}}^2}$$
$$= 0.0257\,\text{Ln}\,\frac{2\times 10^{14}\times 5\times 10^{18}}{1.2\times 1.2\times 10^{20}} = 0.76$$

For 10 V of reverse bias, $V = -10$:

$$X_{\text{mL}} = \left(\frac{2\times 11.7\times 8.85\times 10^{-14}\times 10.76}{1.6\times 10^{-19}\times 2\times 10^{14}}\right)^{1/2}$$
$$X_{\text{mL}} = 8.34\times 10^{-4}\ \text{cm} = 8.34\ \mu\text{m}$$

Then:

$$I = \frac{n_{\text{i}}}{2\tau}\,qX_{\text{mL}}\times 6.45\times 10^{-6}$$
$$= \frac{1.2\times 10^{10}\times 1.6\times 10^{-19}\times 8.34\times 10^{-4}\times 6.45\times 10^{-6}}{2\times 10^{-7}}$$
$$I = 5.1\times 10^{-11}\ \text{A/mil}^2$$

Also:

$$I_{125°} = I_{25°}\frac{n_{\text{i}}(125°)}{n_{\text{i}}(25°)} = 5.1\times 10^{-11}\times\frac{6.8\times 10^{12}}{1.2\times 10^{10}} = \underline{2.89\times 10^{-8}\ \text{A}}$$

This is a 600-fold increase!

Such current levels as well as their dependence on bias level and temperature are generally observed in practice: they describe *well-behaved p–n junctions.* Occasional exceptions, exhibiting far higher leakage, constitute a yield-loss mechanism in LSI circuits.

CHAPTER 2

MOSFETs: A Qualitative View

This chapter presents a simple physical picture of the MOS transistor action without attempting to quantify it in any detailed way. It goes on to discuss the basic features of a MOS *inverter* since this is the elementary 'building-block' for all logic circuits. A few circuit examples are also included. The examples are designed to give some early appreciation of the kind of circuits that are utilized in LSI designs.

The term 'MOS' is descriptive of a Metal gate terminal over an Oxide (SiO_2) dielectic over a Silicon bulk material. The SiO_2 is *silicon dioxide* and provides insulation between the gate and the silicon. Figure 2.1 shows a p-channel MOSFET. The *bulk* is a silicon wafer with light n-type doping. *Source* (S) and *drain* (D) are p+ diffused layers with metal (usually aluminium) contacts. These form p–n junctions with the bulk. The p+ is symbolic of a heavily doped p-type region. (The diffusion process by which the junctions are formed is discussed in the next chapter.) The junctions are never forward biased in normal operation. The *gate* (G) is simply an isolated metal contact separated from the silicon surface by a very thin insulating layer of silicon dioxide. The thickness of this dielectric is typically about 1000 Å (angstrom units) or 0.1 μm. This gives very high gate input resistance (in the region of $10^{15}\,\Omega$) but also close capacitive coupling with the silicon surface. The action is such that current flow between the source and drain is modulated by the gate voltage (V_G). V_G controls the 'thickness' and hence the resistance of a conducting channel linking the source to the drain.

As a matter of convention, we refer all voltages to the source and, for now, assume that the bulk is at the source potential, $V_{BS}=0$. Note that we are free to take the bulk to a positive potential, $V_{BS}>0$, but not negative, thus avoiding forward bias. Similarly, drain voltage must be negative, $V_{DS}<0$; accordingly, p-channel circuits use a negative supply. Gate voltage V_{GS} can be of either polarity but is negative in the usual 'enhancement' mode of operation.

When $V_{GS}=0$, conduction from source to drain is not normally possible since one junction will be reverse biased. (It is V_{GB} that matters here but since $V_{BS}=0$, then $V_{GB}=V_{GS}$). A positive V_{GS} acts capacitively to place equal charges $+Q$ and $-Q$ on the gate metal and on the silicon surface respectively. Thus, an *accumulation* of electrons occurs at the silicon surface under the gate. This does nothing to encourage current flow between source and drain. A negative V_{GS} has the opposite effect—a *depletion* of electron population at the silicon surface. As V_{GS} becomes more negative, electrons are pushed down out of a narrow region near the surface to expose a

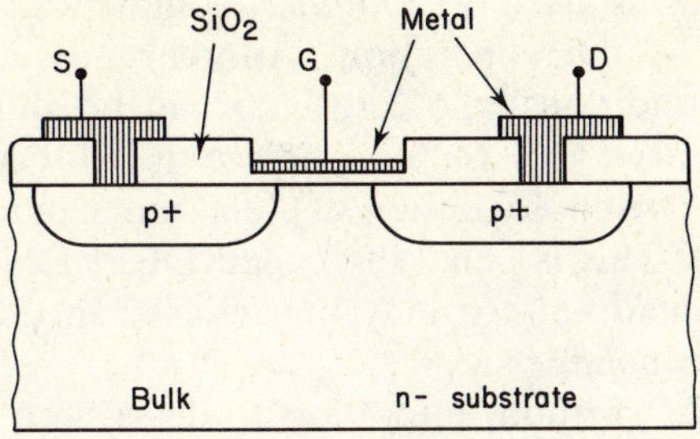

Fig. 2.1. A p-channel MOSFET.

positive ionic charge Q_B coulombs (C) per square centimetre. When V_{GS} goes beyond a certain *threshold voltage*, V_T, Q_B ceases to grow larger and additional positive surface charge is induced, not by repelling 'free' electrons but *by hole accumulation at the surface.* (Remember, a 'hole' describes a particular restricted kind of valence electron motion.) The reason for the change in behaviour will become apparent in Chapter 5 when we examine the thresholding effect in more detail. The *inversion* of surface properties provides a narrow p-type conducting *channel* from source to drain and, for non-zero V_{DS}, marks the beginning of current flow, I_{DS}, from source to drain. The voltage component $(V_{GS}-V_T)$ is dedicated to maintaining the channel; channel depth is very small but increases in proportion to $(V_{GS}-V_T)$ and its resistance decreases accordingly.

An exactly similar story can be told to describe an n-channel MOSFET. Here we have n+ diffusions into a p-type wafer; V_{DS} and V_{GS} will be positive in this instance and the channel will consist of electrons rather than holes. Positive V_{GS} first repels holes from the silicon surface and channel inception is at $V_{GS}=V_T$.

Figure 2.2 illustrates gate control of I_{DS} (assuming a large positive V_{DS}) for the n-channel device. V_T is typically a few volts positive. As shown in curve (a), the device is NORMALLY OFF, i.e. $I_{DS}=0$ when $V_{GS}=0$. Then

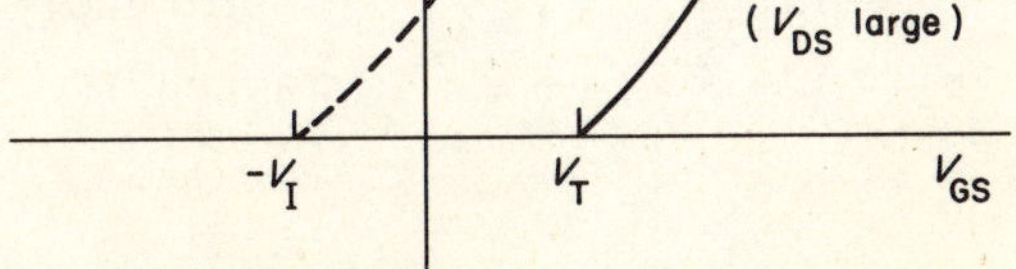

Fig. 2.2. MOSFET (I_{DS} versus V_{GS}).

a (positive) gate bias is used to *enhance* conduction; this is the normal ENHANCEMENT MODE operation. Another possibility is that (under certain processing conditions) the threshold will be shifted to the left and become negative—curve (b). Such a device is NORMALLY ON. Gate voltage can then be used to either deplete the channel ($V_{GS}<0$) or to enhance it ($V_{GS}>0$). This is generally called DEPLETION MODE operation. The term *threshold voltage* may be replaced in this case by the more appropriate *inversion voltage* ($V_I=-V_T$).

We can use Fig. 2.2 without alteration to describe a p-channel device if we are prepared to view I_{DS}, V_{GS}, V_{DS}, and V_T as the *magnitudes only* of quantities which are negative for p-channel working. This approach allows us to use descriptions which are valid for both channel types.

Transistor symbols in Fig. 2.3 show normal supply polarities and current directions. The bulk (B) is either tied to S or given a supply *opposite* in polarity to the normal drain supply.

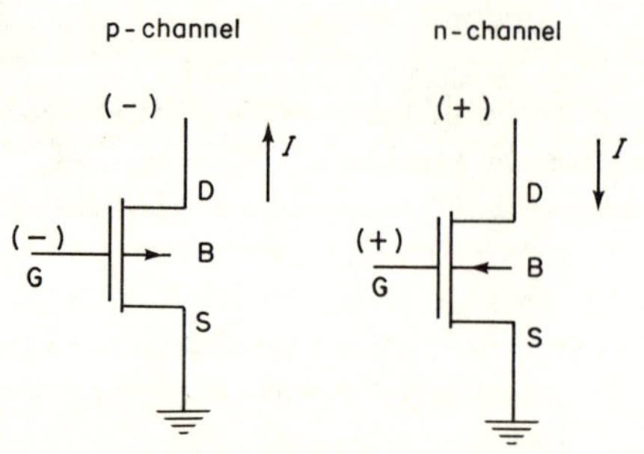

Fig. 2.3. Transistor symbols.

MODES OF CONDUCTION

The example in Fig. 2.4 illustrates the variation in channel depth from source to drain. Channel potential goes from 0 V at S to +5 V at D. With $V_T=+2$ V, the *channel support voltage* is:

$$V_{CS}=V_{GCH}-V_T=V_{GS}-V_{CHS}-V_T \qquad (2.1)$$

$$V_{CS}=12-0-2=10\text{ V} \qquad \text{(source end)}$$

$$v_{CS}=12-5-2=5\text{ V} \qquad \text{(drain end)}$$

where:

V_{GCH} = gate-to-channel voltage

V_{CHS} = channel-to-source voltage

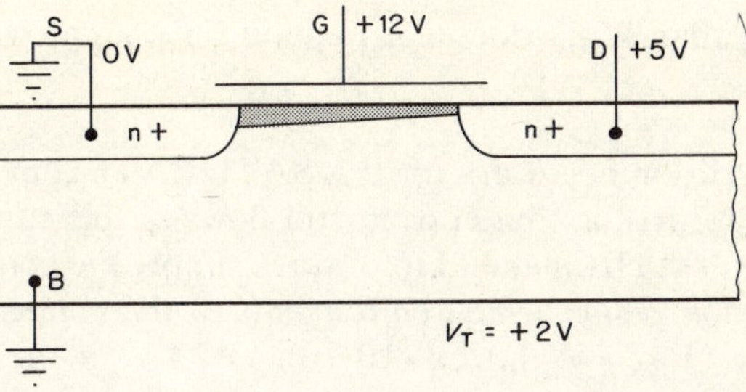

Fig. 2.4. N-channel, non-saturated conduction.

at the point of interest. The concept of channel support voltage V_{CS} as that voltage component responsible for the existence—and the strength—of the channel, is of great assistance in visualizing MOSFET action. While channel potential varies continuously along its length, the statement $V_{CS} = (V_{GCH} - V_T)$ must hold at all points. For n-channel circuits, the channel exists at all points where V_{CS} is positive. The p-channel circuits require negative V_{CS}.

The results above describe a channel which grows narrower toward the drain end. Since $V_{CS} > 0$, even at the drain, the channel exists at all points. This is called NON-SATURATED operation. In this mode, I_{DS} *is controlled by both* V_{GS} *and* V_{DS}. Increase of either voltage causes the current to increase.

For the conditions in Fig. 2.5, $V_{CS} = 8 - 2 = 6$ V at the source but the large V_D makes channel support impossible at the drain end. The channel must

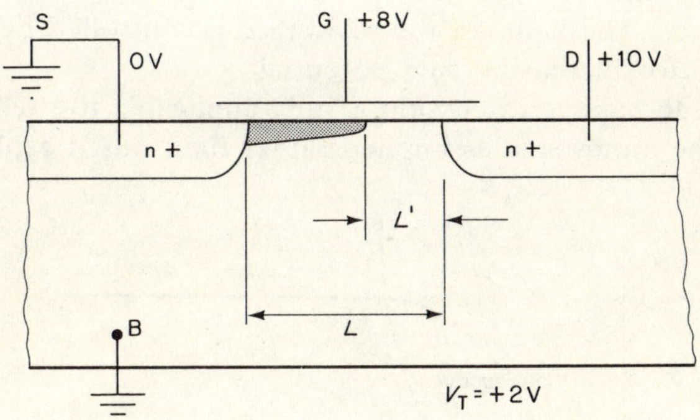

Fig. 2.5. N-channel, saturated conduction.

peter out somewhere between S and D. At the channel end, V_{CS} approaches zero, indicating a channel end voltage V_{CHE} (with respect to S) of:

$$
\begin{aligned}
V_{CHE} &= V_{GS} - V_T \qquad (2.2)\\
&= \text{voltage drop along channel}\\
&= 8 - 2 = 6\ \text{V}
\end{aligned}
$$

Note especially that V_{CHE} does *not* depend on V_{DS} provided that

$$V_{DS} > (V_{GS} - V_T) \tag{2.3}$$

which is the condition necessary for this SATURATED mode of operation.

It may seem surprising that current still flows in this condition. In fact the distance L' from the channel end to drain is simply *the portion of the width of the drain depletion region which corresponds to the voltage span that maintains it,* namely $(V_{DS} - V_{CHE})$ or $(10-6)$, i.e. 4 V in the present example. Accordingly, L' varies in response to changes in $(V_{DS} - V_{CHE})$. The channel injects carriers (electrons for the n-channel device) into the L' region *which are swept across this depletion region to the drain.* This feature is the same as in bipolar transistors where carriers are injected from an emitter and subsequently swept across a collector depletion layer. Both are examples of carriers 'falling downhill', the point being that current is determined by the injection level and not by the drain (or collector) potential.

It follows that, *in the saturated mode,* V_{DS} *modulates channel length but not channel voltage drop or current.* An increase in V_{DS} causes L' to increase and channel length $(L-L')$ to decrease but I_{DS} is unchanged. (A more detailed account shows that I_{DS} changes a little also.) *Gate voltage* V_{GS} *controls both channel voltage drop, its current* I_{DS}, *and length.* An increase in V_{GS} increases V_{CHE} and I_{DS} and channel length and reduces L'.

Figure 2.6 incorporates both modes of operation. Curves are shown for two values of gate voltage. The horizontal segments illustrate how large V_{DS} causes saturated operation and consequent loss of control over I_{DS}. For low V_{DS}, on the other hand, the (I_{DS}, V_{DS}) graph is linear for a given value of V_{GS}. It follows that, in these circumstances, the channel is behaving as a simple resistance. The value of the resistance is controlled by V_{GS} such that resistance falls for increasing gate potential.

Once again, the above discussion is fully applicable to p-channel working if we regard the minus sign as implied rather than stated explicitly. There is

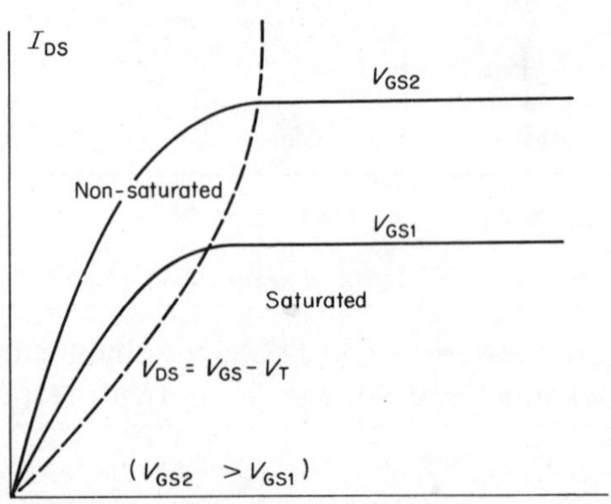

Fig. 2.6. Drain characteristics.

one real difference: n-type channels benefit from the higher mobility of electrons as compared with holes. This increases device transconductance.

THE INVERTER CIRCUIT

MOS circuits are used mostly as digital elements of which the basic building block is the inverter circuit B = A′ (Fig. 2.7).

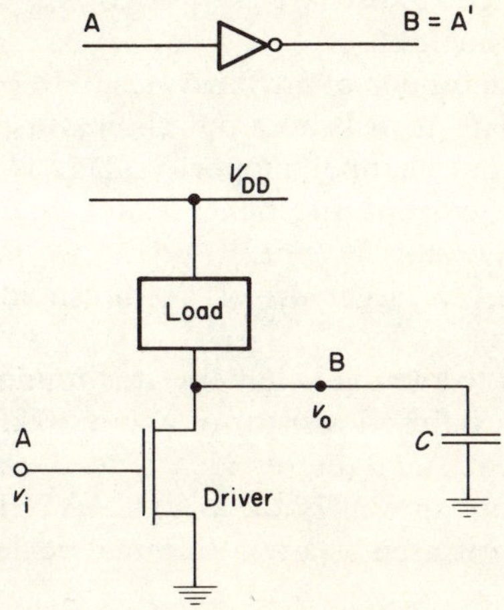

Fig. 2.7. An inverter.

The LOAD device is, in effect, a resistive pull-up. For the moment, assume that it is a diffused resistor of value R ohms. Note that there is no d.c. input current; the coupling is purely capacitive. This is a unique feature of MOS circuitry and it allows modes of operation which are not possible with bipolar devices. Since, in nearly all cases, the inverter drives another similar logic element, it follows that the load connected to the output point B is a capacitive one. In fact, the major component of capacitance C is the gate capacitance at the input to the next circuit.

When v_i is HIGH (at or close to V_{DD}), the DRIVER device is turned hard ON and v_o is at a low level V_{OL} close to ground. If the driver ON resistance is R_{on}, then $V_{OL} = V_{DD}R_{on}/(R_{on}+R)$. This is generally less than $0.2V_{DD}$.

If the input is suddenly switched to a LOW state, the driver turns OFF and the output node capacitance C charges through the load device R with a time constant RC. To ensure rapid switching, the value of R must be kept low. There is a conflicting requirement, however; when the driver is ON, a large potential $(V_{DD}-V_{OL})$ appears across R and causes a steady-state power dissipation of approximately V_{DD}^2/R watts (assuming V_{OL} is small). Thus, a high R value is required for low dissipation. A suitable compromise position yields R values of around 50 kΩ.

The driver, too, must meet certain requirements. In particular, the LOW output level V_{OL} must be sufficiently low to ensure that the driver of the next stage is turned OFF, i.e. $V_{OL} < V_T$, the threshold voltage. If we take it that $V_T = 0.2V_{DD}$, then it is necessary that $R_{on} < R/4$. Thus, the current-carrying ability of the driver in the ON state must be very much greater (by at least a factor of 4) than that of the load element. This may require a rather large driver. Because of its relatively large size, the driver is not a critical element with regard to switching speed. When v_i goes HIGH, the output-node capacitance C discharges through R_{on} about four times faster than it can charge through R!

Modern processing methods allow fairly accurate control of the threshold V_T and it is appropriate to ask how the choice of $V_T = 0.2V_{DD}$ (approximately) arises. Note that channel support voltage is $(V_{GS} - V_T)$ and thus a low V_T leads to more current and hence faster operation for given supply levels. This favours lowering V_T but, if we do, we must lower V_{OL} as well and that calls for an even larger driver. A threshold of about $0.2V_{DD}$ is a good compromise here.

Returning attention to the load element, it is found to be rather impractical to attempt to use a diffused resistor of about 50 kΩ because it takes up a great deal of chip area. A far better idea is to use another transistor as a load device. A frequent option is the SATURATED LOAD as shown in Fig. 2.8. The transistor is used as a two-terminal device with $V_{GS} = V_{DS}$. The

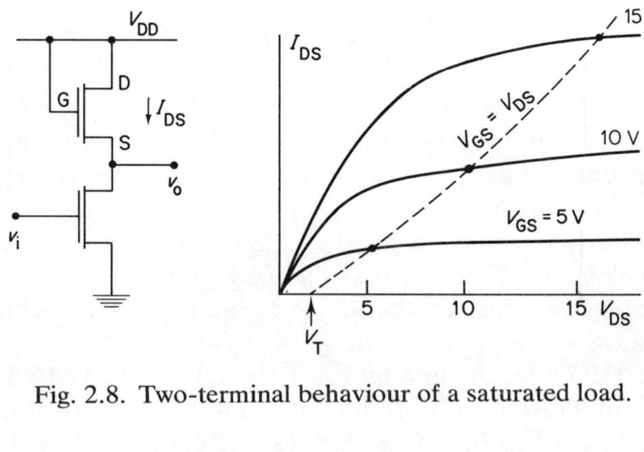

Fig. 2.8. Two-terminal behaviour of a saturated load.

resulting two-terminal characteristics are not very unlike those of a resistor. Non-linearity is not important. High values of effective load resistance are easily built on a small chip area in this manner. The method has an important limitation: v_o reaches a HIGH value of $(V_{DD} - V_T)$ rather than V_{DD} when the driver is OFF. This is because load operation is inherently saturated (see Fig. 2.9):

$$V_{CHE} = V_G - V_T = V_{DD} - V_T \tag{2.4}$$

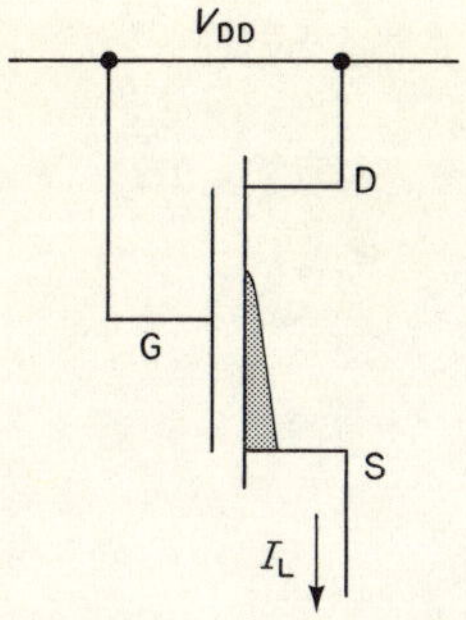

Fig. 2.9. Saturated load.

using ground reference. This gives a drop of V_T volts from drain to channel end for any value of I_L! Thus, the load operates with a fixed channel length. Channel thickness and voltage drop in the channel vary with I_L which also depends on the driver transistor.

The manner in which the saturated load reaches a maximum of $(V_{DD} - V_T)$ at the source node can be visualized as follows: the potential at S rises while the device supplies a current I_L to charge the node capacitance; when the voltage at S reaches $(V_{DD} - V_T)$, there is no channel support voltage left, the channel disappears, I_L becomes zero, and so the node voltage can climb no further.

The upper limit on v_o of $(V_{DD} - V_T)$ calls for a higher supply voltage V_{DD}. This may be undesirable per se (e.g. in wrist-watch applications) and it also increases ON-state dissipation (V_{DD}^2/R). These difficulties are overcome by biasing the load device in the NON-SATURATED condition, and, as a bonus, this also results in faster transient response. The price paid is a separate gate supply V_{GG} and, for non-saturated operation:

$$(V_{GG} - V_T) > V_{DD} \tag{2.5}$$

When this condition is met, v_o has a HIGH value of V_{DD}. The LOW value of v_o is set, as before, by the ratio of driver ON resistance to load resistance.

A better option than either of the above is to use a depletion mode device as the load element. In earlier technologies this was not possible because it called for different thresholds for different devices and, in those times, thresholds were set uniformly for all devices by the choice of process constants. More recently, the method of ion implantation (to be described later) allowed the thresholds of selected devices to be adjusted while others were held constant. The inverter with depletion mode load is shown in Fig. 2.10).

The load device is connected with $V_{GS} = 0$. Let us suppose that the circuits are n channel with a normal threshold V_T of +2 V and that the load devices were implanted so as to shift their thresholds by 8 V in the negative direction. This gives rise to depletion mode loads having $V_T = +2 - 8 = -6$ V. This means that, even when $V_{GS} = 0$, they are turned ON and have the equivalent

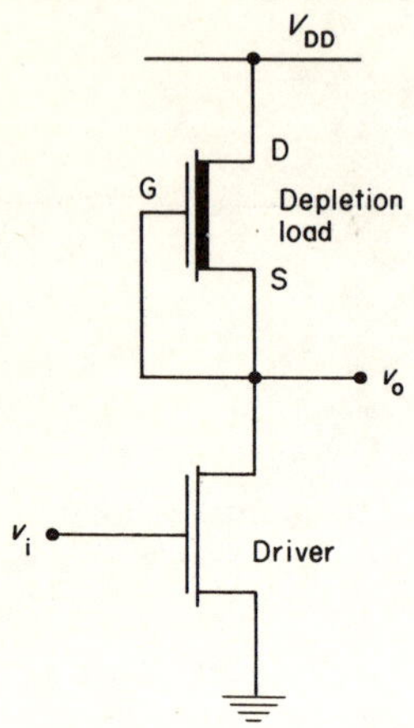

Fig. 2.10. Depletion load.

of 6 V of channel support. The result is that the depletion load is able to supply an almost constant current (as a function of v_o) while charging the output node. This was not true for the saturated load situation; as v_o rose, channel support diminished and the ability to supply current fell accordingly. Also in favour of the depletion load is the fact that v_o can rise as far as V_{DD} without the necessity of a separate gate supply.

Depletion mode devices will be distinguished, as in Fig. 2.10, by the heavy black line linking the source and drain and symbolizing the NORMALLY ON condition.

BUFFER CIRCUITS

The inverter circuits just discussed were seen to be quite asymmetric in their action: a LOW-to-HIGH transition on the output took many times longer than a HIGH-to-LOW transition because of the very different resistances of the driver and load. Also, the transition times are directly proportional to the size of the capacitive load. This means that if two devices are driven by an inverter rather than one (a fan-out of two), the delay is doubled. In practice, large fan-outs are quite common and the corresponding delays for a single inverter stage are unacceptably large. In such situations, special buffer stages are required.

The double-buffer circuit shown in Fig. 2.11 is conducive to much more symmetric switching times. The first inverter stage is a conventional depletion load type; the second stage differs in that the gate of the load device is driven from the first inverter. When v_i is HIGH, v_o is LOW and so also is the gate of T_2 so that V_{GS} for T_2 is close to zero. This is the normal condition for a depletion load. When v_i goes LOW, the gate of T_2 quickly rises towards V_{DD} (since it represents a small load on the first inverter) and so the V_{GS} of T_2 goes strongly positive at first. This greatly increases the current-sourcing ability of T_2 so that, with proper design, its switching

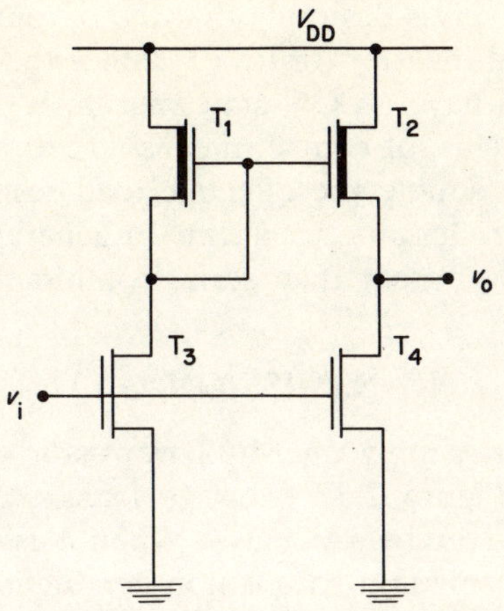

Fig. 2.11. Buffer with active pull-up.

performance approaches that of T_4. This kind of buffer is always desirable where large loads are encountered.

The buffer shown in Fig. 2.11 is an inverting type. It becomes a non-inverting buffer by simply interchanging the gate connections to T_2 and T_4.

COMBINATIONAL LOGIC CIRCUITS

Logic functions can be implemented very readily by replacing the driver transistor by series and parallel transistor combinations. This is illustrated in Fig. 2.12 where:

$$F = (A \cdot B + C)'$$

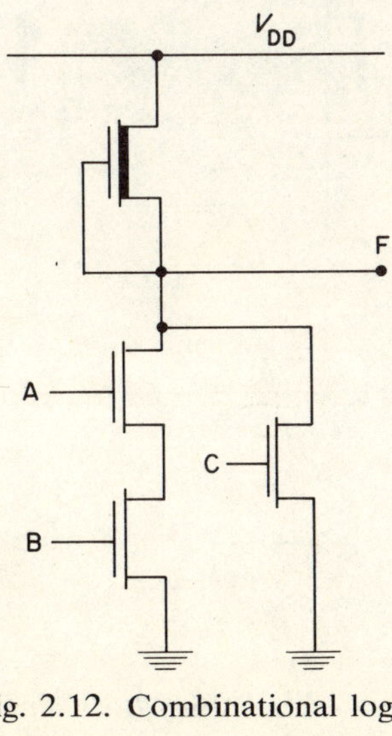

Fig. 2.12. Combinational logic.

In a circuit such as this, it is desirable that the output low voltage V_{OL} be as low as for the standard inverter (or lower). Since the A and B devices are in series, they must each have an ON-state resistance of $R_{on}/2$, half the value of a standard driver. This, of course, means larger devices. Alternatively, it may be acceptable to double the effective load resistance instead, but this option means a longer charging transient. In general, it is preferred, where possible, to use parallel rather than series combinations of driver devices.

TRANSMISSION GATES

A transmission gate is simply an MOS transistor used as a series switch linking two circuits. Figure 2.13 shows a transmission gate between two inverters. Its gate is driven by a signal ϕ. When ϕ is LOW (below the driver threshold V_T) the transmission gate is non-conducting and the inverters are isolated from one another.

When ϕ is HIGH (usually equal to V_{DD}), the transmission gate is conducting and behaves very much like an ideal closed switch because its resistance is low compared to the input impedance of the second inverter. There is one difference, however: for a high value of ϕ equal to V_{DD}, the gate of the second inverter driver can rise no higher than $(V_{DD} - V_T)$. This is not a problem as long as the available gate drive is sufficient to take the output V_{OL} below V_T. It is often convenient to simplify the schematic as shown by replacing each inverter with its logic symbol.

MOS/LSI circuits frequently use an arrangement of two or more *clock*

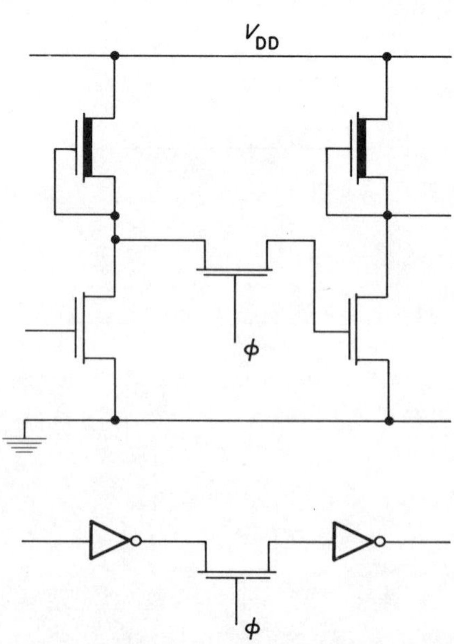

Fig. 2.13. A transmission gate.

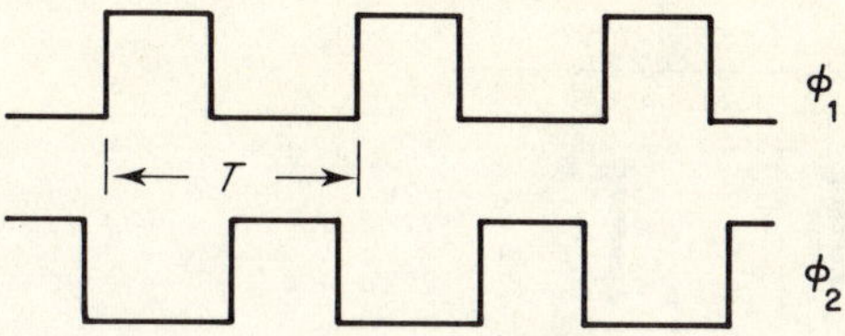

Fig. 2.14. Two-phase clocks.

phases to provide system timing. There are many such schemes and they will be described in detail later. In this section, we will look at some of the possibilities that arise when we use a 'two-phase non-overlapping' clock scheme. The phases are shown in Fig. 2.14 and their LOW and HIGH levels are 0 and V_{DD} respectively. The non-overlapping condition means that the phases are never high simultaneously.

Now consider the circuit of Fig. 2.15. When ϕ_1 is high (or 'during ϕ_1 time'), the logic level at A is inverted and moved to B. Subsequently, during ϕ_2 time, the level at B is inverted and moved to C. Although ϕ_1 is off during this interval, the input voltage on the gate of the central inverter is maintained capacitively for the duration until ϕ_1 again goes high. In this way, the information originally appearing at A is transferred to the right and appears at C after one period, *T* seconds, of the clock has elapsed. There is an information delay of one period or 'one bit-time' from A to C and a delay of *T*/2 or 'one-half bit-time' across each inverter. It is clear that a long string of inverters, with alternate ϕ_1 and ϕ_2 driven gates separating them, will function as a shift register. This is the first step on the road to designing synchronous digital circuitry.

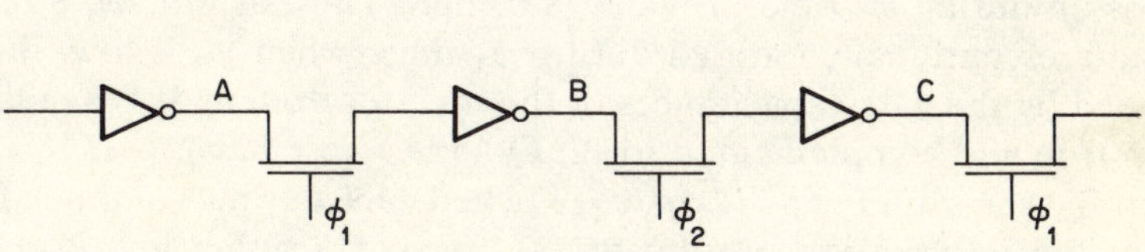

Fig. 2.15. A shift register.

Notice that the term 'HIGH' generally refers to a level at or close to V_{DD}. In the case of n-channel circuitry this will be a positive voltage whereas for p-channel circuits it will be negative. Actual levels vary from about 1.5 V to over 20 V.

MEMORY ELEMENTS

It is entirely possible to construct an MOS memory element in the form of two cross-coupled inverters, just as is done in bipolar memory circuits. Figure 2.16 shows such an element. This kind of memory is often chosen because it has the advantage of being *static*, i.e. it retains its present state

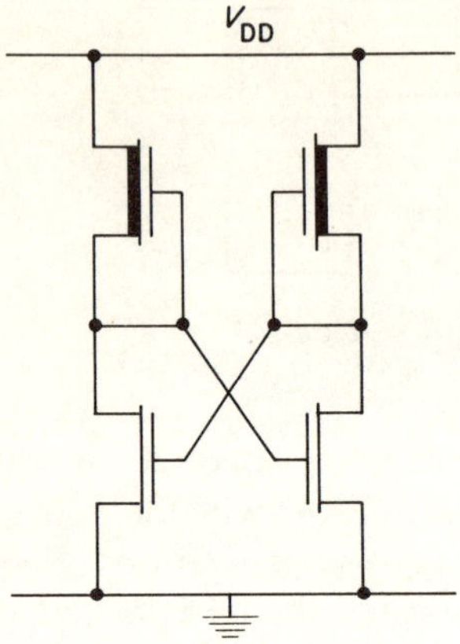

Fig. 2.16. Static memory element.

indefinitely without intervention, as long as power is on. It has the disadvantage that it takes up a considerable chip area.

MOS technologies permit a much simpler concept of information storage to be implemented: because the transistor gates are isolated electrically, they can retain a stored charge for an appreciable time (several milliseconds). The stored charge is used to represent the logic state and the method works very well provided that any charge which leaks away can be periodically replaced. This operation is known as 'memory refresh' and is characteristic of *dynamic memory circuits.*

We can use our two-phase logic mechanism to build a dynamic memory element. An example is shown in Fig. 2.17. Data entry is controlled by the strobe S. The transmission gate T_1 is controlled by ($\phi_1 \cdot S$) meaning ϕ_1 AND S. It turns on during ϕ_1 time provided S is high. Thus, as long as S remains high, data are continually entered during ϕ_1 time; when ϕ_1 is low, the data are retained by the gate capacitances of the inverter drivers. It is available at output in true and complemented form, Q and Q′ respectively. If S is taken low, then T_1 stays OFF and T_2 gets switched ON during ϕ_1 time. In this condition, the information present at the time of S going low is retained

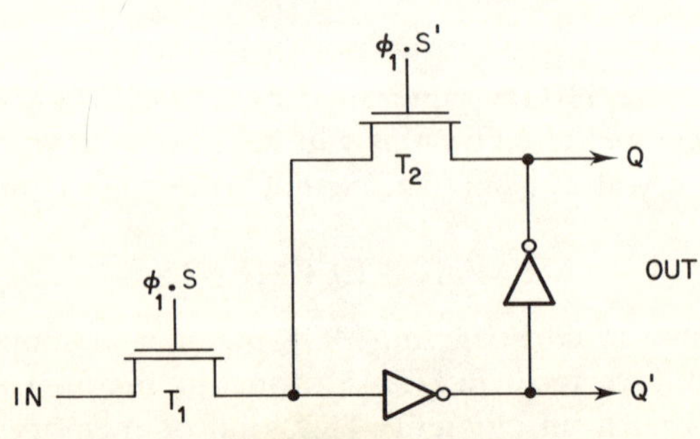

Fig. 2.17. Dynamic memory element.

indefinitely. This is because the input state is re-defined by the output Q every time that T_2 turns on. Thus, if a '1' is stored at the first inverter input, the gate voltage will tend to fall because of leakage but it will get 'topped-up' whenever T_2 goes on. It is intuitively clear that if the refreshing does not occur at sufficiently frequent intervals, then the element will switch to the '0' state and stay there. Note that a '0' state at an inverter input can remain indefinitely because the capacitance is then uncharged.

The output of the dynamic memory element is such that it may change state during ϕ_1 time but not during ϕ_2 time. This is generally stated as 'data is valid during ϕ_2 time'. The circuitry immediately following this would normally be clocked by ϕ_2 for just that reason. In fact it is normal for data to pass through logic elements which are alternately gated by ϕ_1 and then ϕ_2 just as was done for the simple shift register.

The space-saving feature of dynamic memory is not very clear from Figs. 2.16 and 2.17. It will become clear when memories are discussed in detail in Chapter 11 and can be anticipated from the simple fact that the gate capacitance *is* the memory element and all other circuitry is of a peripheral nature.

RATIOED VERSUS RATIOLESS LOGIC

All of the circuits presented so far are of the ratioed type. This means that the output LOW level of an inverter is determined by the *ratio* of the resistances of the driver and load device when the input is HIGH. It was seen previously that ratios R/R_{on} of 4:1 or higher are desirable and that large ratios call for more chip area per inverter. This provides an incentive to devise a ratioless logic scheme which could use minimum-size devices for both driver and load. Such schemes are indeed possible and one is illustrated in Fig. 2.18. It shows inverters in cascade, coupled by transmission gates.

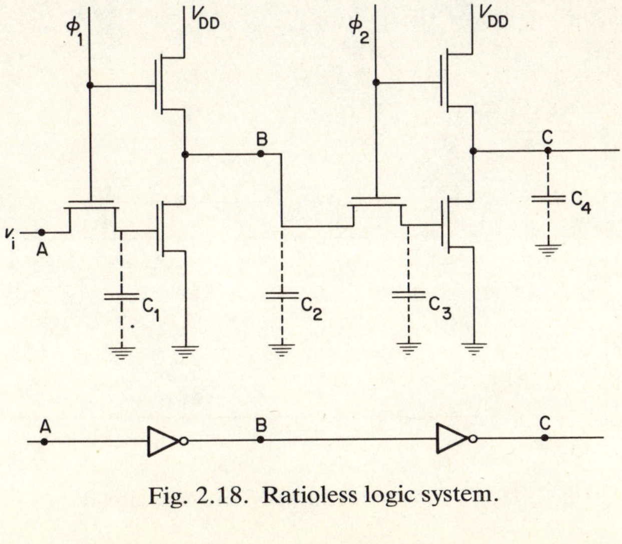

Fig. 2.18. Ratioless logic system.

The difference is that the loads are enhancement mode devices which are switched ON and OFF by the two-phase clock lines. Let us suppose that point A supports a '1' (or a HIGH) level just prior to ϕ_1 time. During ϕ_1 time, the HIGH level is admitted to turn on the first driver and discharge C_2. The load also turns ON. Because these are minimum-geometry devices, the voltage at B falls but not as low as is necessary to represent a logic '0'. It is after ϕ_1 goes low and the load device is switched OFF that point B is free to fall to a '0' level. In fact, B discharges fully to ground level. During ϕ_2 time, the '0' on B is inverted and transferred to C. Because this driver is OFF, the output is free to rise as far as $(\phi_2 - V_T)$ or as far as V_{DD}, whichever is the lower.

This ratioless system offers another benefit. Because the load transistors are OFF for over 50 per cent of the time, the overall power dissipation is greatly reduced.

THE CMOS INVERTER

The conventional MOS inverter dissipates substantial load power when the driver is ON. The complementary MOS process uses p-channel and n-channel devices on the same wafer, as Fig. 2.19 illustrates. Note the grounded p-tub and the bulk connected to the positive V_{CC} supply. All potentials (including v_i and v_o) are *negative* with respect to V_{cc} which can be thought of as the reference level for p-channel transistors. Similarly, the GND line serves as reference for the n-channel transistors.

A 'high' v_i simultaneously turns T_1 OFF and T_2 ON giving a low v_o. With v_i low, T_1 is ON and T_2 is OFF giving high v_o. Since one device is always OFF except during the transitions, the steady-state power dissipation is zero. Thus, except at frequencies over approx. 100 kHz, CMOS offers low power dissipation advantages. A disadvantage is lower packing density due to the large space requirements of the p-tub.

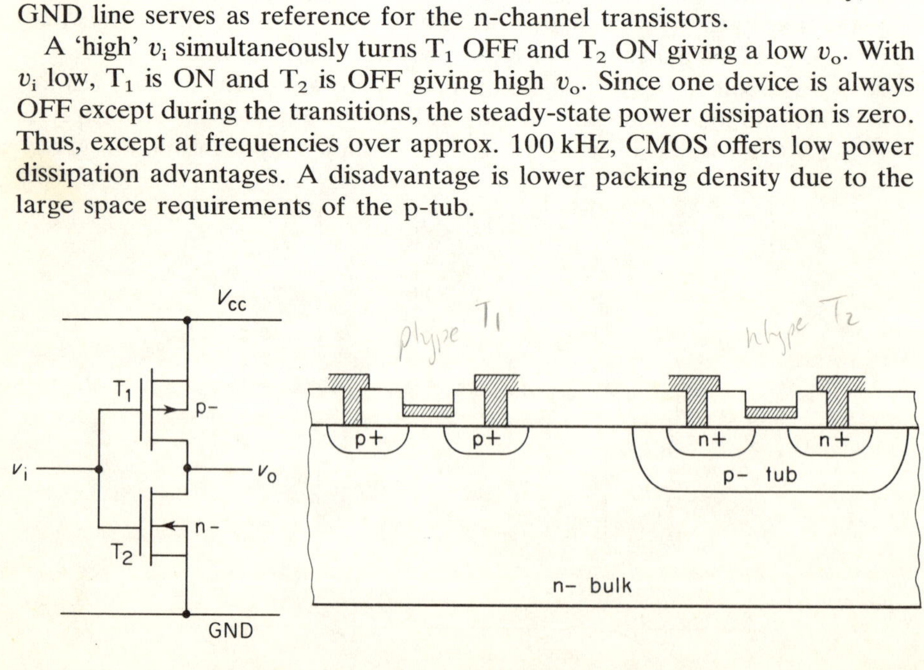

Fig. 2.19. Complementary MOS inverter circuit.

It is hoped that this chapter has helped to demonstrate the suitability of MOS circuits for implementing logical systems. The problems of building a high-performance reliable technology are nonetheless considerable and the next few chapters are devoted to this objective. In Chapters 3 and 4, the relationship between device properties and fabrication steps is thoroughly explored and this is followed by in-depth characterization of circuits fabricated using the well-established fabrication technologies. After device characterization is complete, there is a return in Chapter 10 to application topics. These include some treatment of analog MOS circuits as well as digitally oriented work on the design of memories and programmable logic arrays and other systems-oriented topics.

EXAMPLE 2.1

Are the devices shown saturated or otherwise?

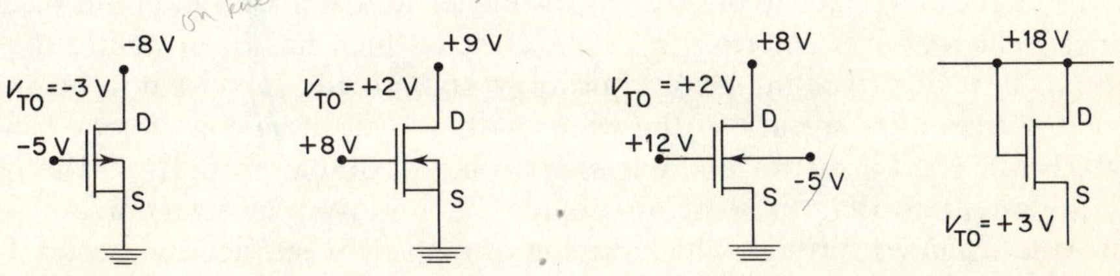

Device A   Device B   Device C   Device D

Solution

A: p channel with grounded substrate. Saturated since $8>(5-3)$; no channel support near drain.

B: n channel with grounded substrate. Saturated since $9>(8-2)$.

C: n channel with substrate bias. Non-saturated since $(12-2)>8$ giving 2 V of channel support at drain.

D: inherently saturated with channel end voltage fixed at $(18-3)$ or 15 V.

CHAPTER 3

Device Fabrication I: *Chemical Processes*

The planar process of integrated circuit fabrication was introduced in 1960. It was the beginning of a major revolution of the electronics industry and the basis for today's large-scale integrated (LSI) circuits. After a preliminary sketch of planar processing steps, we will look at the chemical processes on which fabrication is based.

Circuits are fabricated on one surface of a circular silicon wafer. Dimensions of the wafer are, typically, 3 or 4 inches in diameter and about 0.01 in thick. The wafer is cut from a single crystal which has been lightly doped with n-type or p-type impurities during crystal growth. It is usual to begin by coating the entire surface of the wafer with a layer of silicon dioxide (SiO_2) which is a good insulator and has many other desirable properties. The next requirement is that the silicon dioxide be removed in selected areas to provide *windows* through which doping of the silicon surface can occur. This selective removal of the silicon dioxide is a *patterning* process and depends on photographic methods as will now be described.

The desired pattern is defined on a glass *masking plate* which has the size and shape of the wafer surface and fits snugly against it when in use. The masking plate is opaque in those areas where windows are to occur so that light is prevented from reaching the wafer surface. The wafer is first coated with a thin layer of *photoresist* material on top of the silicon dioxide. After a drying process, the masking plate is placed over the surface and this is followed by exposure to ultraviolet light for a short interval. The photoresist is an emulsion which polymerizes and hardens on exposure, and this happens in all but the window areas, which have been protected by opaque portions of the mask. In the *developing* process which follows, the photoresist in window areas is washed away to expose the silicon dioxide surface. Now the wafer is placed in an acid *etching solution* such as hydrofluoric acid (HF). The acid attacks and gradually removes the silicon dioxide in window areas; the photoresist is able to protect the silicon dioxide in other regions. After etching has proceeded to the full depth of the silicon dioxide layer, the wafer is transferred to a chemical bath which strips away the polymerized photoresist. The patterning process is now complete with windows existing in the silicon dioxide layer at the desired locations. This sequence of events completes the patterning of the silicon dioxide layer and is known as *photolithography*.

The semi-conductor p–n junctions are formed in window areas by heavy

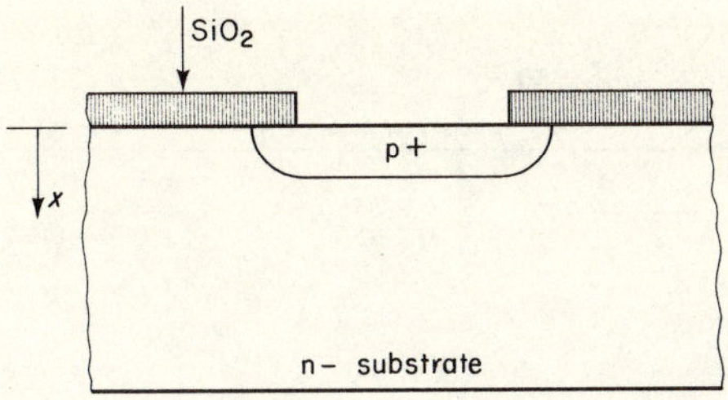

Fig. 3.1. A p diffusion.

doping of a type opposite to wafer doping. The doping ions are furnished from a carrier gas over the wafer's surface and penetrate the silicon by *diffusion.* The diffusion occurs only in window areas because the silicon dioxide (fortunately) serves as a mask against diffusing ions elsewhere (see Fig. 3.1).

This 'diffusion' is fundamentally similar to that described earlier in reference to electron and hole conduction. In this case, however, we are talking of atoms diffusing through a crystal structure. Since they move far less freely than electrons or holes (surely an understatement!), this kind of diffusion has a negligible effect at normal temperatures. At temperatures around 1000 °C, the lattice atoms are under intense thermal motion and diffusion of impurity atoms into the surface can proceed at a significant rate. Accordingly, the diffusion step occurs at precisely controlled high temperatures in special furnaces and causes simultaneous formation of p–n junctions all over the wafer's surface wherever windows have been opened. In fact, the desired integrated circuit is normally reproduced many times on the wafer's surface and this property of simultaneous fabrication is the major economic strength of planar processing.

These diffused layers of dopant material form the source and drain regions of MOS devices. Later, aluminium contacts will link diffused regions to one another and to the bonding pads. Gate contacts are simply aluminium overlay separated from the silicon surface by a very thin silicon dioxide layer. This layer is the dielectric material of the capacitance which links the gate to channel and controls current flow. Its thickness is critical and is usually about 1000 angstrom units, Å for short (which is 0.1 μm or 10^{-5} cm). The only way to achieve this kind of thickness with accuracy is to first remove all the oxide in the gate region completely and then to build up a new layer using a slow process, typically a growth rate of 1000 Å per hour. The result can be seen (somewhat simplified) in Fig. 3.2. Note that this required a second masking operation to define all the gate areas. Note, too, that gates must extend fully from the source to drain in all cases and a small overlap is used to allow for errors.

The entire surface of the wafer is now covered with a layer of aluminium. This is called metallization. Then another photolithographic operation

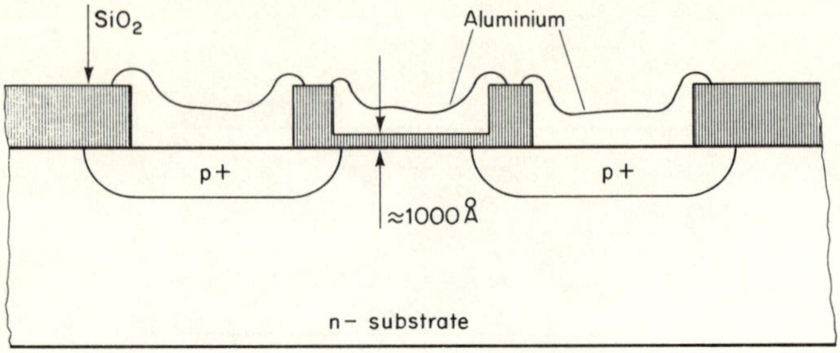

Fig. 3.2. A p-channel device.

(another masking step) is used to remove the aluminium by etching wherever it is not required. It is protected from the etch by photoresist over all device contacts and interconnects. The photoresist is now stripped away and a layer of oxide (i.e. silicon dioxide) is deposited over the entire surface. Its purpose, on this occasion, is simply to protect the wafer from contaminants. Finally, another masking operation is required to remove the oxide over the metal bonding pads by which each integrated circuit is connected, via fine gold wires, to the external package pins.

What has just been presented is a very much simplified picture of 'metal gate thick oxide' fabrication. It has served to introduce the purpose of several important processes which will now be described more fully.

WAFER PREPARATION

The basic raw material for extraction of silicon is silicon dioxide (sand). Initial separation is by reaction with carbon in:

$$SiO_2 + C \rightarrow Si + CO_2$$

After further purification steps, the silicon is available as polycrystalline rods. It is now necessary to heat these rods to their melting point (which is 1415 °C) to provide a molten reservoir of silicon in a quartz crucible. A seed crystal is brought in contact with the melt and gradually withdrawn, rotating all the time. As it is withdrawn, molten silicon cools and 'freezes' on to it while automatically assuming the structure of the seed crystal. This crystal-pulling operation ensures that, as the silicon is drawn from the melt and solidifies, its crystal structure is an extension of that which has already cooled. The final result is a cylindrical silicon bar which has taken its structure from the seed crystal at one end and maintained it throughout. The bar is now sawn into thin slices which, after polishing, become the wafers on which integrated circuits are formed.

Crystal *orientation* is an important parameter for both mechanical and electrical reasons. Using Miller indices, the (111) plane has an orientation such as to pass through coordinates (1, 1, 1) in Fig. 3.3. Cuts along this plane

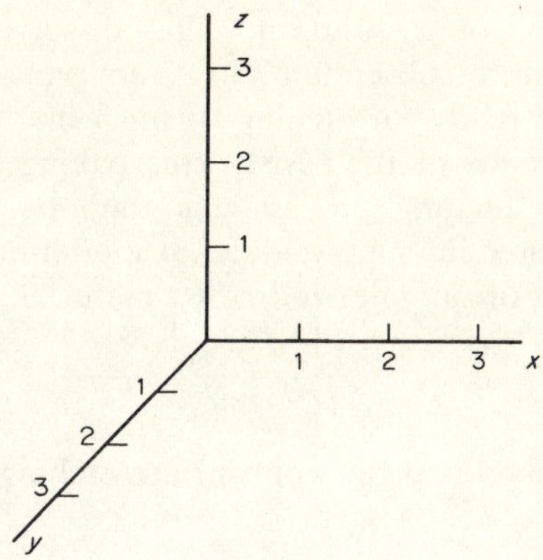

Fig. 3.3. Miller indices.

expose a triangular surface form. The (100) plane is parallel to the yz plane and intersects the x axis at $x = 1$. A rectangular pattern characterizes cuts along this plane. The plane presented on the wafer surface was determined by the seed crystal and both (111) and (100) are of importance in present practice.

The electrical properties of the wafer were determined while the silicon was in the molten state by adding suitable dopants to the melt. It can be considered as a substrate (or 'bulk') material of uniform doping concentration, N_B atoms per cubic centimetre.

PHOTOMASKING

The photomasking operation is more complex than was previously described. The photoresist itself must adhere well to the surface, must serve as an effective mask against the etching solution, and must provide clean, sharp boundaries so that the resolution, the minimum usable line width, is acceptable.

Photoresist is usually applied by a *spinning* process. A small amount is placed on the centre of the wafer and then the wafer is caused to spin at several thousand revolutions per minute. Because it is fairly viscous, the resist moves outwards in a controlled manner and forms a uniform coating on the wafer. Correct spin speed is of critical importance. This operation is followed by a *soft-bake* to remove excess solvent. The bake may involve passing hot air over the wafer or the use of an infrared heat source.

After baking, the wafer is ready to meet the masking plate. A contact printing method is employed. Precise alignment of the masking plate and wafer is achieved in a mechanical assembly with the aid of a microscope and then the exposure (using a high-intensity ultraviolet source) takes place with the mask pressed against the wafer surface. The wafer is now 'developed' by immersion in the developer solution. This dissolves the resist in unpolymerized regions and a subsequent rinse removes residual matter.

The wafer is not yet ready for etching. It must first undergo a *hard-bake* which improves adherence of the resist. This baking step is similar to the earlier soft-bake. The etching process can then be used to remove the surface layer (silicon dioxide or polysilicon or aluminium) wherever it is not protected by a coating of polymerized resist material.

ETCHING

The material to be etched and the appropriate etching solutions are mainly these:

Silicon dioxide, SiO_2	– buffered hydrofluoric acid, HF
Polysilicon	– HF
Silicon nitride, Si_3N_4	– boiling phosphoric acid
Aluminium	– phosphoric acid

The process is simply one of immersion for a suitable interval at an appropriate temperature. Polysilicon is polycrystalline silicon and serves as a conducting medium in some MOS processes. Silicon nitride has properties similar in many ways to silicon dioxide and offers superior passivation properties, i.e. better protection against contaminating ions. It often appears as a passivating layer, sandwiched between two layers of silicon dioxide. The top layer of silicon dioxide has a purpose relating to patterning operations as follows: silicon nitride can only be etched using boiling phosphoric acid. Ordinary photo-resist does not protect against this but silicon dioxide does. Therefore the nitride layer is covered with a thin layer (about 1500 Å) of silicon dioxide. A photoresist mask is used to make a window in the silicon dioxide in the usual way. Then the silicon dioxide serves as a mask while the phosphoric acid etches the nitride layer under the window.

OXIDATION

The protective layer of silicon dioxide is normally *grown* on the silicon surface by oxidation at high temperature in one of the following reactions:

$$Si + O_2 \rightarrow SiO_2 \qquad \text{(dry oxidation)}$$

$$Si + 2H_2O \rightarrow SiO_2 + 2H_2 \qquad \text{(wet oxidation)}$$

The oxidizing ambient is therefore either dry oxygen or water vapour. In both cases, the oxidizing species must pass through any existing silicon dioxide layer and react with silicon at the silicon dioxide–silicon interface. Thus, the silicon for the reaction comes from the wafer itself and, as the reaction proceeds, the silicion–silicon dioxide interface moves down into the wafer surface. About 42 per cent of the final layer thickness is below the original silicon surface.

The growth of silicon dioxide as described above is referred to as 'thermal oxidation' and takes place in a furnace with temperatures in the range 900–1200 °C and with very tight temperature control (less than 1° variation is typical). The furnace comprises a quartz tube surrounded by heating coils. In a dry oxidation, oxygen of very high purity is passed over the wafers. In the case of wet oxidation, the water vapour comes from a 'bubbler'. This is simply a water vessel maintained at a temperature just below boiling point. A stream of oxygen is admitted at the bottom of the vessel and it bubbles up through the water to become saturated with water vapour before it enters the oxidation furnace.

Both oxidation methods have their features and both are used in different phases of a process. Their growth rates are given in graphical form in Appendix I and it is evident at a glance that wet oxidation is several times the faster of the two. It has a disadvantage in that it releases hydrogen gas which quickly diffuses away through the oxide and, in so doing, makes the oxide somewhat weaker and more porous. Dry oxidation produces a denser oxide. Very often, an oxide growth phase begins with a dry oxidation, transfers to a wet process for the sake of rapid growth, and finishes with another dry oxidation. In the growth of a gate dielectric, the need for a thin layer, with precise thickness control, clearly favours a dry oxidation process.

DIFFUSION

Diffusion takes place in a furnace similar to the oxidation furnace and at similar temperatures. A two-step diffusion is usual. The first step is called *predeposition.* Its function is to introduce a precise amount Q atoms per square centimetre of dopant into the silicon surface but, usually, with only slight penetration into the wafer. The second step is called the *drive-in* and it serves to redistribute the dopant Q; no new dopant is introduced but the existing impurities penetrate to a greater depth and the surface concentration falls. The drive-in is performed with temperature and duration as the controlling parameters so as to achieve the desired electrical characteristics. Electrical performance is heavily dependent on the diffusion profile and a full quantitative treatment is undertaken in the next chapter. The term 'profile' refers to the variation of doping ion concentration, $N(x)$ atoms per cubic centimetre, as a function of depth x into the semi-conductor. Two key parameters of the final junction profile are the surface concentration N_s and the junction depth x_j (in micrometres). Values of x_j in the range 1–10 μm are typical.

The term 'infinite source' is often used in reference to the predeposition step and means simply that the supply of doping material on the surface is sufficiently 'in excess' so that its availability does not limit the diffusion process. The dopant reaches the wafers in a carrier gas at high temperature and forms a deposit on the surface. There is a limit on maximum concentration (atoms/cm^3) that the dopant atoms can achieve in the silicon. This is known as the *solid solubility limit* and, for a given dopant species, depends only on temperature. The solid solubility limit versus temperature is shown in Appendix E for several dopant types. This limit is utilized during predeposition as a convenient and predictable constant concentration at the silicon surface. The concentration, of course, falls with depth and, in general, penetration increases with time, as suggested by Fig. 3.4. The dopant source is often a powder which is heated and transported in the carrier gas (usually nitrogen) to the furnace. Alternatively, a liquid source and a bubbler may be used (as in the case of wet oxidation).

Sometimes a rather different method is used: the wafer is coated with a doped layer of silicon dioxide and, during 'predeposition', the dopant diffuses into the silicon. The coating method is either by chemical vapour deposition (CVD) or by a spinning procedure like that used with photoresist. Whatever the method employed, the predeposition is always followed by an etching operation which removes all surplus dopant from the wafer surface.

The drive-in which follows is no more than a spreading of the dopant from regions of high to regions of low concentration at a suitably high temperature. However, it is usual to allow this to happen while simultaneously re-growing a new layer of silicon dioxide over the wafer. This seals the diffused regions from contamination as well as providing an insulating layer. Note, too, that any subsequent high-temperature process will contribute to the drive-in, regardless of its primary purpose, and it must be considered when the diffusion parameters are first chosen. The rate of drive-in varies a lot from one dopant species to another as determined by the relevant diffusion coefficient D. Appendix G shows D (square centimetres per second) versus temperature for common dopants. As would be expected, the diffusion coefficient D increases rapidly with temperature.

The final diffused layer has an impurity concentration which varies by several orders of magnitude from top to bottom. As a result, the concept of a uniform resistivity is completely inappropriate. We do, however, speak of an average resistivity, $\bar{\rho}$, for the layer. Now consider a square section of diffused layer with side length l and depth x_j.

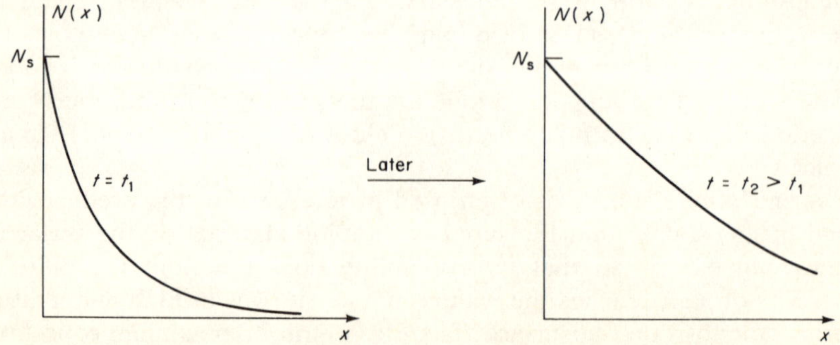

Fig. 3.4. Penetration of impurities during predeposition.

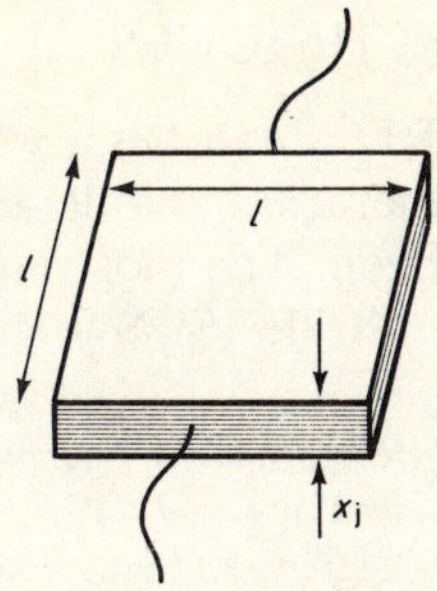

Fig. 3.5. Sheet resistance.

The resistance between end faces, as suggested by Fig. 3.5, is:

$$R = \bar{\rho}\frac{\text{length}}{\text{area}} = \bar{\rho}\frac{l}{lx_j} = \frac{\bar{\rho}}{x_j}$$

Note that it is independent of l; it is the same for any square and is known as the *sheet resistance* $R_{\square}$. Its dimensions are simply ohms, usually described as 'ohms per square' or simply $\Omega/\square$.

ION IMPLANTATION

Ion implantation is another method by which impurity atoms can be introduced to the silicon surface and is a very different approach from that of high-temperature diffusion. Surface penetration is achieved by accelerating the ions to very high energies, typically through voltages of 50 to 150 kV, before they pass, via suitable focusing and deflection systems, to impact the wafer surface at high velocities. The ions are extracted from a gaseous source and separated on the basis of their weight by magnetic deflection. The average depth of penetration is a function of the accelerating potential and the ion mass. Penetration depth varies in a statistical manner with a profile which is roughly Gaussian. It shows maximum concentration of implanted ions at the nominal penetration depth while concentration diminishes at higher and at lower depths. Because a certain amount of crystal damage occurs during implantation, it is followed by a high-temperature anneal to restore the original structure.

Ions can be implanted, to selected regions only, by the use of a mask. Silicon dioxide is an effective mask provided its thickness is several times the penetration depth. Metal overlay and even photoresist can also serve as effective masking materials. Ion implantation offers the great advantage of allowing small well-controlled doses to be applied to the surface of the wafer. It is also possible to implant through a thin layer of silicon dioxide as, for example, when used in gate regions for threshold adjustment. In spite of high equipment costs, ion implantation has become an extremely valuable tool of modern wafer processing. It can be used to achieve a variety of goals of which several examples appear in later chapters.

DEPOSITION TECHNIQUES

The term chemical vapour deposition (CVD) describes a variety of processes wherein a material is reacted at high temperature and deposited with the aid of a carrier gas such as hydrogen or nitrogen. The most important examples are deposition of silicon dioxide, silicon nitride (Si_3N_4), and polycrystalline silicon.

Silicon dioxide can be deposited via the reaction involving silane (SiH_4) and oxygen thus:

$$SiH_4 + 2O_2 \rightarrow SiO_2 + 2H_2O$$

using N_2 as a carrier gas at temperatures in the 200–500 °C range. Notice that this provides the option of depositing SiO_2 after metallization because the required temperature is sufficiently low to avoid melting of the aluminium (aluminium melts at 660 °C). Impurities can be included to produce a doped oxide if required, e.g. as a predeposition mechanism.

Deposition of silicon nitride can occur around 700 °C via the reaction:

$$3SiH_4 + 4NH_3 \rightarrow Si_3N_4 + 12H_2$$

to provide a dense dielectric offering excellent passivation properties.

Polysilicon is deposited via the reaction:

$$SiH_4 \rightarrow Si + 2H_2$$

at around 1000 °C in hydrogen. This is an important step in *silicon gate* technology where the gate metal is a polysilicon layer. In this context, the deposition is made on top of a thin oxide layer. The same reaction, performed on the silicon surface (suitably prepared), can bring about *epitaxial growth.* This means that the deposited silicon takes on the structure of the crystal. The method can accommodate uniform doping with p-type or n-type impurities introduced during the deposition phase. It is widely used in the fabrication of bipolar circuits.

The metallization step, i.e. the deposition of an aluminium layer, is normally performed by vacuum evaporation. A high vacuum provides a long mean-free path so that vaporized aluminium can radiate freely to provide the desired deposit on the wafer surface. Evaporation is accomplished either by filament heating or by electron-beam bombardment. The deposited aluminum must meet exacting requirements. A particular problem is *electromigration.* High current densities cause migration of the atoms of the metal and can cause open-circuit conditions and, hence, circuit failure. This problem gets more acute as device size gets smaller. The current density limit imposed by electromigration is a few milliamperes per square micrometer.

One other deposition technique that deserves mention is *cathode sputtering.* The apparatus uses high voltage to accelerate positive ions, generated in a glow discharge, towards the cathode which has a surface coating of the required deposition material. The ion bombardment causes release, or 'sputtering', of this material, some of which ends up as a deposit on the wafer. The rate of film growth by this method is rather low. Having reviewed the individual process steps, it is now appropriate to examine a few of the fabrication processes which have become important in modern device manufacture.

A THICK OXIDE ALUMINIUM GATE PROCESS

This kind of process achieved major importance in the seventies. While there are several variations, the basic process steps are shown in Fig. 3.6 and are as follows:

(a) The wafer surface is cleaned and a few thousand angstroms of silicon dioxide is grown over the entire surface to serve as a diffusion mask.
(b) The FIRST MASKING operation uses photoresist in a standard patterning procedure to define areas in which the oxide is to be etched. Silicon dioxide is then etched away wherever source or drain diffusions are to occur and where a diffused layer is to serve as a sub-metal interconnect medium.
(c) Junctions are formed by diffusion. A p-channel process is used for this illustration. Some re-oxidation may occur during drive-in.
(d) A thick oxide layer is grown or deposited. This layer needs to be around 15,000 Å in thickness. The reasons are electrical and will be discussed in Chapter 5 in reference to thresholding.
(e) The SECOND MASKING step is for etching down to the surface at terminal locations.
(f) A thin oxide is re-grown over the entire wafer but its intended use is in gate regions, to serve as a dielectric. Thickness is typically 1000–1300 Å. No mask is needed here.
(g) The THIRD MASKING step is for the removal of the thin oxide over the diffused layers wherever metal contacts are to be applied.
(h) Aluminium is deposited by vacuum evaporation over the entire surface. Then, the FOURTH MASKING step defines the regions where unwanted metal is to be etched. The metal must completely cover thin oxide areas in gate regions.

It is common practice to seal the surface with a passivating layer of silicon dioxide which is deposited at low temperature (so as to avoid melting the aluminium). When this is done, a FIFTH MASKING step is required to create windows in the oxide over the bonding pads. The processing of the wafer is then complete.

In the process just described, it is essential that the gate extends fully as far as the source and drain. Note that the gate is defined by a separate masking step after source and drain diffusions have been made and, therefore, errors in mask alignment must be considered. To allow for alignment

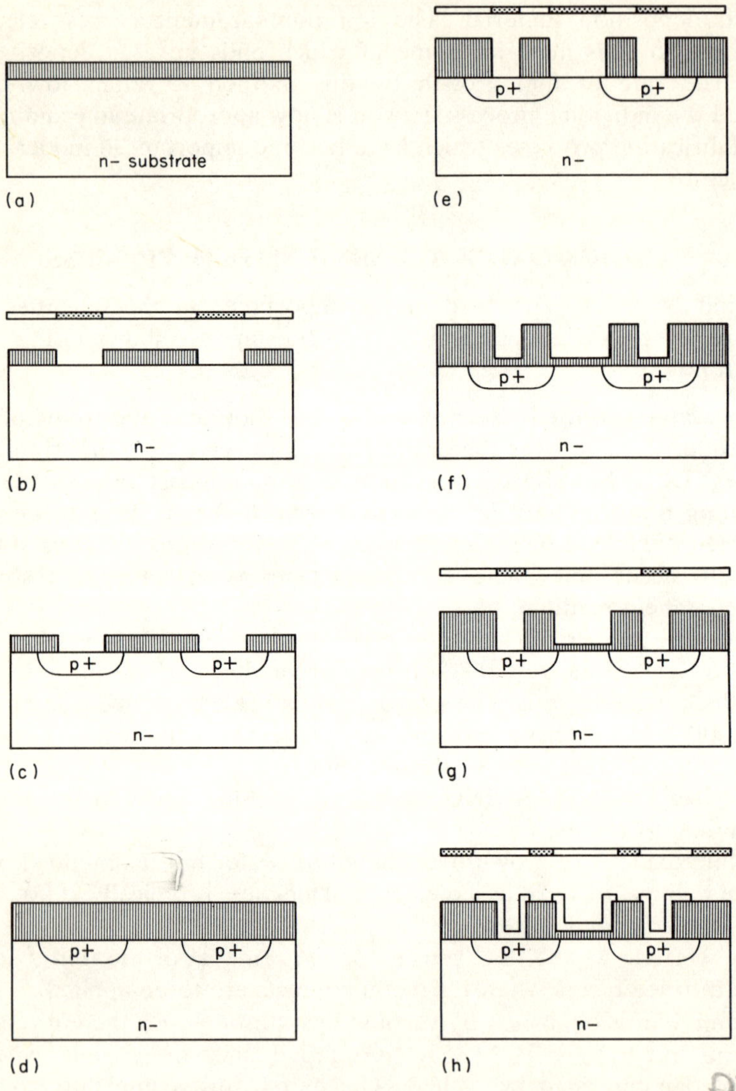

Fig. 3.6. Thick oxide metal gate processing.

error, the gate must deliberately overlap the diffused areas. We shall see that this is undesirable electrically and it also increases the area requirements a little.

A SILICON GATE PROCESS

This process has come into widespread use in more recent times and offers several advantages over the aluminium gate process. Some of the advantages relate to electrical performance and will be discussed in a later chapter. Figure 3.7 shows an n-channel process at various stages of completion and can be documented as follows:

(a) A silicon nitride (Si_3N_4) layer is grown on the silicon surface and selectively etched as defined by the FIRST MASK. In fact, the nitride remains in those areas which will later contain transistors. Two separate transistors are shown in the illustrations.

(b) The silicon is etched to about 1 μm depth in areas not covered by nitride. A boron deposition is performed to make the silicon surface

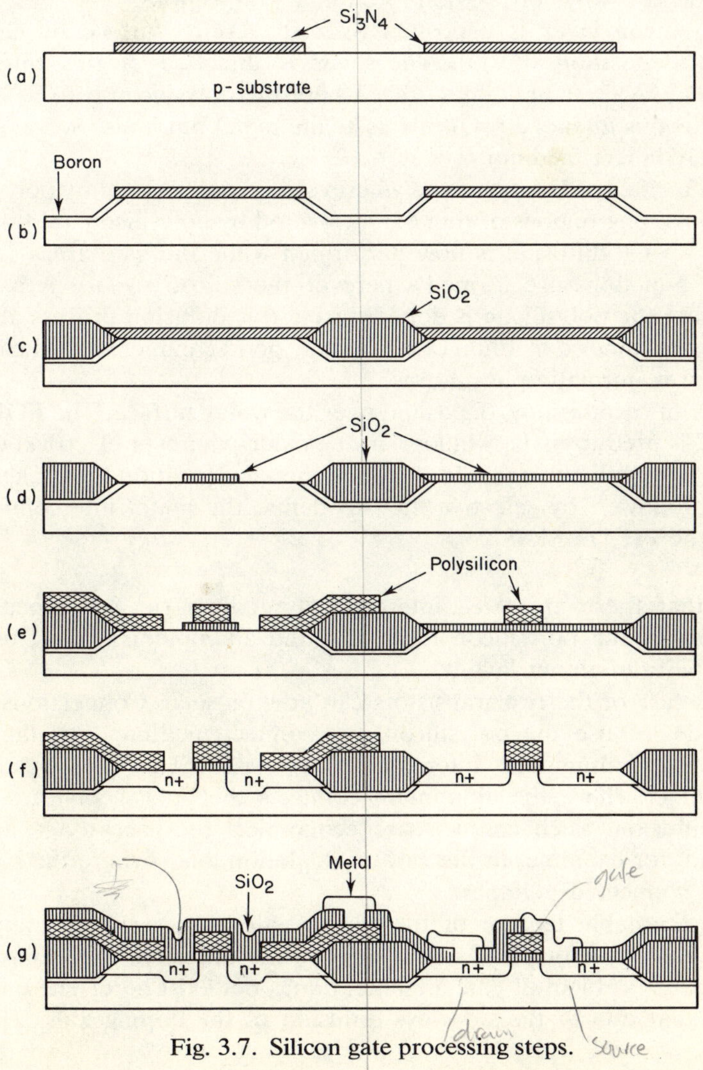

Fig. 3.7. Silicon gate processing steps.

more strongly p-type. (This is to prevent *channelling* and will be discussed in Chapter 5).

(c) A thick oxide (about 15,000 Å) is grown selectively in areas not covered with nitride. This will serve to separate adjacent devices.

(d) The nitride is now stripped and a thin gate oxide is grown over the entire surface. (Ion implant through the thin oxide can be used at this stage to achieve threshold voltage adjustment). The SECOND MASK defines areas where the thin oxide is to be stripped. This is done at locations where polysilicon will soon be deposited with the intention that it should be in electrical contact with the silicon below. From this point forward, the processing of the two transistors is different so as to illustrate two ways of providing terminal connections.

(e) A polysilicon layer is deposited over the entire surface by chemical vapour deposition (CVD). The layer is undoped at this time. The THIRD MASK then defines areas where the polysilicon is to be etched. Polysilicon is intended primarily as a gate metal but it also serves well as an interconnect medium.

(f) The thin gate oxide is removed wherever it is not covered by polysilicon. The remaining regions of thin oxide covered by polysilicon are transistor gates. An n+ diffusion is now performed while the gate areas serve as masks. Junctions are formed wherever the silicon surface is exposed. Note that the polysilicon is doped during this diffusion process also and thereby becomes a medium of reasonably good conductivity. The dopant material is normally phosphorus.

(g) A layer of oxide is now deposited over the wafer surface. The FOURTH MASK is used to make windows in the oxide wherever electrical contact with the polysilicon is required. Aluminium evaporation then takes place and is followed by selective etch to define the metal interconnections using the FIFTH MASK.

Notice that there are *three* interconnect mediums in this process: the diffused layers, the polysilicon level, and the aluminium level. This gives great flexibility in circuit layout.

The structure of the two transistors can now be seen. Connections to the left-hand device use the polysilicon interconnect medium. For the right-hand transistor, aluminium interconnects are used. This also illustrates a *butting contact* where the aluminium connects the polysilicon gate to the (source) diffusion. Such contacts are economical in space usage and are widely used, for example, in the case of depletion loads where the gate and source are connected together.

The most notable feature of the silicon gate process is the manner in which the gate structure serves as a mask while diffusions are being made. Therefore, the problem of gate alignment does not exist here. The only gate overlap is that due to the sideways diffusion of the doping ions. This is a self-aligning process. It results in higher packing density (more circuits per square millimetre) and superior electrical performance.

Some of the links between fabrication parameters and subsequent electrical properties have been hinted at in this chapter. The next chapter provides a full quantitative study of these links and thereby enables the designer to tailor the fabrication process to meet desired performance criteria.

CHAPTER 4

Device Fabrication II: *Engineering Aspects*

The final electrical properties of integrated devices are determined, in one way or another, during fabrication. This chapter provides a quantitative basis for control of fabrication processes to achieve desired results. Much of the information is in graphical form and is largely the result of investigations conducted in the early sixties at various locations, but notably at Bell Laboratories. The discussion includes oxide growth, the evaluation of diffused layers, and certain measurement techniques appropriate to this work.

GROWTH OF OXIDE LAYERS

The information necessary for growth rate calculations is, as already mentioned, contained in a pair of graphs in Appendix I. The calculations are quite straightforward in the case of single-layer growth. For example, a wet oxidation for 1 hour at 1200 °C grows 0.8 μm, i.e. 8000 Å of oxide. It should be noted, however, that growth does not proceed in direct proportion to the time available. This is because the reaction happens at the silicon–silicon dioxide interface. For thin layers, the oxygen or water molecules can diffuse rapidly through the oxide and the growth proceeds at a rate which is 'reaction limited'. For thick layers, the growth rate is limited by the rate of movement of oxidizing species through the oxide and the problem becomes increasingly more acute for thicker oxides. This effect renders it impractical to grow oxides to thicknesses in excess of about 1.5 μm (15,000 Å).

It was mentioned earlier that oxides are often grown in a few steps such as a dry/wet/dry sequence, for example. In such cases, the calculations for a growth phase over *existing* oxide must be made to take account of that oxide. This is done simply by treating the existing oxide as though it had just been deposited by the method now contemplated. An example will illustrate. Consider a two-step process:

1. Dry oxidation, 2 hours, 1100°
2. Wet oxidation, 0.1 hour, 1200°

Taken separately, these would grow 2000 and 2300 Å respectively, a total of 4300 Å but this result is incorrect. The first figure of 2000 Å is valid. If it had been grown in a wet phase at 1200°, it would have taken 0.07 hours. This is now followed by 0.1 hours, a total equivalent 'wet' time of 0.17

hours. From the graph, 0.17 hours at 1200 °C yields about 3200 Å of oxide. This is the correct result and is considerably less than the 4300 Å figure which ignores the influence of existing oxide on the growth rate. The 'correct' method relies on the assumption that O_2 or H_2O molecules pass through the oxide at a rate which depends only on the oxide thickness and not on how it was grown. This seems a reasonable assumption and has good support experimentally.

MODELLING THE DIFFUSION PROCESS

Theoretical modelling of particle diffusion, based on laws attributed to Fick, is quite successful in predicting the properties of diffused layers in silicon. Attention is drawn to the diffusion coefficient, D square centimetres per second, which is given in Appendix G for popular dopants. As already mentioned, D increases strongly with temperature. Referring to Fig. 4.1, we will treat the one-dimensional situation where particles diffuse in the positive x direction only. The dopant concentration is $N(x, t)$ atoms per cubic

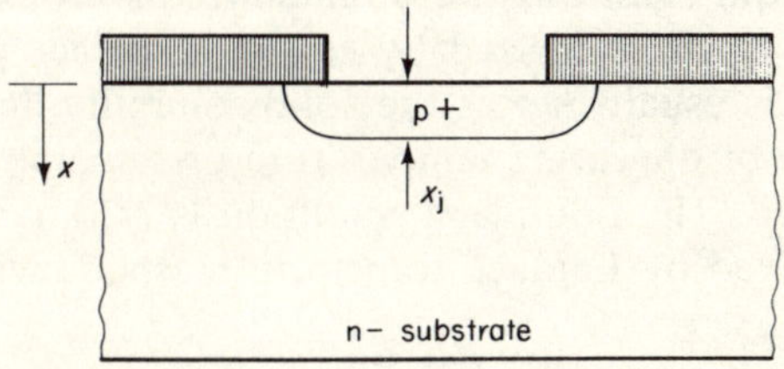

Fig. 4.1.

centimetre and, because N changes with depth, there is a particle flux of F atoms per square centimetre per second in the x direction:

$$F = -D\frac{\delta N}{\delta x} \qquad \text{(Fick's first law)} \qquad (4.1)$$

This is entirely analogous to the diffusion of electrons and holes as discussed in Chapter 1. In this context, however, the time scale is very much longer and we can 'freeze' the action by simply cooling the wafer.

In Fig. 4.2, the nett flow rate into a volume $A\,\mathrm{d}x$ is seen to be:

$$FA - \left(F + \frac{\delta F}{\delta x}\mathrm{d}x\right)A = -\frac{\delta F}{\delta x}\mathrm{d}xA$$

and it must equal the rate of particle accumulation in the volume which is $\mathrm{d}(NA\,\mathrm{d}x)/\mathrm{d}t$. Thus:

$$\frac{\mathrm{d}N}{\mathrm{d}t}\mathrm{d}xA = -\frac{\delta F}{\delta x}\mathrm{d}xA$$

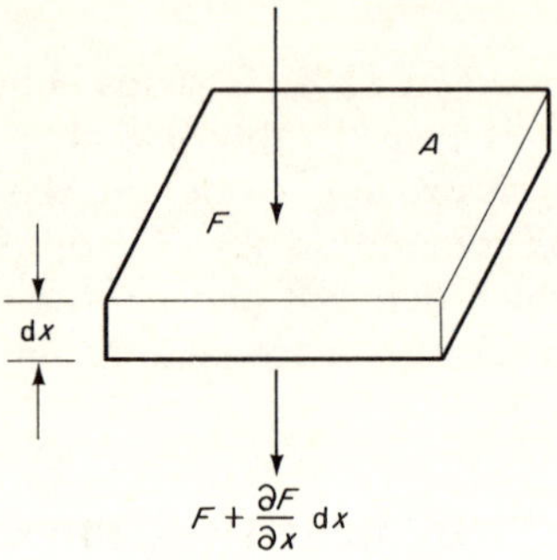

Fig. 4.2.

and using Eqn. (4.1) for F in the right-hand term we get:

$$\frac{\delta N}{\delta t} = D\frac{\delta^2 N}{\delta x^2} \qquad \text{(Fick's second law)} \tag{4.2}$$

When this equation is solved subject to the relevant boundary conditions, an expression for $N(x, t)$ results. In mathematical terms, the predeposition and drive-in both obey Eqn. (4.2) but the boundary conditions are different.

The predeposition is characterized by constant surface concentration N_s and we saw that this is usually set as the solid solubility limit of the dopant in silicon. The choice of operating temperature gives some control over the value of N_s if required. The boundary condition is $N(0, t) = N_s = \text{const.}$ and Eqn. (4.2) can be solved by Laplace transform methods with this condition to give:

$$N(x, t) = N_s \operatorname{erfc} \frac{x}{2\sqrt{(Dt)}} \tag{4.3}$$

Note that erfc(z) is the *complementary error function* and is not unlike the *Gaussian function* $\exp(-z^2)$. Both of these functions are plotted in Appendix F. A tabulation of erfc(z) is also included.

The graphs in Fig. 4.3 are sketches of Eqn. (4.3) evaluated at two points in time, t_1 and the later time t_2. They show how the concentration at a given depth increases progressively with time. Their rather odd shape looks contrary to what is intuitively expected. This is quickly explained by observing that $N(x, t)$ is shown on a log scale; on a linear scale it would be apparent that N changes more slowly at greater depth.

The concentration $N(x, t)$ refers to only those atoms which have diffused into the surface. There exists also a uniform concentration, N_B atoms per cubic centimetre, of doping ions of opposite type. N_B represents the lightly doped condition of the wafer itself and is indicated by the dotted line in Fig. 4.3. Because $N(x, t)$ and N_B are dopants of opposite type, the *nett* or *effective* doping level for purposes of conduction is $N(x, t) - N_B$. The depth x_j at which $N(x_j) - N_B = 0$ marks the semi-conductor junction. The conduction mechanism changes from p-type to n-type on passing through x_j. In Fig. 4.3 the junction depth is x_1 at time t_1 and increases to x_2 at time t_2.

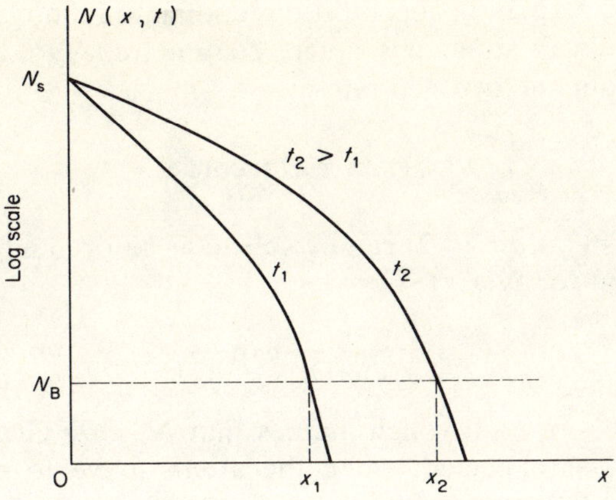

Fig. 4.3. Predeposition.

If we set $N(x, t) = N_B$ with $x = x_j$ in Eqn. (4.3) and rearrange, we get:

$$\frac{N_B}{N_s} = \operatorname{erfc} \frac{x_j}{2\sqrt{(D_p t_p)}} \tag{4.4}$$

where D_p and t_p specify the diffusion coefficient during predeposition and its duration respectively. The total dose (atoms/cm^2) of diffused atoms, when deposition is complete, can be evaluated as:

$$\begin{aligned} Q &= \int_0^\infty N(x, t_p)\, dx = N_s \int_0^\infty \operatorname{erfc}\left[\frac{x}{2\sqrt{(D_p t_p)}}\right] dx \\ &= 2\sqrt{(D_p t_p)} N_s \int_0^\infty \operatorname{erfc}\left[\frac{x}{2\sqrt{(D_p t_p)}}\right] d\left[\frac{x}{2\sqrt{(D_p t_p)}}\right] \\ &= 2\sqrt{(D_p t_p)} \frac{N_s}{\pi^{1/2}} \\ &= 2N_s\left(\frac{D_p t_p}{\pi}\right)^{1/2} \end{aligned} \tag{4.5}$$

The predeposition is a method of introducing an exact amount of dopant into the surface. In general, its penetration is not great and the resulting junction properties are unsuitable in many situations. A subsequent drive-in causes much deeper penetration while the surface concentration falls substantially.

The boundary conditions for the drive-in are quite complicated if one considers the distribution of the atoms that constitute the predeposition Q. Fortunately, Q is very close to the surface, compared to the final depth after drive-in and this permits a simplification of the boundary condition which works well in most circumstances. Q is viewed as a 'very thin' layer located at the surface and is modelled as an impulse function, i.e. a layer of thickness tending to zero and concentration tending to infinity but with finite total 'charge', namely Q atoms per square centimetre. Also, this 'charge' is conserved throughout the drive-in phase:

$$\int_0^\infty N(x, t)\,dx = Q = \text{const.}$$

With these constraints, Eqn. (4.2) can be solved as before and the result is a gaussian distribution for $N(x, t)$:

$$N(x, t) = \frac{Q}{\sqrt{(\pi D t)}} \exp \frac{-x^2}{4Dt} \tag{4.6}$$

At the start of drive-in, $t = 0$ which implies that $N_s = Q/(\pi D t)^{1/2}$ is initially infinite but it falls, as expected, when the atoms move to greater depth. Evaluating Eqn. (4.6) when drive-in is complete ($t = t_d$) and substituting Eqn. (4.5) for Q, we get:

$$N(x, t_d) = \frac{2N_s}{\pi} \left(\frac{D_p t_p}{D_d t_d}\right)^{1/2} \exp \frac{-x^2}{4D_d t_d} \tag{4.7}$$

The drive-in profiles in Fig. 4.4 are immediately distinguished by their declining surface concentrations. Wafer fabrication may involve a number of diffusion steps and every high-temperature treatment promotes even those (existing) diffusions for which it is not primarily intended. In general, the effect of all these is that of a (Dt) value such that:

$$(Dt) = D_1 t_1 + D_2 t_2 + D_3 t_3 + \cdots \tag{4.8}$$

The foregoing results make it possible to select suitable temperatures and diffusion times so as to obtain a desirable profile, expressed in terms of N_s, N_B, x_j, and the shape (erfc or Gaussian). The next section explains how the profile description relates to the electrical properties of the diffused layer.

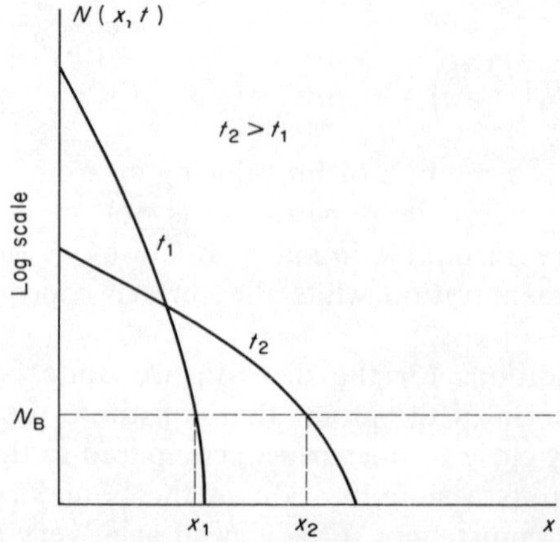

Fig. 4.4. Drive-in.

EVALUATION OF DIFFUSED LAYERS

Sheet resistance $R_s(\Omega/\square)$ is a key parameter of diffused layers. It was expressed in Chapter 3 in terms of junction depth x_j and average resistivity $\bar{\rho}$ as:

$$R_s = \frac{\bar{\rho}}{x_j} \quad \Omega/\square \tag{4.9}$$

The 'average conductivity', $\bar{\sigma} = 1/\bar{\rho}$, is also widely used:

$$R_s = \frac{1}{\bar{\sigma} x_j} \tag{4.10}$$

The actual variation of conductivity $\bar{\sigma}$ with depth reflects the impurity profile (erfc or Gaussian) and can therefore be predicted with accuracy. It follows that mean conductivity values ($\bar{\sigma}$) can be calculated; such calculations were undertaken by J. C. Irvin at Bell Laboratories and the results published in 1962. The results are applicable to layers which have had their upper surface removed to a depth x. This is facilitated by giving average conductivity figures for a layer extending from x to x_j. When $x/x_j = 0$, $\bar{\sigma}$ is the mean conductivity for a full layer.

The shaded area in Fig. 4.5 is representative of those impurities that contribute to $\bar{\sigma}$ and the expression for sheet resistance in the general case becomes:

$$R_s = \frac{1}{\bar{\sigma}(x_j - x)} \quad \Omega/\square \tag{4.11}$$

Irvin's results are presented in Appendix A, for both erfc and Gaussian profiles and for four values of N_B (spanning three decades) in each case. Results are given for p-type and for n-type layers. The appendix includes a convenient guide to utilization. Most often, the user specifies (x/x_j) and N_s

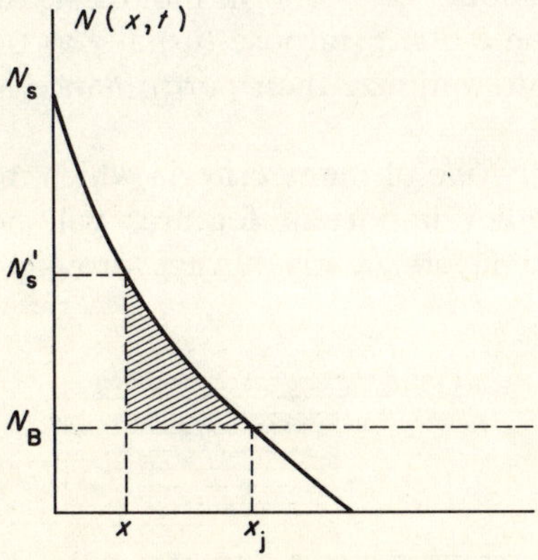

Fig. 4.5.

and obtains a value of $\bar{\sigma}$ from the graphs. This is readily converted to sheet resistance using Eqn. (4.11). As an example, suppose we want to find the sheet resistance of a layer, having a surface concentration $N_s = 10^{20}$, n-type, in a p-type substrate uniformly doped to a level $N_B = 10^{16}$ atoms/cm^3, and the profile specified as Gaussian. We obtain $\bar{\sigma}$ from Fig. A4(c) of Appendix A as $\bar{\sigma} = 3.7 \times 10^2 = 370\ (\Omega \cdot \text{cm})^{-1}$ for a full layer ($x/x_j = 0$). Then the sheet resistance is:

$$R_s = \frac{1}{\bar{\sigma}(x_j - x)} = \frac{1}{\bar{\sigma} x_j} = \frac{1}{370 x_j}$$

For a layer of depth $x_j = 3\ \mu\text{m} = 3 \times 10^{-4}$ cm:

$$R_s = \frac{10^4}{3 \times 370} = 9.01\ \Omega/\square$$

If this layer had been partially etched such that $x/x_j = 0.3$, we would have

$$R_s = \frac{1}{\bar{\sigma}(x_j - x)} = \frac{1}{130(3 - 0.9) \times 10^{-4}} = 36.6\ \Omega/\square$$

where $\bar{\sigma}$ was read off the ($x/x_j = 0.3$) line intersection with $N_s = 10^{20}$ on the same graph as before. Notice the very much reduced value of $\bar{\sigma}$ for the partial layer; it reflects the fact that doping concentration is a lot higher near the surface.

Diffused layers are frequently made to serve as resistors. We might, for example, wish to fabricate a 500 Ω resistor using the n-type diffusion just described with an R_s value of 9.01 Ω/□ (full depth). This would call for 500/9.01 = 55 squares (approx.) laid end to end. We would make the side length of a square as small as the photolithography would allow (typically 0.3 mil where a 'mil' is 1/1000 of an inch), and then lay out the resistor in a compact manner as suggested in Fig. 4.6. The example serves to indicate that such a resistor could easily be objectionable on the grounds of excessive space requirements. It should be borne in mind that the diffused layers are generally tailored for some other purpose (usually as transistor regions) and the designer is not free to optimize their performance in the role of diffused resistors.

Sheet resistance is only one of many criteria which may be important in a diffused layer. Some other important features will now be discussed. In MOS circuitry, diffused layers nearly always provide electrical insulation

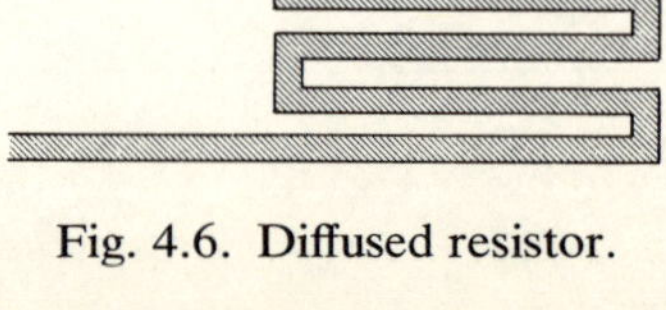

Fig. 4.6. Diffused resistor.

from the underlying substrate by ensuring that the semi-conductor junction is always reverse-biased. It was seen in Chapter 1 that large reverse bias can lead to avalanche breakdown:

$$BV = 2.7 \times 10^{12} N_B^{-2/3} \tag{4.12}$$

This approximate expression gives warning that a high bulk doping level N_B can result in low breakdown voltage (BV) and, in practice, it places an upper limit on the allowable N_B levels.

Chapter 1 also dealt with the subject of depletion layer width for p–n junctions. This places another constraint on N_B as follows. From Eqn. (1.20), the depletion layer width is:

$$X_{mL} = \left[\frac{2k_{si}e_0(\psi_0 - V)}{qN_B}\right]^{1/2} \tag{4.13}$$

and X_{mL} extends, almost in its entirety, below the junction into the substrate, being largest for maximum reverse bias. It must never be so large that the depletion layer extends all the way from the drain diffusion to the source diffusion because this would result in *punch-through* and consequent currents which completely alter the device behaviour (Fig. 4.7). The source to drain spacing is about 0.2 mil (5 μm) minimum and X_{mL} must be less than this.

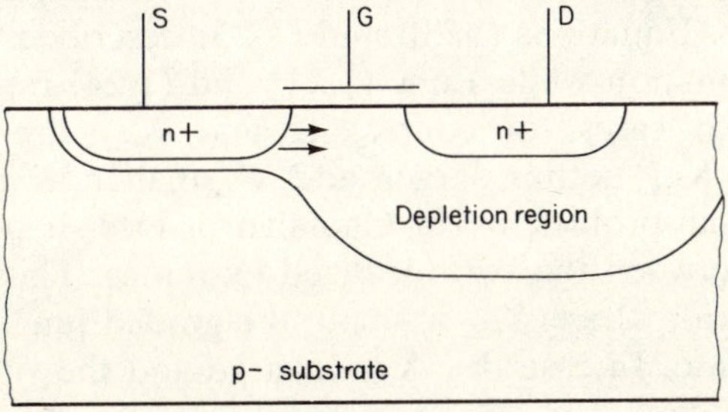

Fig. 4.7. Punch-through.

The upper and lower limits on N_B are evident from the table below which uses Eqns (4.12) and (4.13) and is valid for $(\psi_0 - V) = 20$ V of reverse bias. The resistivity figures are determined from N_B alone using Appendix D and are included for convenience since wafers are usually specified in this way.

N_B, atoms/cm^3	$\bar{\rho}$, $\Omega \cdot$ cm p-channel	n-channel	BV, V	X_{mL}, μm
5×10^{14}	9.0	26.0	429	7.2
1×10^{15}	5.0	13.0	270	5.1
3×10^{15}	2.0	4.0	130	2.9
1×10^{16}	0.6	1.5	58	1.6

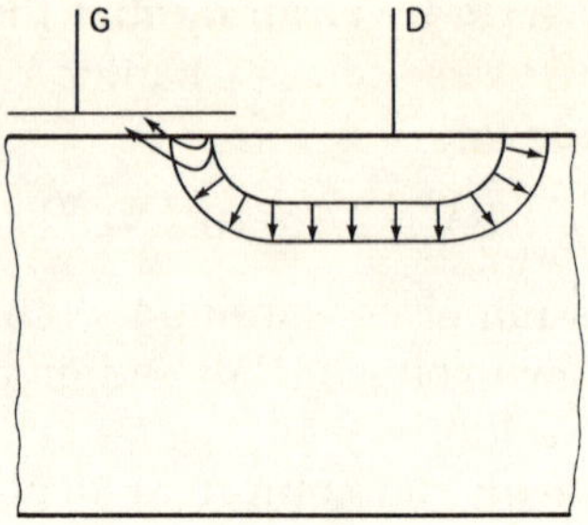

Fig. 4.8. Gate-modulated breakdown.

It is evident that breakdown voltage becomes a problem for $N_B > 10^{16}$. In fact, the true situation is worse than this because the close proximity of a gate metal can cause avalanche breakdown at a lower voltage by distortion of the electric field. This *gate-modulated breakdown* mechanism is illustrated in Fig. 4.8. The table also shows that punch-through may occur for approx. $N_B < 10^{15}$. There remains only a narrow range, about one decade, of usable N_B values.

The most important properties of the semi-conductor junction formed by the diffused layer are its capacitance C per unit area and the value of X_{mL}. Recall that, in Chapter 1, it was possible to write approximate expressions for both of these. Equations (1.20) and (1.23) described the one-sided step junction approximation while Eqns (1.21) and (1.24) represented a graded junction. In both cases, of course, C and X_{mL} are strongly voltage-dependent with X_{mL} getting larger and C smaller as reverse bias is increased. The actual profiles are of Gaussian or erfc shape and tend to give results that lie between the two idealized extremes. This can be seen from Fig. 4.9. In case (a), where X_{mL} is small, the graded junction approximation is quite appropriate. In case (b), X_{mL} is large and the situation approaches the step-junction case.

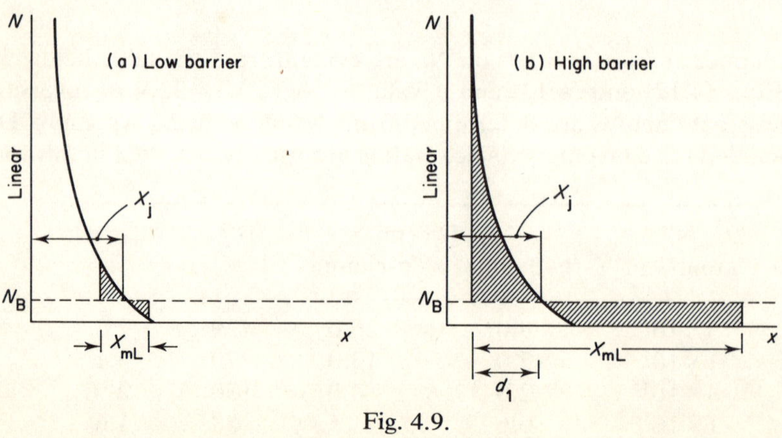

Fig. 4.9.

Because the profiles are precisely known, it is only a matter of computation to obtain precise figures for C and X_{mL} in all situations. This was done by Lawrence and Warner of Bell Laboratories who integrated Poisson's equation analytically for both Gaussian and erfc layers and generated the results which are presented graphically in Appendix B. Graphs are given for various values of N_B/N_s and, in normal usage, X_{mL} and C can be read after V/N_B and x_j have been specified. The occurrence of V/N_B as a parameter is not surprising since it also appears in Eqn. (1.20). The original publication gave information on d_1 (see Fig. 4.9) as well as X_{mL}. For most purposes, we can assume that $d_1 \ll X_{mL}$. It is worth noting that results become relatively independent of x_j for large values of V/N_{BC} and this characteristic marks the approach to step-junction behaviour. The appendix includes a utilization guide on page 216. While these graphs should always be used for precise results, Eqns (1.20) to (1.24) provide useful insights to the interdependence of the parameters.

The use of Lawrence and Warner's results is now illustrated by an example. The profile is Gaussian, the diffusion is n-type, $N_B = 1.3 \times 10^{15}$, $N_s = 2.5 \times 10^{19}$, $x_j = 1.7$ μm. Results are given for external applied potentials of 0 and 10 V. Because $N_B/N_s = 5 \times 10^{-5}$, we take the nearest available graphs, those for $N_B/N_s = 10^{-4}$ in Fig. B.12. We also assume a junction barrier ψ_0 of 0.6 V. The results can then be read off and tabulated as follows:

Bias	V	V/N_B	x_j	X_{mL}	C, pF/cm^2	C, fF/mil^2
0	0.6	4.6×10^{-16}	1.7	0.85	1.2×10^4	77.4
10	10.6	8.1×10^{-15}	1.7	3.2	3.1×10^3	20.0

On these graphs, V refers to total barrier height, i.e. $(\psi_0 - V)$ in our notation, x_j and X_{mL} are in micrometres, C is read in picofarads per square centimetre but is readily converted to femtofarads per square mil by:

$$C(\text{fF/mil}^2) = C(\text{pF/cm}^2) \times 6.45 \times 10^{-6} \times 10^3$$

because 1 mil$^2 = 6.45 \times 10^{-6}$ cm^2 and 1 pF $= 10^3$ fF. It is worth remembering the 'order of magnitude' of the above results because of their critical influence on the performance parameters of MOS circuits.

The importance of X_{mL} has already been noted. The importance of C is mainly with regard to switching speed for MOS inverter circuitry. This is a considerable part of the total capacitive load on a typical inverter output node. Notice that it is greatly reduced by reverse bias. This implies that 'substrate bias' could improve speed performance appreciably.

The influence of junction depth is evident from the graphical data. Firstly, notice that the average conductivity $\bar{\sigma}$ is not dependent on x_j but the resulting sheet resistance R_s falls with increasing junction depth, as would be expected. The influence of x_j on X_{mL} and C depends on the level of applied voltage. For sufficiently high voltage, the contours on Appendix B, representing various x_j values, approach one another. Where this occurs, junction depth has little influence on electrical parameters. At lower voltages, x_j is significant and the graded junction approximation of eqn (1.21) shows why. A deep junction will have a smaller slope $|dN/dx|$ at x_j than does a shallow one. This means that the effective doping level $N(x)-N_B$ is low over a considerable distance on both sides of the junction and therefore a large depletion width X_{mL} results. This, in turn, means that the deep junction has lower capacitance than a shallow one, a significant advantage. Examination of the graphs will show, however, that junctions come close to step-junction behaviour for voltages which are well within normal operating ranges.

The penetration of doping ions through a window during diffusion proceeds horizontally as well as vertically. Thus, the final diffused layer extends under the window edges for a short distance. This horizontal penetration is frequently estimated as $0.75x_j$ and the estimate is adequate for many purposes. More precise information was published by the IBM authors Kennedy and O'Brien and their findings are given graphically in Appendix H. The contours are lines of constant doping concentration with normalized parameter $C=N(x)/N_s$ and the results are given for both erfc and Gaussian profiles. Sideways diffusion is significant in calculating the total junction capacitance of a diffused layer and also, of special importance, in calculating the capacitance between the gate metal and diffused layer due to gate overlap. This capacitance can be quite large because the gate overlaps the source and drain diffusions with only a very thin oxide dielectric to separate them. The extent of the overlap can be severe in the case of metal gate processing because of the need to allow for alignment errors in gate-mask placement. In silicon gate processing the problem is less severe because the only overlap is that resulting from sideways diffusion with the polysilicon gate itself assuming the role of a gate mask.

MOS DEVICE CAPACITANCES

Several sources of capacitance are inherent to the MOS structure and it is important to have an idea of their relative importance. The dielectric is frequently silicon dioxide and, if the oxide is bounded above and below by conducting planes, capacitance estimation is easy:

$$C=\frac{k_{ox}e_0}{t_{ox}} \quad \text{F/cm}^2 \tag{4.14}$$

where $k_{ox}=3.9$ is the relative dielectric constant of the oxide (SiO_2), $e_0=8.85\times10^{-14}$ F/cm is the free space permittivity, and t_{ox} is the oxide thickness in centimetres. Thus, for a 1000 Å gate oxide:

$$C = \frac{3.9 \times 8.85 \times 10^{-14}}{10^{-5}} = 3.4 \times 10^{-8}\ \mathrm{F/cm^2} = 222\ \mathrm{fF/mil^2}$$

For a 15,000 Å field oxide (Fig. 4.10), capacitance is proportionately reduced giving 222 (1000/15,000) = 15 fF/mil^2. Also, the surface of a diffused layer qualifies quite well as a 'conducting plane' for purposes of capacitance estimation.

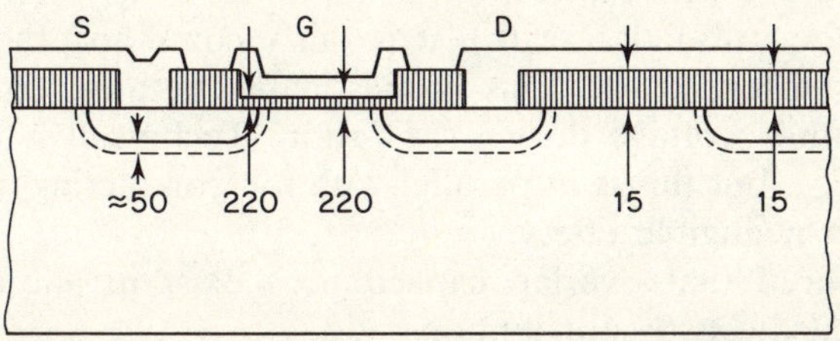

Fig. 4.10. Some typical MOS capacitances in femtofarads per square mil.

Figure 4.10 offers a pictorial summary of capacitance types and values. Because of voltage dependence, p–n junction capacitance varies widely around the given value of 50 fF/mil^2. Capacitance due to gate overlap of the source and drain diffusions is important to device performance and is relatively large due to the thin oxide. Reduction of this capacitance is a prime concern for fast operation.

The interterminal capacitances C_{GS} and C_{GD} of a transistor are not limited to overlap capacitance alone. To illustrate, we consider the case of a device *with no overlap*. It is at once clear from Fig. 4.11 that *the channel* is instrumental in furnishing finite values of C_{GS} and C_{GD} because it is, after all, a physical extension of these conducting bodies. Both C_{GS} and C_{GD} have

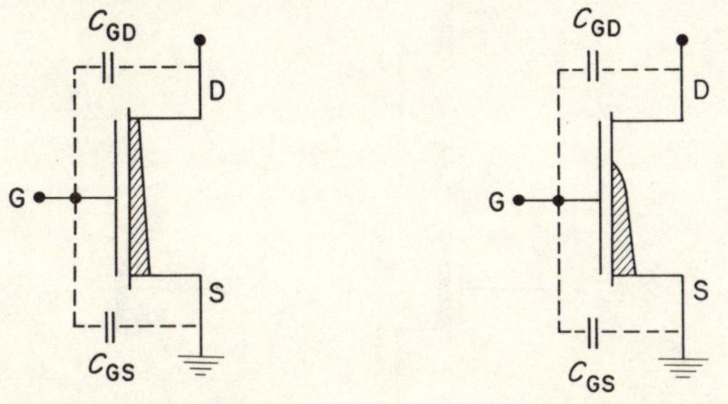

Fig. 4.11. Channel influence on C_{GS} and C_{GD}.

their origins in the gate-to-channel capacitance C_G which we will define as:

$$C_G = \frac{k_{ox}e_0}{t_{ox}} A \qquad \text{F} \tag{4.15}$$

where A is the gate area. In the non-saturated case (a), one can reason that the average voltage in the bottom half of the channel is controlled largely by the source while that in the upper half is under drain control. By this reasoning, $C_{GS} = C_{GD} = 0.5C_G$ seems reasonable. But, since the channel grows narrower towards the drain end, the source tends to dominate in determining channel voltage with the result that estimates of C_{GS} may be as high as $(2/3)C_G$. In the saturated case (b), it is fairly common practice to choose $C_{GS} = (2/3)C_G$ with $C_{GD} = 0$. This is reasonable because the drain does not influence channel voltage during saturation. Technically, there is another capacitance, C_{SD}, but this is in parallel with the conducting channel and consequently has a negligible effect.

It must be remembered that overlap capacitances exist in addition to those just discussed and the two contributions add.

It is worth giving a moment's thought to the various capacitances that contribute to form the capacitive load C_L on the inverter in Fig. 4.12. The largest component, probably, is the gate capacitance of the next stage driver. The diffusion-to-substrate capacitance of the output node comes next. (Note that a single diffusion serves as the drain for the driver and the source for the load device.) With a little thought, it becomes clear that both transistor channels tend to enlarge this contribution! C_L will vary during a switching excursion due to changes in channel lengths and due to junction capacitance variations with bias level. These factors combine to make a precise analysis very difficult indeed. The usual engineering answer to this situation is to apply a constant but pessimistic value to C_L and to perform transient

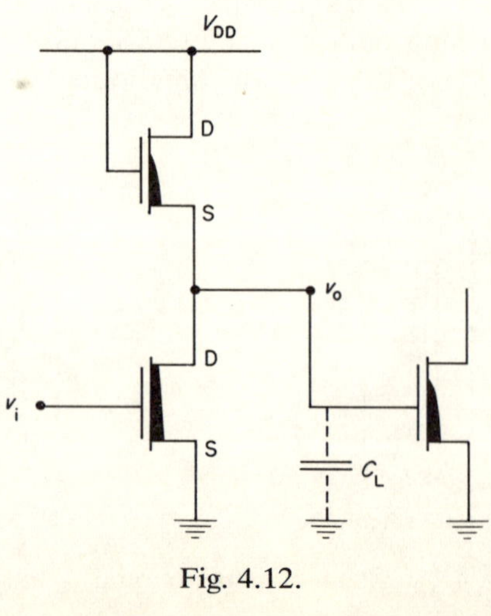

Fig. 4.12.

analysis on that basis. This approach can work quite well as long as the value for C_L is chosen carefully and with good judgement.

MEASUREMENT TECHNIQUES

A chemical staining technique can be used to measure junction depth. The material is first lapped at a small angle ϕ to the surface (Fig. 4.13) in order to enlarge the depth dimension for inspection purposes. The stain is then applied and tends to darken the p-type region without affecting the n-type region. The depth can then be measured by microscope or interference methods.

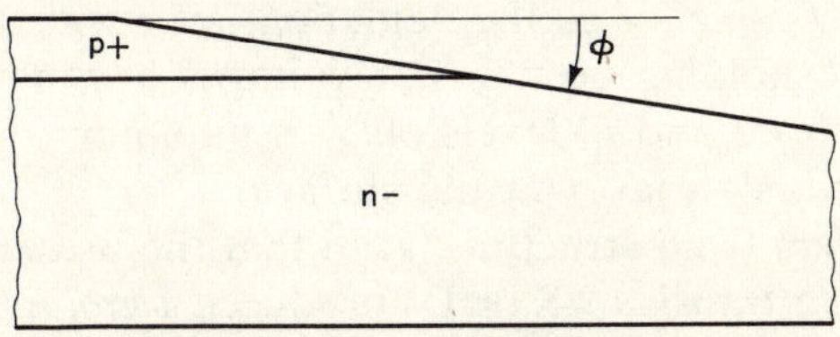

Fig. 4.13. Junction depth measurement.

The average resistivity $\bar{\rho}$ of a surface layer can be measured by the four-point probe method (Fig. 4.14). Probes have equal spacing d. A current

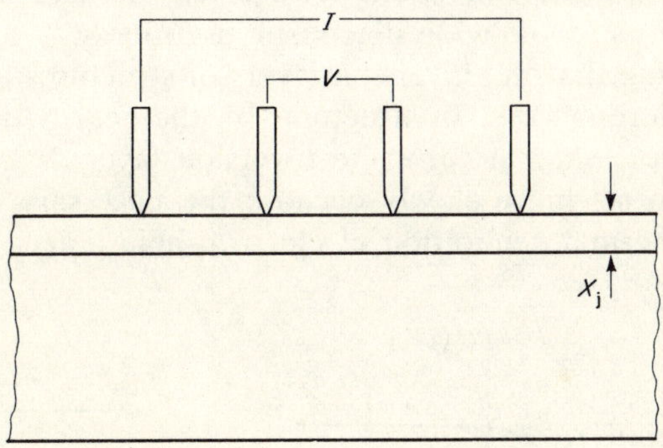

Fig. 4.14. Measuring resistivity.

I is circulated through the outer probes and the resulting voltage V between the inner probes is measured. For thin layers ($x_j \ll d$), a theoretical analysis yields:

$$\bar{\rho} = 4.53 \frac{V}{I} x_j \qquad \Omega \cdot \text{cm} \tag{4.16}$$

independent of the spacing d. Therefore, the sheet resistance is simply:

$$R_S = \frac{\bar{\rho}}{x_j} = 4.53 \frac{V}{I} \qquad \Omega/\square \tag{4.17}$$

C–V testing is a method of measuring (small-signal) capacitance as a function of applied bias potential and displaying the result as a capacitance versus voltage plot. It is easily automated and non-destructive and produces data quickly and accurately. This makes it convenient for checking at various steps of fabrication which greatly facilitates process control. Figure 4.15 shows *C–V* plots for two types of structure. Structure (a) is simply a metal oxide–substrate capacitance with no diffusion present. For positive voltages the surface is accumulated and the capacitance is simply $C_{ox} = k_{ox}e_0A/t_{ox}$ where A is the area of the metal overlay. With negative applied voltage the depletion region (shown dotted) grows and effectively places a depletion capacitance C_d in series with C_{ox}. The nett capacitance $C = C_dC_{ox}/(C_d + C_{ox})$ gets smaller as the depletion layer widens and this process continues until thresholding occurs. For voltages beyond threshold, depletion layer width is fixed and C levels off at a minimum. It is not influenced by the 'floating' inversion layer at the surface.

Structure (b) differs from structure (a) in that the presence of an adjacent grounded p region (a transistor S or D) produces a *grounded* inversion layer rather than a floating one. In this case, therefore, for voltages beyond threshold, the nett capacitance C rapidly returns to its original value C_{ox} since there is no depletion layer influence remaining. This behaviour is typical of transistor gate capacitance C_G. In estimating C_G, it is best to take $C_G = C_{ox}$ ignoring the dip in the *C–V* plot. The *C–V* plot can obviously be used to monitor changing threshold voltages as well as changes in t_{ox}, N_B, etc., which accounts for its wide diagnostic usefulness.

It is worth noting that the inversion layers of structures (a) and (b) above arise in quite different ways. In structure (b), the nearby diffusion provides all the carriers necessary to form the inversion layer. In structure (a), the inversion layer forms more slowly because the necessary carriers must be furnished from thermal generation of electron-hole pairs.

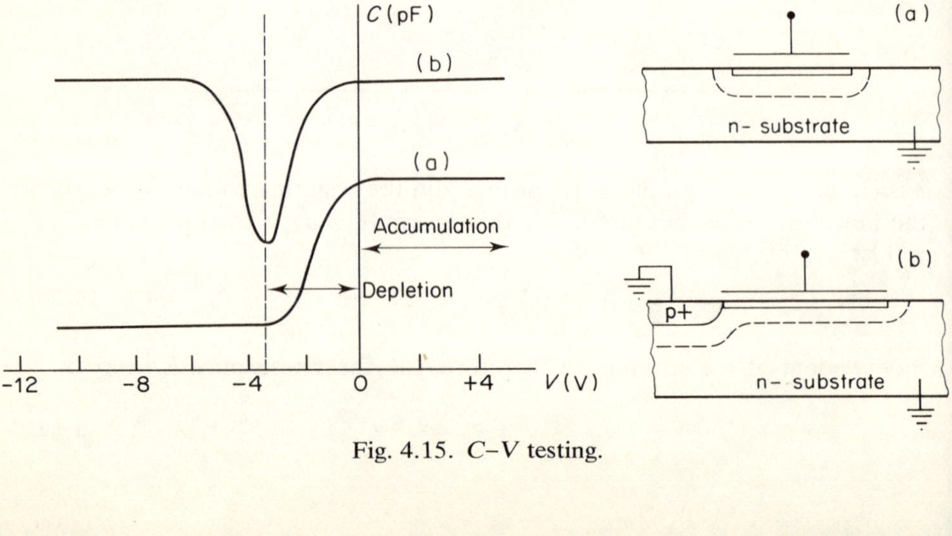

Fig. 4.15. *C–V* testing.

EXAMPLE 4.1

A constant-source boron diffusion (p-type) is made on an n-type wafer of $0.5\ \Omega \cdot$ cm resistivity. Surface concentration of boron is kept at the solid solubility limit. Diffusion proceeds for 1 hour at 1200 °C. Find the junction depth x_j and the lateral diffusion width.

Solution

From Appendix D, $N_B = 1.3 \times 10^{16}$. From Appendix E, $N_s = 5.3 \times 10^{20}$. The diffusion coefficient at 1200 °C is, from Appendix G, $D = 1.4 \times 10^{-12}$ cm^2/sec. At $x = x_j$, $N(x, t) = N_B$ and:

$$\frac{N(x, t)}{N_s} = \frac{N_B}{N_s} = \text{erfc}\,\frac{x_j}{2\sqrt{(Dt)}} = \frac{1.3 \times 10^{16}}{5.3 \times 10^{20}} = 2.45 \times 10^{-5}$$

From Appendix F, $x_j/2\sqrt{(Dt)} = 3.0$ or:

$$x_j = 3.0 \times 2(1.4 \times 10^{-12} \times 3600)^{1/2} = 4.2 \times 10^{-4}$$

$$\underline{x_j = 4.2\ \mu\text{m}}$$

From Appendix H, $x_L/x_j = 2.5/3.0 = 0.833$ and the lateral diffusion width $=$ $4.2 \times 0.833 = \underline{3.5\ \mu\text{m} = x_L}$.

EXAMPLE 4.2

In a p-type Gaussian diffusion process, the predeposition level is 10^{16} atoms/cm^2 of boron. The n-type substrate has $1\ \Omega \cdot$ cm resistivity. Find the junction depth after a diffusion for 80 minutes at 1000 °C.

Solution

$$N(x, t) = \frac{Q}{(\pi Dt)^{1/2}}\, e^{-x^2}/4Dt = N_B \qquad \text{at } x = x_j$$

For $1\ \Omega \cdot$ cm, Appendix D gives $N_B = 5 \times 10^{15}$ atoms/cm^3. From Appendix G, $D = 3 \times 10^{-14}$ at 1000 °C so that:

$$N_s = \frac{Q}{(\pi Dt)^{1/2}} = \frac{10^{16}}{(\pi \times 3 \times 10^{-14} \times 80 \times 60)^{1/2}} = 4.70 \times 10^{20}$$

Rearranging $e^{x_j^2/4Dt} = N_s/N_B$; $x_j^2 = 4Dt\ \text{Ln}(N_s/N_B)$:

$$x_j = \left(4Dt\ \text{Ln}\,\frac{N_s}{N_B}\right)^{1/2} = [4 \times 3 \times 10^{-14} \times 80 \times 60\ \text{Ln}(94{,}000)]^{1/2}$$

$$\underline{x_j = 0.81\ \mu\text{m}}$$

EXAMPLE 4.3

A boron diffusion to a $2\,\Omega \cdot$ cm n-type substrate is found by direct measurement to have $x_j = 2.5\ \mu$m and sheet resistance $R_s = 120\ \Omega/\square$. If the profile is Gaussian, find N_s. Find also the contact potential at 27 °C and at 75 °C. Find the depletion layer width X_{mL} and capacitance C in femtofarads per square mil for 0 V and for 20 V of reverse bias. Compare results with the step-junction approximation. For a minimum width of 0.3 mil, what diffusion length is necessary to provide a 50 kΩ resistor?

Solution

For $2\,\Omega \cdot$ cm, substrate $N_B = 2.8 \times 10^{15}$. Since $R_s = 1/(\bar{\sigma} x_j)$:

$$\bar{\sigma} = \frac{1}{R_s x_j} = \frac{1}{120 \times 2.5 \times 10^{-4}}$$
$$= 33.33\ (\Omega \cdot \text{cm})^{-1}$$

From Appendix A, Fig. A.5(b) $\underline{N_s = 1.0 \times 10^{19}}$
The contact potential $\psi_0 = (kT/q)\text{Ln}(N_A N_D/n_i^2)$ and, near the junction, $N_A \to N_B$ while $N_D = N_B$, so that:

$$\psi_0 = \frac{kT}{q}\text{Ln}\left(\frac{N_B}{n_i}\right)^2 = \frac{2kT}{q}\text{Ln}\,\frac{N_B}{n_i}$$

At 27 °C: $$\psi_0 = \frac{2 \times 1.38 \times 10^{-23} \times 300}{1.6 \times 10^{-19}}\text{Ln}\,\frac{2.8 \times 10^{15}}{1.4 \times 10^{10}} = 0.63\ \text{V}$$

At 75 °C: $$\psi_0 = \frac{2 \times 1.38 \times 10^{-23} \times 348}{1.6 \times 10^{-19}}\text{Ln}\,\frac{2.8 \times 10^{15}}{4.4 \times 10^{11}} = 0.52\ \text{V}$$

$N_B/N_s = 2.8 \times 10^{15}/10^{19} = 2.8 \times 10^{-4}$. We use the 10^{-4} Lawrence and Warner graph of Fig. B.12 in Appendix B:

0 V applied: $$\frac{V}{N_{BC}} = \frac{0.63}{2.8 \times 10^{15}} = 2.25 \times 10^{-16}, \qquad x_j = 2.5 \times 10^{-4}\ \text{cm}$$

so that

$$X_{mL} = 0.7 \times 10^{-4}\ \text{cm}, \qquad C = 1.5 \times 10^4\ \text{pF/cm}^2$$

20 V applied: $$\frac{V}{N_{BC}} = \frac{20.63}{2.8 \times 10^{15}} = 7.3 \times 10^{-15}$$

and

$$X_{mL} = 3 \times 10^{-4}\ \text{cm}, \qquad C = 3.3 \times 10^3\ \text{pF/cm}^2$$

Tabulating results:

Applied voltage, V	V/N_{BC}	$X_{mL}(\mu m)$	$C(pF/cm^2)$	$C(fF/mil^2)$
0	2.25×10^{-16}	0.7	1.5×10^4	96.7
20	7.3×10^{-15}	3.0	3.3×10^3	21.3

Note that $C(fF/mil^2) = C(pF/cm^2)\times 6.45\times10^{-6}\times10^3$. Using $X_{mL} = [2k_{si}e_0(\psi_0 - V)/qN_B]^{1/2}$ (step-junction approximation) gives:

At 0 V: $\underline{X_{mL} = 0.54\ \mu m}$

At 20 V: $\underline{X_{mL} = 3.1\ \mu m}$

The approximation is better for higher V, as expected.

A 50 kΩ resistor requires 50,000/120 or 416 squares, a total length of (416×0.3) or <u>125 mils.</u>

CHAPTER 5

The Threshold Voltage in MOS Transistors

The threshold voltage, and the stability thereof, depend on many factors, the most important of which are treated here. Expressions for the threshold voltage (V_T) will be obtained which accurately reflect observed performance.

The threshold voltage V_T was a very troublesome parameter in the early days of MOS circuitry. It was difficult to control the value of V_T during processing and, in addition, the threshold suffered from drift mechanisms which were only poorly understood at that time and which sometimes led to unexpected failures in service. In more recent times, ion implantation has become available as a method of threshold adjustment which permits accurate control of the final value. It is, nevertheless, important to understand the influences that contribute to V_T because they are intimately related to the choice of technology and to quality control considerations and also because V_T is by no means constant but is voltage-dependent via a mechanism known as *body-effect.*

The contact potential ψ_0 plays a major role in the discussion of the threshold voltage. From Chapter 1 (Eqn. 1.16):

$$\psi_0 = \frac{kT}{q} \mathrm{Ln} \frac{N_A}{n_i} + \frac{kT}{q} \mathrm{Ln} \frac{N_D}{n_i} = \psi_p + \psi_n \tag{5.1}$$

Some simplification is possible.

As Fig. 5.1 illustrates, N_A approaches N_B (the bulk doping level) at a junction x_j and $N_B = N_D$, the donor level. With no great error:

$$\psi_0 = \frac{kT}{q} \mathrm{Ln} \frac{N_B}{n_i} + \frac{kT}{q} \mathrm{Ln} \frac{N_B}{n_i}$$

$$= 2\phi_F \quad \text{with} \quad \phi_F = \frac{kT}{q} \mathrm{Ln} \frac{N_B}{n_i} \tag{5.2}$$

This is applicable to both diffusion types and to depletion regions under gates, as will be discussed. ϕ_F is usually called the *Fermi potential.* The threshold voltage can be viewed as the sum of several contributions which will be dealt with in turn in the discussion that follows.

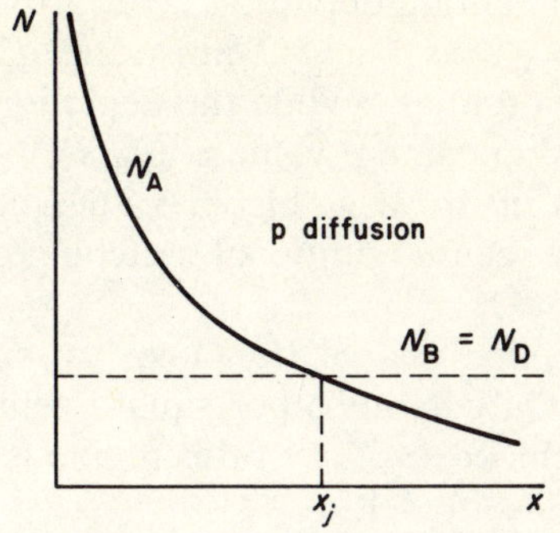

Fig. 5.1. Diffusion profile.

THE CONTACT POTENTIAL CONTRIBUTION

We begin with the simplest bias conditions such that the drain and bulk are both at the source potential $v_{DS} = v_{BS} = 0$ (Fig. 5.2). In the electric circuit linking the gate and bulk, the full external potential appears across the gate dielectric and the top of the substrate, implying a full charge balance between the gate metal and the silicon–silicon dioxide interface. This is entirely analogous to the equal and opposite charges $+Q$ and $-Q$ on the plates of a capacitor.

In the discussion that follows, the symbol V_{TO} will be used to describe the 'basic' threshold voltage under nominal bias conditions, i.e. a value which does not include the voltage-dependent 'body-effect'. The threshold voltage V_{TO} is the gate-to-source voltage (or, in this case, gate-to-bulk) which initiates a conducting channel between the source and drain. Thus, with V_{TO} volts applied ($V_{GS} = V_{TO}$), the line AB in Fig. 5.2 becomes an equipotential since $V_{DS} = 0$. Further, both diffusions have depleted regions with a contact potential $2\phi_F$ to the bulk. It follows that, with V_{TO} applied, the substrate under the gate must have a depletion layer *with the same potential difference* $2\phi_F$ *from the channel to the bulk.* As the gate voltage increases from zero towards the threshold value, mobile charge is pushed away from the surface

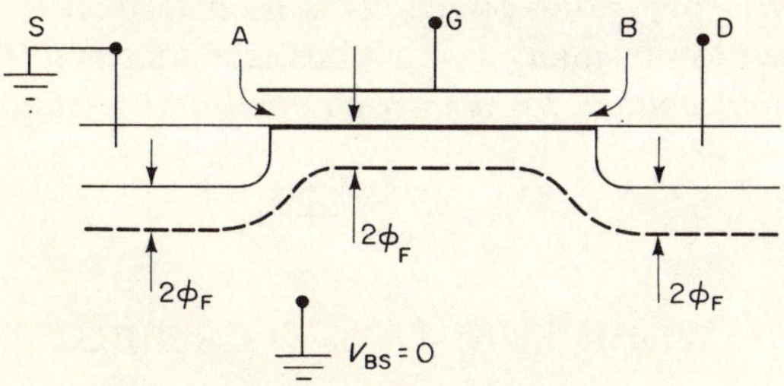

Fig. 5.2. Thresholding.

under the gate and the resulting depletion layer grows wider. Its maximum width occurs at $V_{GS}=V_{TO}$, as Fig. 5.2 illustrates. (The dotted line is the surface of the bulk equipotential.) While the depletion layer incorporates an 'uncovered' ionic charge (maximum value is Q_B at $V_{GS}=V_{TO}$) to match the gate charge for voltages up to V_{TO}, higher values of the gate voltage and gate terminal charge will secure additional matching charge *by enlarging the channel thickness.*

The total depletion region charge (for 1 cm^2 of surface) under the gate reaches a maximum of Q_B coulombs per square centimetre with a drop of $2\phi_F$ volts across it. The thickness X_{mL} of this region is predicted very well by the *one-sided step-junction* model:

$$X_{mL}=\left[\frac{2k_{si}e_0(2\phi_F)}{qN_B}\right]^{1/2} \tag{5.3}$$

Then, since Q_B simply represents uncovered doping ions, we get:

$$\begin{aligned} Q_B &= qN_BX_{mL}=[2k_{si}e_0qN_B(2\phi_F)]^{1/2} \\ &= \text{depletion layer charge per square centimetre of surface} \end{aligned} \tag{5.4}$$

This charge Q_B will have an equal and opposite charge on the gate electrode. As in any capacitor, this charge terminates an electric field through the intervening silicon dioxide dielectric. The capacitance C_{ox} per square centimetre from the gate to the channel through the oxide dielectric is:

$$C_{ox}=\frac{k_{ox}e_0}{t_{ox}} \quad \text{F/cm}^2 \tag{5.5}$$

with the relative dielectric constant $k_{ox}=3.9$ for silicon dioxide and the oxide thickness t_{ox} in centimetres. (Note that 1 Å $=10^{-8}$ cm.) The voltage across the dielectric which will support a charge Q_B is Q_B/C_{ox}. *The threshold voltage contribution* (V_{TO1}) *from factors just discussed includes* Q_B/C_{ox} *across the oxide and* $2\phi_F$ *across the deletion layer.*

We will use V_{TO1}, V_{TO2}, etc., to give the *threshold voltage magnitude* without prejudice to the fact that p-channel thresholds have *negative* polarity and n-channel thresholds are *positive.* (In the case of depletion mode operation, the term inversion voltage V_I is used instead.)

With this convention in mind, it must be clear that *both* Q_B/C_{ox} *and* $2\phi_F$ *enlarge the threshold voltage for both p-channel and n-channel devices:*

$$V_{TO1}=\frac{Q_B}{C_{ox}}+2\phi_F \tag{5.6}$$

THE WORK FUNCTION CONTRIBUTION

The *work function* (in electronvolts, eV) is the energy that must be supplied to an electron to take it across the surface barrier. If the work function of the

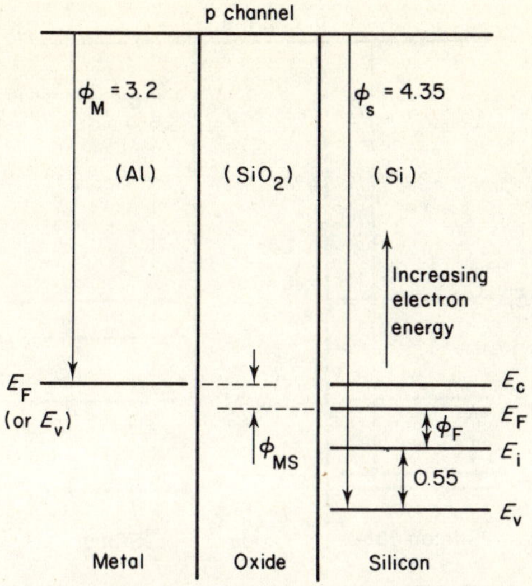

Fig. 5.3. Energy levels (aluminium gate).

gate metal differs from that of the silicon substrate by X electronvolts, this will in turn influence the threshold voltage by X volts.

We take first the case of an aluminium gate p-channel (n-type substrate) device with the gate metal work function ϕ_M and silicon work function ϕ_s stated in volts. The energy level diagram of Fig. 5.3 will help—at least for those familiar with such representations.

The factor influencing the threshold voltage is the distance ϕ_{MS} between Fermi levels (E_F). In intrinsic silicon, E_F coincides with E_i, a level half-way between the valence band upper boundary E_V and the bottom of the conduction band E_c. In n-type silicon as shown, E_F moves up from E_i by the amount ϕ_F. (In p-type silicon, it moves down by the same amount.) Work functions measured to the top of the valence band (as is usual) are 3.2 for aluminium and 4.35 for silicon. Also $(E_c - E_v) = 1.1$ for silicon and $(E_i - E_v) = 1.1/2 = 0.55$. Thus:

$$\phi_{MS} = 3.2 - (4.35 - 0.55 - \phi_F)$$

$$\phi_{MS} = -0.6 + \phi_F \quad \text{(aluminium gate p-channel device)} \qquad (5.7)$$

Since $\phi_F = (kT/q)\text{Ln}(N_B/n_i)$ is typically $= 0.3$ V and never much more, ϕ_{MS} is a negative number. As such *it makes the threshold voltage more negative or less positive.*

If we now consider an aluminium gate n-channel device, the only change will be that in the silicon (p-type) substrate E_F will be *below* E_i by the amount ϕ_F. Then:

$$\phi_{MS} = -0.6 - \phi_F \qquad \text{(aluminium gate n-channel device)} \qquad (5.8)$$

This again influences the threshold in the negative direction.

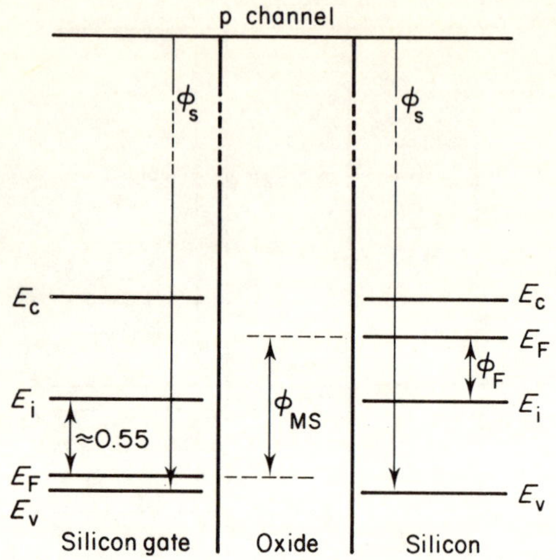

Fig. 5.4. Energy levels (silicon gate).

Another popular technology (silicon gate) uses polycrystalline silicon as a gate metal. It alters V_{TO} and has the capacity to function as a gate mask during high-temperature source and drain diffusions so as to avoid alignment difficulties. Figure 5.4 is the energy level diagram for a silicon gate p-channel structure. The gate material is doped, not during deposition but in the course of the subsequent diffusion operation. Accordingly, it has heavy p-type doping to the extent of lowering E_F approximately as far as E_V. As a result:

$$\phi_{MS} = 0.55 + \phi_F \qquad \text{(silicon gate p-channel device)} \tag{5.9}$$

The component in Eqn. (5.9) is positive and influences the threshold in the positive direction. With silicon gate n-channel devices, the doping type changes in both the gate and substrate material and then:

$$\phi_{MS} = -0.55 - \phi_F \qquad \text{(silicon gate n-channel device)} \tag{5.10}$$

The combined effect of the potential barrier and work function on threshold can now be found. Examination of Eqns (5.7), (5.8), (5.9), and (5.10) shows that *the ϕ_F term everywhere has the effect of reducing the threshold level* (*amplitude*). As such, it always subtracts from the $2\phi_F$ term in Eqn. (5.6) and *the nett effect is a single ϕ_F term which increases the threshold magnitude.* If we now use V_{TO} to mean the *algebraic* sum of all contributions, we get four expressions as follows:

V_{TO}	p-channel	n-channel	
Aluminium gate	$-0.6 - \phi_F - Q_B/C_{ox}$	$-0.6 + \phi_F + Q_B/C_{ox}$	(5.11)
Silicon gate	$+0.55 - \phi_F - Q_B/C_{ox}$	$-0.55 + \phi_F + Q_B/C_{ox}$	

ϕ_F, Q_B, and C_{ox} are positive numbers. Contributions of ϕ_F and Q_B *enlarge threshold magnitude* as long as the enhancement operation applies.

There is one other major contributor to threshold voltage—one which always shifts V_{TO} in the negative direction—to be discussed.

FIXED POSITIVE SURFACE-STATE CHARGE Q_{ss}

The location of positive charge Q_{ss} (per square centimetre of surface) at the silicon dioxide–silicon interface is an experimental fact. The behaviour of Q_{ss} is now well understood—or, at least, predictable—as follows:

(a) Its magnitude is fixed.
(b) Its magnitude is set independently of t_{ox} or N_B.
(c) It lies in the silicon dioxide within 200 Å of the interface.
(d) It is immobile up to about 250 °C.
(e) The observed magnitude of Q_{ss} as a function of silicon substrate orientation decreases in the order (111)>(110)>(100) in ratios of 3:2:1 approximately.
(f) Its magnitude is closely connected to thermal oxidation processes and depends on both the type of oxide growth (wet or dry) and the temperature. It seems that Q_{ss} varies from one high-temperature treatment to another but the most important fact is that *the last high-temperature treatment, and it alone, determines the final value of* Q_{ss}. This Q_{ss} in turn influences V_{TO}, as will be described.

Apart from the fixed Q_{ss}, other positive charges within the silicon dioxide are hydrogen (H^+) and sodium (Na^+) ions. Sodium, in particular, is a contaminant hard to exclude. Unlike Q_{ss}, these ions move very slowly across the dielectric in response to high gate voltages. Positive gate voltage pushes the ions towards the silicon interface; negative voltage draws them towards the gate. This can cause long-term drift of the threshold voltage and unexpected failures. The effect of any positive ionic charge Q^+ at the silicon dioxide–silicon interface is to require a neutralizing *negative* charge (and voltage) at the gate metal in addition to other components of V_{TO} (see Fig. 5.5). This involves a voltage drop across the oxide of $\sum Q/C_{ox}$ where

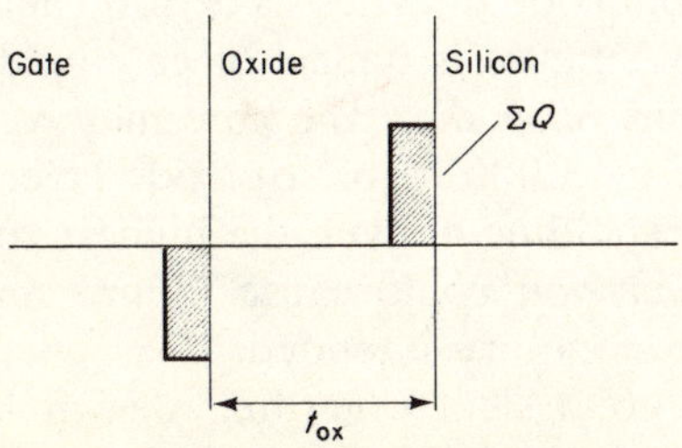

Fig. 5.5.

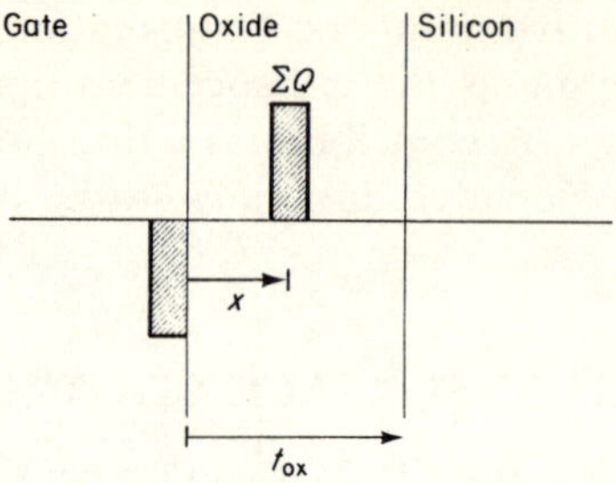

Fig. 5.6.

$\sum Q = Q_{ss} + Q_{Na} + \ldots$, etc., i.e. the total ionic interface charge. A similar charge $\sum Q$ located at a distance $x < t_{ox}$ from the gate metal requires less support voltage since the electric field acts over a shorter distance. In this case (Fig. 5.6), the voltage requirement is:

$$\Delta V_G = -\frac{x}{t_{ox}} \frac{\sum Q}{C_{ox}} \tag{5.12}$$

This indicates that the charge at $x = 0$ (at the gate–oxide interface) has no influence. In conclusion, Q_{ss}, Q_{Na}, *etc., move the threshold voltage in the negative direction*, as given in Eqn. (5.12).

The set of expressions (5.11) are now modified to include a Q_{ss} term at $x = t_{0x}$.

V_{TO}	p-channel	n-channel
Aluminium gate	$-0.6 - \phi_F - Q_B/C_{ox} - Q_{ss}/C_{ox}$	$-0.6 + \phi_F + Q_B/C_{ox} - Q_{ss}/C_{ox}$
Silicon gate	$+0.55 - \phi_F - Q_B/C_{ox} - Q_{ss}/C_{ox}$	$-0.55 + \phi_F + Q_B/C_{ox} - Q_{ss}/C_{ox}$

(5.13)

Appendix J shows the behaviour of theoretical V_{TO} curves versus the substrate doping level for different Q_{ss} values. The curves are for aluminium gate n-channel and p-channel devices with values computed according to the set of expressions (5.13). Oxide thicknesses of 1000 and 15,000 Å are treated in each case. The 1000 Å figure is typical of gate oxide. The 15,000 Å case is shown because regions *other than* the gate may well have an aluminium conductor running over a 15,000 Å (or so) oxide layer over silicon and here one wants to *avoid* thresholding at even the highest voltages likely to occur. Such unwanted *field inversion* could cause failure since it introduces conducting channels where none are intended.

These curves tell a great deal regarding design limitations. Note that p-channel thresholds are essentially negative—enhancement operation only.

On the other hand, n-channel enhancement operation (i.e. positive thresholds) calls for high N_B and low Q_{ss}. In all cases, field inversion thresholds must be acceptably high (greater than the highest supply potential by a safe margin). For $N_B > 10^{15}$ (approx.), Q_B plays a dominant role in setting the threshold. Much more could be said about the implications for the designer and about the trade-offs which must be made. In fact, threshold voltage is a major distinguishing factor between different technologies and is the subject of further discussion in Chapter 9.

The set of expressions also show the implications of a switch from aluminium to silicon gate processing:

For p channel:	1.15 V shift in positive direction
For n channel:	0.05 V shift in positive direction

For p-channel devices a large drop in V_{TO} occurs and could even lead to depletion mode operation. Lower V_{TO} can be very desirable for low-voltage wrist-watch circuits and the like. With n-channel devices, the change is small.

BODY-EFFECT ON THRESHOLD VOLTAGE

Thresholds have been calculated for the situation involving a $2\phi_F$ volt drop from the channel to the substrate. A much more general case is depicted in Fig. 5.7. Again, V_{TO} is measured *using source reference*—not the bulk. A p-channel transistor is depicted with *substrate bias* $V_{BS} = +5$ V. Also, channel potential falls from 0 to -8 V between the source and drain. A selected point X along the channel may be at, say, -6 V. This implies a depletion layer wide enough to support $(6+5+2\phi_F)$ volts at that point. V_{TO}, measured from the source, does not include V_{BS}. What does happen is that the now wider depletion layer uncovers a larger Q_B charge and this acts directly *to enlarge* V_T. The result is a threshold voltage which varies with bulk bias and from point to point along the channel.

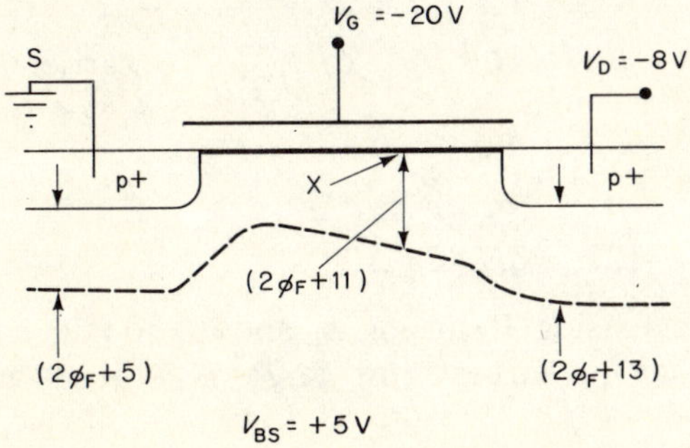

Fig. 5.7. Bulk-effect.

If V_{CB} describes the *magnitude* of the channel to bulk voltage at point X, the added ΔQ_B, over and above zero bias conditions, is, from Eqn. (5.4):

$$\Delta Q_B = [2k_{si}e_0 qN_B(V_{CB}+2\phi_F)]^{1/2} - [2k_{si}e_0 qN_B(2\phi_F)]^{1/2}$$
$$= (2k_{si}e_0 qN_B)^{1/2}[(V_{CB}+2\phi_F)^{1/2} - (2\phi_F)^{1/2}]$$

The addition to the threshold voltage is $\Delta V_T = \Delta Q_B / C_{ox}$, i.e. $\Delta V_T = \Delta Q_B t_{ox}/k_{ox}e_0$, which gives:

$$\Delta V_T = \frac{t_{ox}}{k_{ox}e_o}(2k_{si}e_0 qN_B)^{1/2}[(V_{CB}+2\phi_F)^{1/2} - (2\phi_F)^{1/2}]$$

$$= \frac{t_{ox}}{k_{ox}}\left(2k_{si}\frac{qN_B}{e_0}\right)^{1/2}[(V_{CB}+2\phi_F)^{1/2} - (2\phi_F)^{1/2}]$$

Introducing the *body-effect factor M:*

$$M = \frac{t_{ox}}{k_{ox}}\left(\frac{2k_{si}qN_B}{e_0}\right)^{1/2} \tag{5.14}$$

The total threshold voltage in the channel at point X (an arbitrary point) is:

$$V_{TCH} = V_{TO} + M[(V_{CB}+2\phi_F)^{1/2} - (2\phi_F)^{1/2}] \tag{5.15}$$

with V_{CB} = *channel to bulk voltage magnitude. M* ranges typically between 0.5 and 10. As it acts to *enlarge threshold magnitude*, it should appear with a negative sign for p-channel and a positive sign for n-channel devices. Body-effect is generally quite large and tends to complicate MOSFET analysis. The difficulty arises from the fact that, with body-effect, V_T is no longer constant. It varies from point to point along the channel and it varies with time whenever circuit conditions are changing.

TEMPERATURE DRIFT OF THRESHOLD VOLTAGE

The temperature-variable components of threshold voltage are ($\phi_F + Q_B/C_{Ox}$), and both of these parameters tend to enlarge V_T. Both ϕ_F and Q_B decrease with increasing T as indicated by:

$$n_i(T) = 3.87 \times 10^{16} T^{3/2} \exp\frac{-1.21}{2kT/q} \qquad \text{carriers/cm}^3$$

$$\phi_F = \frac{kT}{q}\text{Ln}\frac{N_B}{n_i}$$

$$Q_B = [2k_{si}e_0 qN_B(2\phi_F)]^{1/2}$$

n_i increases with increasing T causing ϕ_F and Q_B to fall. The overall effect is fairly small and, at its largest (for large N_B), it is around the figure -5.0 mV/°C.

The examples which follow are designed to illustrate the threshold voltage levels and the associated difficulties which are characteristic of p-channel and n-channel circuits and of the fabrication processes employed. The subject arises again in Chapter 9 where a number of technologies are evaluated on their various performance characteristics.

EXAMPLE 5.1

A typical p-channel process uses 5 Ω · cm n-type wafers with 1100 Å gate oxide and 14,000 Å oxide thickness in field regions. The typical Q_{ss}/q is 3.5×10^{11} charges/cm². Find the gate and field threshold voltages for aluminium gate technology. Find the new gate threshold if a silicon gate is used.

Solution

First determine:

$$\phi_F = \frac{kT}{q} \text{Ln} \frac{N_B}{n_i}$$

Assume:

$$T = 300\text{ K}; \qquad n_i = 1.41\times 10^{10}$$

Appendix D gives:

$$N_B = 10^{15}\text{ atoms/cm}^3$$

$$\phi_F = 0.0258\text{ Ln}\frac{10^{15}}{1.41\times 10^{10}} = \underline{0.29\text{ V}}$$

Then:

$$\begin{aligned} Q_B &= [2k_{si}e_0 q N_B(2\phi_F)]^{1/2} \\ &= (2\times 11.7\times 8.85\times 10^{-14}\times 1.6\times 10^{-19}\times 10^{15}\times 2\times 0.29)^{1/2} \\ &= \underline{1.38\times 10^{-8}\text{ C/cm}^2} \end{aligned}$$

$$C_{ox(G)} = \frac{k_{ox}e_0}{t_{oxG}} = \frac{3.9\times 8.85\times 10^{-14}}{1100\times 10^{-8}} = 3.14\times 10^{-8}\text{ F/cm}^2$$

$$C_{ox(F)} = \frac{3.9\times 8.85\times 10^{-14}}{14{,}000\times 10^{-8}} = 2.46\times 10^{-9}\text{ F/cm}^2$$

Aluminium gate:

$$V_{TO(G)} = -0.6 - \phi_F - \frac{1}{C_{ox}}(Q_B + Q_{ss})$$

$$= -0.6 - 0.29 - \frac{1.38\times 10^{-8} + 3.5\times 10^{11}\times 1.6\times 10^{-19}}{3.14\times 10^{-8}}$$

$$= 0.6 - 0.29 - 0.44 - 1.78 = \underline{-3.11\text{ V}}$$

$$V_{TO(F)} = -0.6 - 0.29 - \frac{1.38\times10^{-8} + 5.6\times10^{-8}}{2.46\times10^{-9}}$$

$$= -0.6 - 0.29 - 5.61 - 22.76 = \underline{-29.2\text{ V}}$$

In changing to silicon gate, add +1.15 giving:

Silicon gate:

$$V_{TO(G)} = -3.11 + 1.15 = \underline{-1.96\text{ V}}$$

EXAMPLE 5.2

An n-channel process uses a low resistivity p-type wafer ($1\,\Omega\cdot$cm) with (100) orientation which gives a low valued $Q_{ss}/q = 1.0\times10^{11}$ and gate oxide thickness 1000 Å. These measures are intended to ensure enhancement mode operation. Find V_{TO} from the equations and check the result against Appendix J. If sodium ions can contribute an additional $Q_{Na^+} = 2.9\times10^{-8}$ C/cm^2, what are the likely consequences of long-term sodium drift? Determine the field thresholds, with and without sodium effects, for 15,000 Å of field oxide.

Solution

From Appendix D:

$$N_B = 1.6\times10^{16}$$

$$\phi_F = \frac{kT}{q}\text{Ln}\frac{N_B}{n_i} = 0.0258\text{ Ln}\frac{1.6\times10^{16}}{1.4\times10^{10}}$$

$$= \underline{0.36\text{ V}}$$

$$Q_B = (2\times11.7\times8.85\times10^{-14}\times1.6\times10^{-19}\times1.6\times10^{16}\times2\times0.36)^{1/2}$$

$$= \underline{6.18\times10^{-8}\text{ C/cm}^2}$$

$$C_{ox(G)} = \frac{3.9\times8.85\times10^{-14}}{1000\times10^{-8}} = 3.45\times10^{-8}\text{ F/cm}^2$$

$$C_{ox(F)} = \frac{3.9\times8.85\times10^{-14}}{15{,}000\times10^{-8}} = 2.3\times10^{-9}\text{ F/cm}^2$$

$$Q_{ss} = 1.0\times10^{11}\times1.6\times10^{-19} = 1.6\times10^{-8}\text{ C/cm}^2$$

$$Q_{Na^+} = 2.9\times10^{-8}$$

Without Na^+:

$$V_{TO} = -0.6 + \phi_F + \frac{Q_B}{C_{ox}} - \frac{Q_{ss}}{C_{ox}}$$

$$= -0.6 + 0.36 + 1.8 - 0.46$$

$$= \underline{+1.1\text{ V}}$$

Normal gate voltage is positive and sodium ions will tend to drift towards the silicon–silicon dioxide interface where they have maximum effect. They reduce V_{TO} by:

$$\frac{Q_{Na^+}}{C_{ox}}=\frac{2.9\times10^{-8}}{3.45\times10^{-8}}=0.84\ \text{V}$$

With full Na^+ influences:

$$V_{TO}=1.1-0.84=\underline{+0.26\ \text{V}}$$

A slightly worse situation would lead to depletion mode operation as a result of long-term drift.

Field threshold is as follows:

$$V_{TF}=-0.6+0.36+\frac{6.18\times10^{-8}}{2.3\times10^{-9}}-\frac{1.6\times10^{-8}}{2.3\times10^{-9}}-\frac{2.9\times10^{-8}}{2.3\times10^{-9}}$$

$$=-0.6+0.36+26.8-6.95-12.6$$

$$=19.6-12.6$$

i.e. 19.6 V without Na^+ effect and 7.0 with Na^+ drift (a possible failure mode). The results show agreement with the graphical data in Fig. J1.

EXAMPLE 5.3

Calculate the body-effect factor M for the processes of Examples 5.1 and 5.2. Assume the aluminium gate process and ignore Na^+ effects; then calculate the 'output high' voltage for a saturated load ($V_{GG}=V_{DD}$) if supplies are −18 V (p channel) and +10 V (n channel) respectively.

Solution

$$M=\frac{t_{ox}}{k_{ox}}\left(\frac{2k_{si}qN_B}{e_0}\right)^{1/2}$$

For Example 5.1:

$$M=\frac{1100\times10^{-8}}{3.9}\left(\frac{2\times11.7\times1.6\times10^{-19}\times10^{15}}{8.85\times10^{-14}}\right)^{1/2}=0.58$$

For Example 5.2:

$$M=\frac{1000\times10^{-8}}{3.9}\left(\frac{2\times11.7\times1.6\times10^{-19}\times1.6\times10^{16}}{8.85\times10^{-14}}\right)^{1/2}=2.1$$

$$V_{TCE}=V_{TO}+M[(V_{CB}+2\phi_F)^{1/2}-(2\phi_F)^{1/2}]$$

$$V_{TCE}=V_{TO}+M[(V_{DD}-V_{TCE}+2\phi_F)^{1/2}-(2\phi_F)^{1/2}]$$

since the channel end voltage is ($V_{DD}-V_{TCE}$) with respect to GND or bulk.

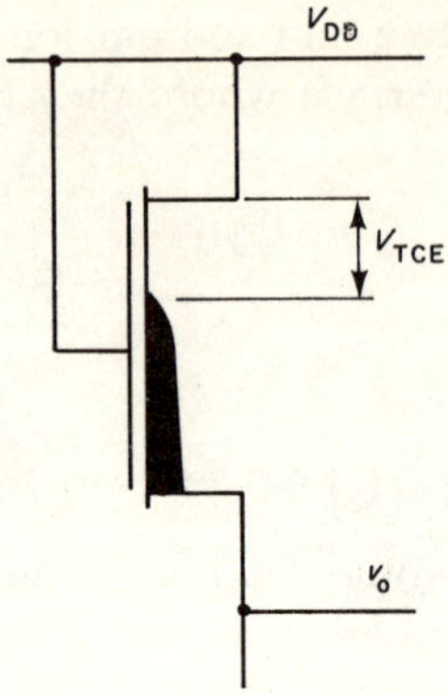

A convenient solution is to *iterate:*
For Example 5.1:

$$V_{TCE} = 3.11 + 0.58[(18 - V_{TCE} + 2 \times 0.29)^{1/2} - (2 \times 0.29)^{1/2}]$$

$$V_{TCE} = 3.11 + 0.58[(18.58 - V_{TCE})^{1/2} - (0.58)^{1/2}] \qquad \text{(using positive signs)}$$

Try $V_{TCE} = 3.5$ V:

$$3.5 \rightarrow 4.92$$

Try $V_{TCE} = 4.5$ V:

$$4.5 \rightarrow 4.84 \qquad \text{(getting closer!)}$$

Try $V_{TCE} = 4.8$ V:

$$4.8 \rightarrow 4.82 \qquad \text{(acceptable)}$$

Take 4.81 as the final figure.

Taking account of the sign for the p-channel process:

$$\underline{V_{TCE} = -4.81 \text{ V}}$$

$$\underline{V_{OH} = -(18 - 4.81) = -13.19 \text{ V}}$$

For Example 5.2:

$$V_{TCE} = 1.1 + 2.1[(10 - V_{TCE} + 0.72)^{1/2} - (0.72)^{1/2}]$$

Try $V_{TCE} = +5$ V:

$$V_{TCE} \rightarrow 4.34$$

Try 4.5 V:

$$4.5 \rightarrow 4.55$$

Try 4.52 V:

$$4.52 \rightarrow 4.54$$

Try 4.54 V:

$$4.54 \rightarrow 4.54 \qquad \text{(correct)}$$

$$\underline{V_{TCE} = 4.54 \text{ V}}$$

$$\underline{V_{OH} = 10 - 4.54 = 5.46 \text{ V}}$$

Note the large body-effect for the n-channel process.

CHAPTER 6

Current Control in MOS Transistors

The terminal properties of a MOS device are formulated in this chapter. We derive a simple expression for current flow in a MOSFET from first principles. Taking the case of an n-channel device of width W and length L from the source (S) to the drain (D), we determine I_{DS} in terms of V_{GS}, V_{DS} and device constants assuming non-saturated conditions. The derivation is somewhat simplified and fairly brief but, hopefully, offers some insights in to the behaviour of the MOS device. Despite some shortcomings this model is entirely adequate for most purposes. The reader who wishes to avoid the derivation can move directly to the result in Eqn. (6.1).

For a channel with current density $J(x, y)$ as in Fig. 6.1 and depth x_c:

$$I_{DS} = W\int_0^{x_c} J(x, y)\,\mathrm{d}x$$

and since $J = nq\mu E$ where E is the electric field as shown, $E = \delta V(x, y)/\delta y$, we get:

$$I_{DS} = W\int_0^{x_c} nq\mu \frac{\delta V(x, y)}{\delta y}\,\mathrm{d}x$$

where n is the electron concentration in the channel, q is the electron charge, and μ is the mobility of channel electrons. It is known that μ varies across the channel and even channel potential V may depend slightly on x. However, to preserve simplicity we proceed with an average mobility $\bar{\mu}$ and treat V as a function of y only. This gives:

$$I_{DS} = Wq\bar{\mu}\frac{\mathrm{d}V}{\mathrm{d}y}\int_0^{x_c} n\,\mathrm{d}x$$

We identify:

$$\begin{aligned} q\int_0^{x_c} n\,\mathrm{d}x &= \text{channel electron charge per square centimetre} \\ &= \text{gate charge} - V_T \text{ inducing charge} \\ &= C_{ox}[V_{GS} - V(y)] - C_{ox}V_T \\ &= C_{ox}[V_{GS} - V_T - V(y)] \end{aligned}$$

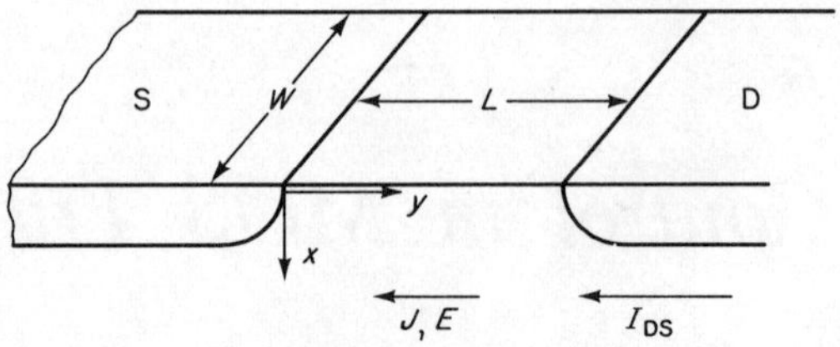

Fig. 6.1. Device conventions.

where C_{ox} is gate dielectric capacitance per square centimetre. Thus:

$$I_{DS} = W\bar{\mu}\frac{dV}{dy}C_{ox}[V_{GS} - V_T - V(y)]$$

$$\int_0^L I_{DS}\,dy = W\bar{\mu}C_{ox}\int_0^{V_{DS}}[V_{GS} - V_T - V(y)]\,dV(y)$$

$$I_{DS}L = W\bar{\mu}C_{ox}[(V_{GS} - V_T)V_{DS} - 0.5V_{DS}^2]$$

Finally:

$$I_{DS} = \beta'\frac{W}{L}[(V_{GS} - V_T)V_{DS} - 0.5V_{DS}^2] \qquad \text{(non-saturated)} \qquad (6.1)$$

where

$\beta' = C_{ox}\bar{\mu} = k_{ox}e_0\bar{\mu}/t_{ox} =$ 'gain' for a square device, A/V^2

$\bar{\mu} =$ effective channel mobility

$V_T =$ threshold voltage

$W =$ channel width

$L =$ channel length

The 'gain' β' is unlike the bipolar transistor β since the MOSFET is voltage-controlled. For device dimensions W and L we define:

$$\beta = \beta'\frac{W}{L} \qquad A/V^2 \quad \text{or} \quad ℧/V \qquad (6.2)$$

Typical β' is equal to 5 $\mu A/V^2$. It is influenced by many factors as discussed later. In general, high W/L gives a high-current device. Very low W/L is suitable for high-resistance load transistors. Note that Eqn. (6.1) takes no account of body-effect on threshold voltage. Because body-effect is largest at the drain end, then, for a conservative estimate, the highest (drain end) threshold could be used.

Equation (6.1) describes operation in a non-saturated mode only, i.e. for $(V_{GS} - V_T) > V_{DS}$. For saturated operation, the channel end voltage is no longer V_{DS} but $(V_{GS} - V_T)$. Making this substitution in Eqn. (6.1) gives

$$I_{DS} = \frac{\beta'}{2}\left(\frac{W}{L - L'}\right)(V_{GS} - V_T)^2 \qquad \text{(saturated)} \qquad (6.3)$$

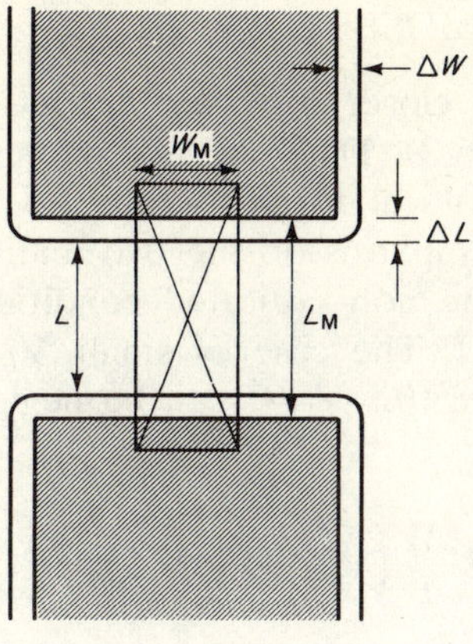

Fig. 6.2.

The actual channel length $(L-L')$ is used in Eqn. (6.3) and thus takes account of channel length modulation. The distance L' across the drain-depletion layer is calculated in the manner of a single-sided step junction:

$$L' = \left\{\frac{2k_{si}e_0[V_{DS}-(V_{GS}-V_{TCE})]}{qN_B}\right\}^{1/2} \tag{6.4}$$

The voltage term is simply the drain-to-channel end voltage where V_{TCE} is the channel end threshold voltage. This is larger than V_{TO} due to body-effect. Body-effect is determined by channel-to-bulk voltage V_{CB} and, at the channel end:

$$V_{CB} = (V_{GS} - V_{TCE}) + V_{SB} \tag{6.5}$$

Then V_{TCE} is evaluated from:

$$V_{TCE} = V_{TO} + M\{[(V_{GS} - V_{TCE}) + V_{SB} + 2\phi_F]^{1/2} - (2\phi_F)^{1/2}\} \tag{6.6}$$

All quantities are positive numbers for n-channel working. For p-channel working, both V_{TO} and the body-effect term are negative. Equation (6.6) is a quadratic in V_{TCE}. It can be solved using the standard formula approach or, just as conveniently, by iteration, using a calculator. The worked examples at the end of the chapter will further clarify the sign convention in use.

The effective value of length L differs from the mask dimension (call it L_M) due to lateral spreading of the source and drain diffusions. As Fig. 6.2 illustrates:

$$\left(\frac{W}{L}\right)_{eff} = \frac{W_M}{L_M - 2\Delta L - L'} \tag{6.7}$$

in the general case and, for junction depth x_j, $\Delta L = 0.75x_j$ (approx.). If the gate metal extends beyond the side walls of the diffusions, then the effective width becomes $(W_M + 2\Delta W)$. These adjustments can greatly alter the effective geometry of the device.

It is worth while to take a closer look at the MOSFET behaviour as described by Eqns (6.1) and (6.3). They describe modes of behaviour which were previously seen in graphical form in Fig. 2.6. The same graph is reproduced as Fig. 6.3 with emphasis on the two main operating zones.

Equation (6.1) describes the non-saturated condition with a full-length channel and $V_{DS}<(V_{GS}-V_T)$. The case of small V_{DS} ($\ll V_{GS}-V_T$) is of interest because then the term $(0.5\,V_{DS})^2$ can be neglected and Eqn. (6.1) reduces to:

$$I_{DS}=\beta'\frac{W}{L}(V_{GS}-V_T)V_{DS} \tag{6.8}$$

or

$$\frac{V_{DS}}{I_{DS}}=R_{on}=\frac{1}{\beta'(W/L)(V_{GS}-V_T)} \tag{6.9}$$

In Eqn. (6.9) the device is seen simply as a linear resistance under gate control. It is widely used in this mode as a switch which is turned ON and OFF by taking the gate to supply potential and to ground respectively. The 'OFF' switch has no channel and only a very small leakage current flows. The 'ON' switch has an ON-resistance R_{on} of a few hundred ohms typically. This may seem high at first glance but if it appears as a series element in a high-impedance circuit (as it usually does) the current is very low and the voltage drop across the switch may be negligible. The obvious example of this application is the 'pass' transistor seen in Chapter 2 where it provided

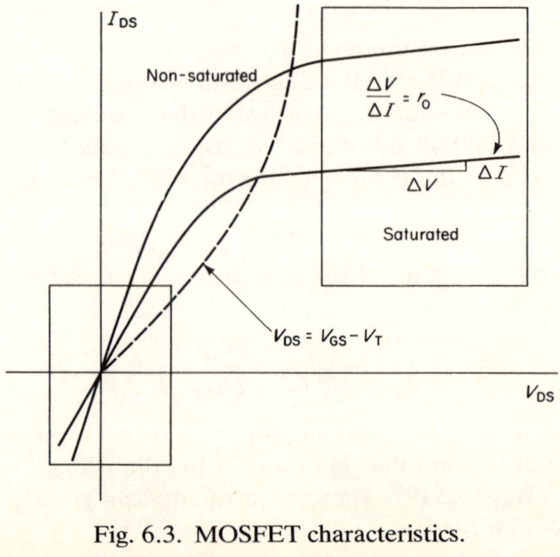

Fig. 6.3. MOSFET characteristics.

isolation, if required, between successive inverter stages. It is also a valid option in analog signal multiplexing and will be discussed in that role in Chapter 10.

When the transistor operates in the non-saturated mode as a switch, there is no real distinction between the roles of the source and drain. Current flow can and does change direction and for that reason the linear part of the characteristic is deliberately extended into the third quadrant in Fig. 6.3.

In the saturated mode described by Eqn. (6.3), there is a clear distinction between the source and drain. The drain is the end nearer the supply rail. The channel extends out from the source but does not reach the drain. At first glance, current is controlled by V_{GS} alone. This would imply that the small-signal resistance, r_0, viewed from the drain is infinite (see Fig. 6.4). Such is not the case because an increase in drain potential causes L' to increase and so the channel $(L-L')$ grows shorter. This, in turn, causes increased current flow. It turns out that r_0 is not very large after all and this appears again as a significant parameter in the discussion of analog MOS circuits in Chapter 10. With reference to Fig. 6.3, the finite value of r_0 means that the graphs in the saturated region are not fully horizontal but have a small positive slope.

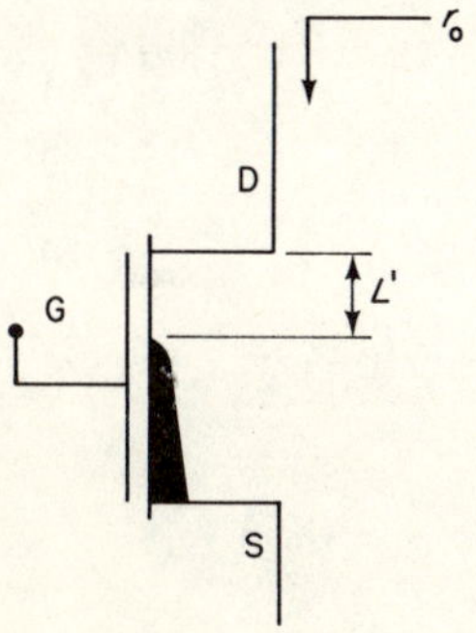

Fig. 6.4. Saturated FET.

LIMITATIONS OF THE MODEL

The model employed to obtain the current-control equations (6.1) and (6.3) is less than precise because of its failure to take account of body effect on threshold voltage. Stated otherwise, it made provision for a constant Q_B term as given by equation (5.4) but failed to recognize that the total Q_B increases as a function of channel-to-bulk potential. The precise solution is a little more complex. It is not strongly favoured in this book because the results are a bit too cumbersome for hand calculation. Nevertheless, it is helpful to know how our model differs from the more exact one and, in particular, to see how it can be made to yield answers which come close to the precise ones even under unfavourable conditions. For this reason, we now state the exact equation for non-saturated operation:

$$I_{DS} = \beta'\left(\frac{W}{L}\right)[V_{GS} - V_{TP})V_{DS} - 0.5V_{DS}^2 - (2M/3)\cdot D] \tag{6.10}$$

with

$$D = (V_{DB} + 2\phi_F)^{3/2} - (V_{SB} + 2\phi_F)^{3/2} \tag{6.11}$$

V_{DB} = drain-to-bulk potential.

V_{SB} = source-to-bulk potential.

$V_{TP} = V_{TO} - Q_B/C_{ox}$.

V_{TP} is the basic threshold voltage excluding the Q_B term. Because it depends only on features of the device fabrication, it is a 'process variable' and is fixed (independent of voltage levels) for a given process. Compatibility with the simpler equation (6.1) can be demonstrated for the case where:

$$V_{DS} \ll V_{SB} + 2\phi_F \tag{6.12}$$

In this situation, the expression for D can be shown to approximate to:

$$D = (3/2)V_{DS}(V_{SB} + 2\phi_F)^{1/2} \tag{6.13}$$

and substituting in (6.10) gives:

$$I_{DS} = \beta'\left(\frac{W}{L}\right)[(V_{GS} - V_{TP})V_{DS} - 0.5V_{DS}^2 - V_{DS}M(V_{SB} + 2\phi_F)^{1/2}] \tag{6.14}$$

For grounded bulk, $V_{SB} = 0$ and $M(2\phi_F)^{1/2} = Q_B/C_{ox}$ so that

$$I_{DS} = \beta'\left(\frac{W}{L}\right)\left[\left(V_{GS} - V_{TP} - \frac{Q_B}{C_{ox}}\right)V_{DS} - 0.5V_{DS}^2\right]$$

$$= \beta'\left(\frac{W}{L}\right)[(V_{GS} - V_{TO})V_{DS} - 0.5V_{DS}^2] \tag{6.15}$$

which is in agreement with Eqn. (6.1). When bulk bias is applied, $V_{SB} \neq 0$, it can be seen that:

$$V_{TP} + M(V_{SB} + 2\phi_F)^{1/2}$$

is in fact the threshold voltage calculated at the source and including body effect. Thus, (6.14) becomes

$$I_{DS} = \beta'\left(\frac{W}{L}\right)[(V_{GS} - V_{TS})V_{DS} - 0.5V_{DS}^2] \tag{6.16}$$

We can conclude that, as long as Eqn. (6.12) is valid, the simpler equation in the form of (6.1) will give good results if the threshold voltage is calculated at the source and takes account of body effect.

Equation (6.10) cannot be further simplified when V_{DS} is large. It can be observed, however, that it yields current values which are smaller than those predicted by (6.16). The reason is easily seen: while (6.16) takes account of body effect at the source, it does not recognize the increased body effect and higher thresholds that occur at other points in the channel. Higher threshold voltage would clearly give reduced I_{DS}. The question therefore arises: What value of threshold voltage might be used in the simple model to give good agreement with (6.10) even under worst-case conditions? An experimental answer to this is illustrated in Fig. 6.5. The case chosen is that of an n-channel transistor with $m = 2.5$, $V_{TP} = -0.4$ V, and $\beta = 170\ \mu A/V^2$ with $V_{GS} = +12$ V. The non-saturated portion of the characteristic is illustrated in Fig. 6.5(a) for the grounded-bulk situation and in Fig. 6.5(b) for 5 V of bulk bias ($V_{SB} = +5$ V). Curves are given for the exact solution of Eqn. (6.10) and for the simpler model using thresholds evaluated at source (V_{TS}), at the drain (V_{TD}) and also using an average figure $V_{TM} = (V_{TS} + V_{TD})/2$. It is evident that, even for high V_{DS}, the use of V_{TM} gives very acceptable results. It is clear too that calculations for case (b), which includes bulk bias, are less error-prone. This can be anticipated from Eqn. (6.12).

Note also that the high value of M enlarges differences between the two models. For p-channel circuits, M is generally much smaller and better agreement can be expected.

The transition to saturated operation occurs for $V_{DS} = V_{GS} - V_{TD}$. Notice that it is the drain-end threshold that marks this transition because it is, by definition, the point at which channel support *at the drain* becomes zero. Using V_{DSAT} to mean the value of V_{DS} at the onset of saturation, it follows that:

$$V_{DSAT} = V_{GS} - V_{TD} \tag{6.17}$$

$$V_{DSAT} = V_{GS} - V_{TP} - M[V_{DSAT} + V_{SB} + 2\phi_F]^{1/2} \tag{6.18}$$

V_{DSAT} is readily determined from (6.18) by iteration. It is important to be aware that this result is in full agreement with the exact model. No approximation is involved. For higher values of V_{DS}, V_{DSAT} is still quite meaningful: It is the channel-end voltage ($V_{GS} - V_{TCE}$) and is in agreement with Eqn. (6.6).

For operation in the saturated region, the channel-end voltage is fixed, independent of V_{DS}. Consequently, the body-effect influence is fixed. In this region, therefore, only channel-length modulation produces further increases in I_{DS}. The current is reasonably well predicted as:

$$I_{DS} = \frac{\beta'}{2}\left(\frac{W}{L - L'}\right)(V_{GS} - V_{TM})^2 \tag{6.19}$$

The following conclusions can now be drawn from our examination of the current model:

(a) Equations (6.1) and (6.3) yield accurate results when V_T is selected as

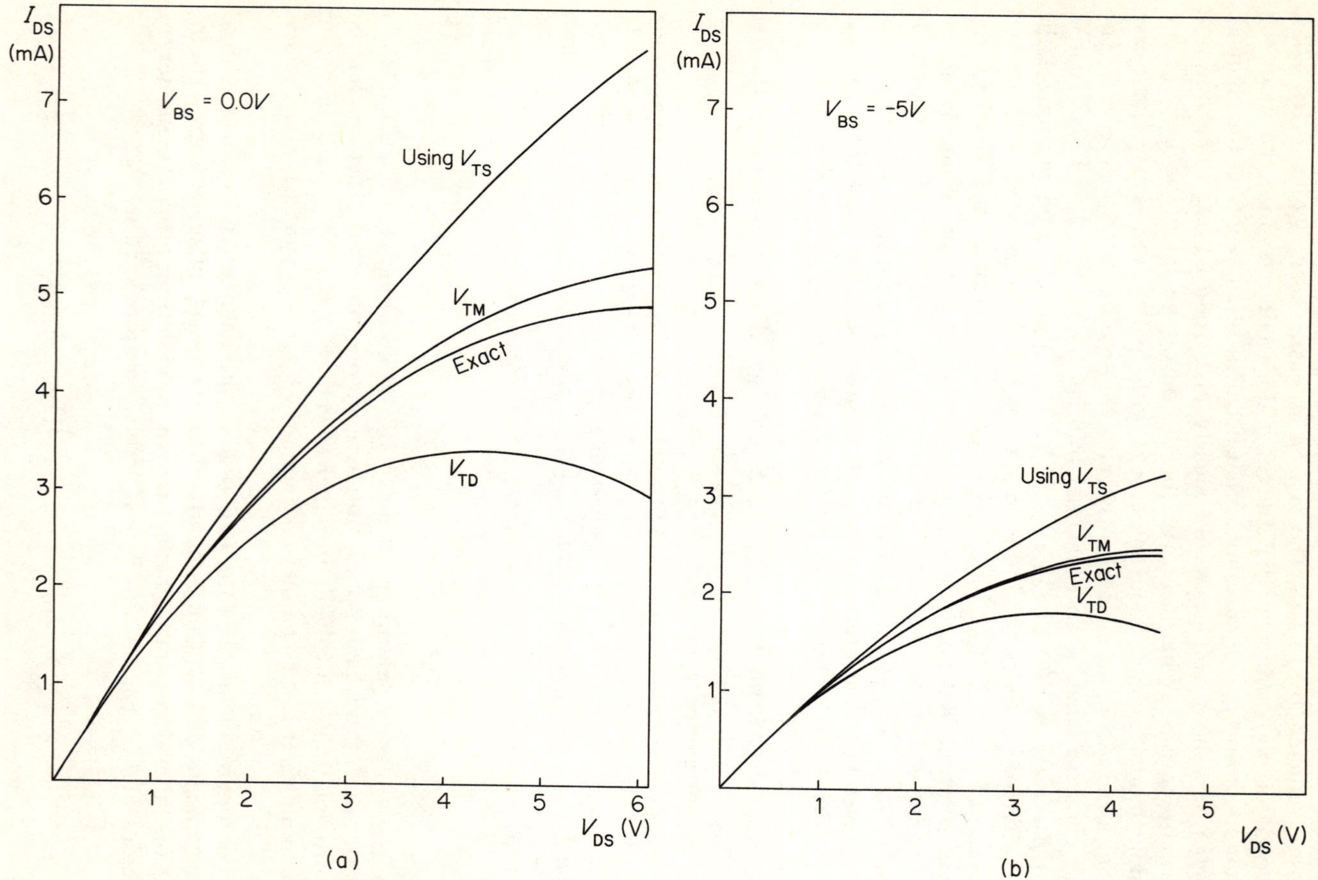

Fig. 6.5. Comparisons with the exact model.

the average of threshold levels at opposite ends of the conducting channel.

(b) The transition to saturated working is determined exactly by Eqn. (6.18) which yields V_{DSAT} and also the channel-end threshold which is simply $V_{TCE} = (V_{GS} - V_{DSAT})$.

VARIATIONS IN β'

It was seen that $\beta' = C_{ox}\bar{\mu} = k_{ox}e_0\bar{\mu}/t_{ox}$. In practice β' is by no means constant, even from point to point in the channel at a given time. It is influenced by many factors which are now listed and separately discussed.

1. Electric fields E_G and E_B above and below the channel
2. Channel type and bulk doping level N_B
3. Wafer orientation
4. Temperature
5. Ion implant
6. Oxide thickness

1. Electric fields

Channel mobility tends to be adversely influenced by the electric field perpendicular to the direction of current flow. Above the channel, the field E_G is determined by the *channel support voltage* V_{CS}:

$$V_{CS} = V_{GS} - V_{CHS} - V_{TCH} \tag{6.20}$$

or, verbally, V_{CS} = gate voltage − channel voltage − threshold voltage at the point of evaluation. V_{CS} normally varies along the channel. The field E_B under the channel is determined by the *channel-to-bulk voltage* V_{CB}:

$$V_{CB} = V_{SB} + V_{CHS} \tag{6.21}$$

In general, β' falls with increasing values of V_{CS} and of V_{CB}. To achieve good accuracy in device characterization, the β' value could be taken as the average of the two estimates for opposite ends of the channel.

2. Channel type and bulk doping level

Surface mobilities are typically only about 50% of mobility values in bulk silicon. They are very sensitive to bulk doping level N_B and the following figures are fairly representative:

N_B(cm^{-3}):		10^{15}	10^{16}
Mobility:	electrons	500	200
(cm^2/v-sec)	holes	300	150

These figures make it clear that high doping levels have a large adverse effect on mobility and, consequently, on device performance. As expected, electrons are more mobile than holes.

3. Wafer orientation

In general, the (111) orientation yields higher mobility figures than (100).

4. Temperature

Experimental data agrees well with the lattice scattering theory which predicts a mobility that varies as $T^{-3/2}$. Thus, for a known value β_0 at temperature T_0 in deg K, the value β at some other temperature T is obtained from:

$$\frac{\beta'}{\beta'_0}=\left(\frac{T}{T_0}\right)^{-3/2} \tag{6.22}$$

For example, a comparision between values at 25 °C and 150 °C yields:

$$(\beta'/\beta'_0)=[(273+150)/(273+25)]^{-3/2}=0.59$$

The β at 150 °C is only 59% of the 25 °C value.

5. Ion implant

Ion implantation changes the effective Q_B (atoms per cm^2) and alters mobility accordingly.

6. Oxide thickness

β' varies in inverse proportion to t_{ox}.

EVALUATION OF PARAMETERS

The device characteristics, as represented graphically in Fig. 6.3, can be displayed very conveniently with the aid of a curve tracer. Fig. 6.6 shows the results obtained in this manner for an n-channel transistor. Comparison of the two photographs (with the same scale factors for both) shows the dramatic fall in current levels due to body effect when 5V of bulk bias are applied.

While device parameters can be deduced from such data, body-effect is a complicating feature. The complication can be avoided in tests where V_{DS} is maintained as a very low and constant voltage level. Such tests are described and documented in Appendix K for p-channel and n-channel specimens. They are well suited to the determination of the basic threshold voltage V_{TO} and, by measuring the higher threshold caused by substrate bias, the

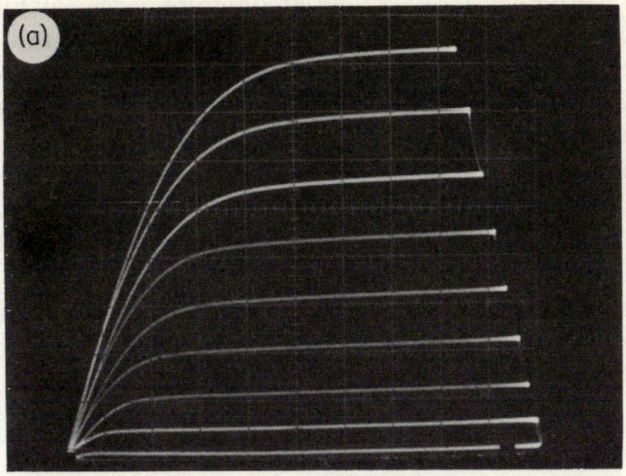

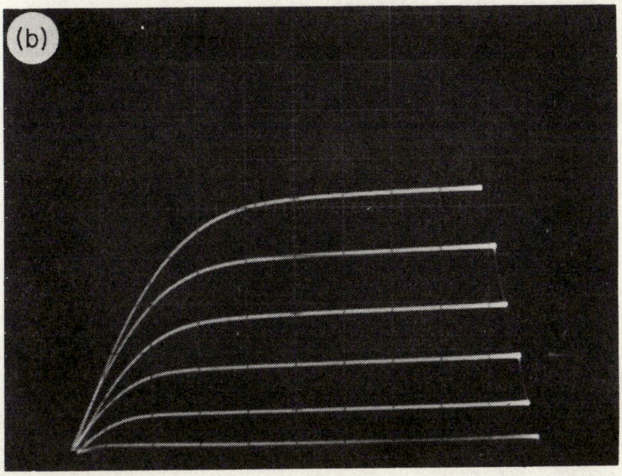

Fig. 6.6. Characteristics of an n-channel FET of mask dimensions $W = 1.5$ mil, $L = 0.4$ mil. Horizontal scale: V_{DS} at 1 V/Div. Vertical scale: I_{DS} at 0.5 mA/Div. Gate voltage applied in 10 steps of 1 V from zero. Photograph (a) uses $V_{SB} = 0$ V and (b) uses $V_{SB} = +5$ V. Body effect reduces currents substantially.

body-effect factor M can be calculated. Note that the (I_{DS} vs. V_{GS}) characteristics should, in theory, be straight lines with an offset corresponding to the threshold voltage. In practice, currents are lower than expected at high gate voltages. This is due to the transverse field effect which causes mobility, and hence β', to decrease. Application of bulk bias increases the transverse field under the channel and similarly reduces β'. The appendix includes plots of β' versus channel-support voltage at different levels of bulk bias. This data is computed from the I_{DS}/V_{GS} graphs using the non-saturated current equation. Note that precise results can be expected from this equation since V_{DS} is very small. It should also be noted that the computed β' figures are global estimates and are not strictly appropriate to small-signal modelling. The information regarding β' obtained in this way can be used to attempt accurate performance predictions. Most often, it permits a well informed decision as to what value of β' might seem appropriate to specified working conditions.

The operations described in Appendix K provide for experimental determination of all parameters required by our model. The particular results obtained have certain features which are normally observed and are worth noting:

(a) The basic threshold V_{TO} tends to be lower (in absolute magnitude) for the n-channel case.
(b) Body-effect is much more pronounced for n-channel. $|M| = 2.1$ compared with 0.4 for the p-channel case.
(c) n-Channel devices have a typical β' of 16 μA/V^2 compared with 6 μA/V^2 for p-channel.
(d) Channel mobility is more heavily influenced by gate voltage than by bulk bias.

EXAMPLE 6.1

A p-channel driver device is formed with a geometric length $L = 0.4$ mils (the minimum recommended for the process) and width $W = 10$ mils on an n-type wafer with a bulk doping level $N_B = 1.3 \times 10^{15}$ atoms/cm^3. Gate oxide thickness is 1200 Å and $V_{TO} = -2.3$ V. Take ΔL as 0.06 mils and $V_{BS} = 0$ V. Find the drain end threshold voltage and the current through the device when $V_{GS} = -12$ V and $V_{DS} = -5$ V for an assumed β' of 5 μA/V^2. Take $2\phi_F$ as 0.6 V.

Repeat the current estimate using the p-channel β' data from Appendix K.

Solution

$$M = \frac{t_{ox}}{k_{ox}} \left(\frac{2k_{si}qN_B}{e_0}\right)^{1/2}$$

$$= \frac{1200 \times 10^{-8}}{3.9} \left(\frac{2 \times 11.7 \times 1.6 \times 10^{-19} \times 1.3 \times 10^{15}}{8.85 \times 10^{-14}}\right)^{1/2}$$

$$= 0.72$$

Assume non-saturated conditions. The drain end threshold becomes:

$$V_{TD} = V_{TO} + M[(V_{DB} + 2\phi_F)^{1/2} - (2\phi_F)^{1/2}]$$
$$= -2.3 - 0.72[(5+0.6)^{1/2} - (0.6)^{1/2}] = -3.45\ \text{V}$$
$$= \underline{\text{channel-end threshold (at drain)} = -3.45\ \text{V}}$$

Note that $V_{GS} - V_{TD} > V_{DS}$ which verifies non-saturated operation. A value for β' of $5\ \mu\text{A/V}^2$ gives:

$$\beta = 5 \times 10^{-6} \frac{10}{0.4 - 2(0.06)} = 1.78 \times 10^{-4}\ \text{A/V}^2$$

Then:

$$I_{DS} = \beta[(V_{GS} - V_{TM})V_{DS} - 0.5V_{DS}^2]$$

and

$$V_{TM} = -(2.3 + 3.45)/2 = -2.87,$$

giving

$$I_{DS} = 1.78 \times 10^{-4}[(12 - 2.87)5 - 0.5(5^2)] = \underline{5.90\ \text{mA}}$$

Using p-channel data from Appendix K, β' can be taken as the average of values at the source and drain ends as follows:
At the source, E_G is set by (12–2.3) or 9.7 V. Also $V_{BS} = 0$ giving $E_B = 0$. From the graph:

$$\underline{\beta'_{(S)} = 4.85 \times 10^{-6}}$$

At the drain, the overhead field E_G is due to $(12 - 5 - 3.45)$ or 3.55 V and E_B is set by $V_{\text{bulk-drain}} = 5$ V. The graph gives:

$$\underline{\beta'_{(D)} = 6.10 \times 10^{-6}}$$

Taking the average:

$$\underline{\beta'_{(\text{ave})} = \frac{4.85 + 6.10}{2} 10^{-6} = 5.47 \times 10^{-6}\ \text{A/V}^2}$$

and

$$\beta = 5.47 \times 10^{-6} \frac{10}{0.4 - 0.12} = 1.95 \times 10^{-4}$$

$$I_{DS} = 1.95 \times 10^{-4}[(12 - 2.87)5 - 0.5(5^2)] = \underline{6.46\ \text{mA}}$$

With a little practice, a suitable value of β' can be judged from the graphs without the detailed averaging process illustrated here.

EXAMPLE 6.2

The n-channel load device shown operates with 5 V of substrate bias. The threshold voltage without substrate bias is -0.3 V (in the depletion mode) and the body factor is $M = 1.2$. Assume $(W/L)_{\text{eff}} = 0.8$ and calculate the

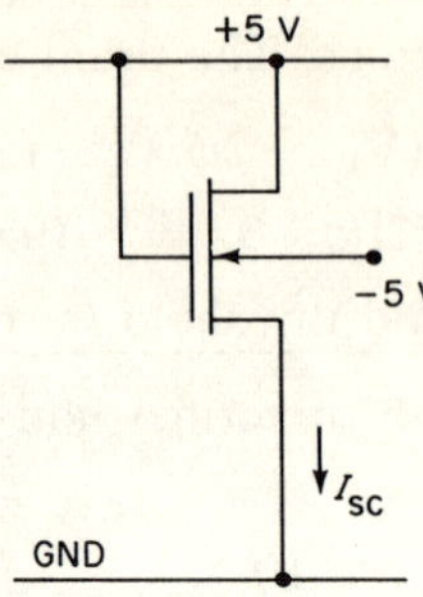

short-circuit current I_{sc} when the source is grounded. Assume that the n-channel β' data of Appendix K is applicable. Take $N_B = 1.3 \times 10^{15}$, $\phi_F = 0.3$, and $L_{eff} = 0.4$ mils for a full channel.

Solution

The source end threshold is:

$$V_{TO} = -0.3 + 1.2[(5 + 0.6)^{1/2} - (0.6)^{1/2}]$$
$$= -0.3 + 1.91 = \underline{+1.61\ V}$$

The operation is inherently saturated with a drop of V_{TCE} from the drain to the channel end.

$$V_{TCE} = -0.3 + 1.2[(5 - V_{TCE} + 5 + 0.6)^{1/2} - (0.6)^{1/2}]$$
$$= -0.3 + 1.2[(10.6 - V_{TCE})^{1/2} - (0.6)^{1/2}]$$

Try $2.5 \rightarrow 2.18$

Try $2.2 \rightarrow 2.24$

Try $2.24 \rightarrow 2.24$

Then: $\underline{V_{TCE} = 2.24\ V}$

$$V_{TM} = (V_{TO} + V_{TCE})/2 = 1.92\ V$$

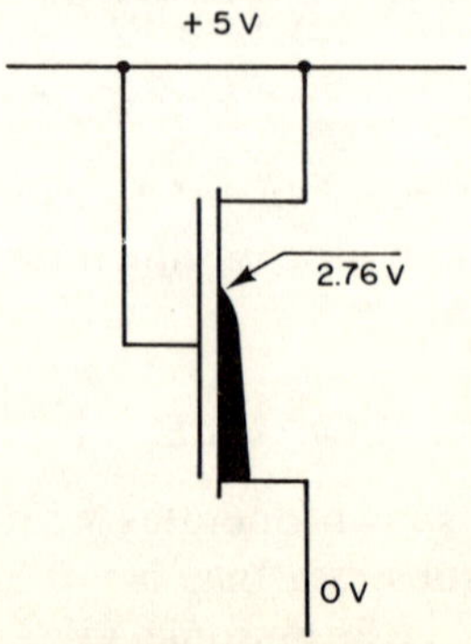

At the source end, β' is set by $(5-1.61)=3.39$ V overhead and by 5 V under the channel. From Appendix K:

$$\underline{\beta'_{(S)}=16.5\times10^{-6}}$$

At the channel end, β' is set by 0 V overhead (since the channel pinches off here) and by $(2.76+5)$ or 7.76 V below. Then:

$$\underline{\beta'_{(D)}=17.4\times10^{-6}} \qquad \text{(approx.)}$$

The average is

$$\beta'_{(ave)}=16.95\times10^{-6}\ \text{A/V}^2$$

Then, the length L' from the drain to channel end is:

$$L'=\left(\frac{2k_{si}e_0V_{TCE}}{qN_B}\right)^{1/2}=\left(\frac{2\times11.7\times8.85\times10^{-14}\times2.24}{1.6\times10^{-19}\times1.3\times10^{15}}\right)^{1/2}$$

$$=\underline{1.5\times10^{-4}\ \text{cm}=1.5\ \mu\text{m}=0.06\ \text{mils}}$$

Then $W=0.4\times0.8=0.32$ mils, $(L-L')=0.4-0.06=0.34$ mils, and

$$I_{sc}=\frac{\beta'_{(ave)}}{2}\left(\frac{W}{L-L'}\right)_{eff}(V_{GS}-V_{TM})^2;$$

$$=0.5\times16.95\times10^{-6}\frac{0.32}{0.34}(5-1.92)^2$$

$$=\underline{75.6\ \mu\text{A}}$$

CHAPTER 7

MOS Inverters: Static Transfer Characteristics

The term 'static transfer characteristic' is used here in reference to the basic inverter configuration shown in Fig. 7.1. It is simply the graph of v_o versus v_i and its general shape is suggested in Fig. 7.1(b). This chapter is concerned with the detailed determination of static transfer characteristics of various inverter types.

The inverter stage was seen in Chapter 2 to be the fundamental building block for digital circuitry. In the digital role, some of the prime *static* considerations were the output HIGH level V_{OH}, the output LOW level V_{OL}, and the power dissipation in the load element in the output LOW state. (There was also some concern over the *dynamic* behaviour, but that is a consequence of the static inverter properties and other factors so it is logically left until later.) The very same circuit is capable of functioning as an analog signal amplifier. The main point of interest in that application is the slope of the static transfer characteristic at the operating point. Accordingly, there are several reasons for investigating the static transfer characteristics of MOS inverters.

There are three main categories of inverter as follows;

1. P-channel and n-channel inverters using enhancement transistors only
2. Inverters using depletion loads
3. CMOS inverters

These will all be examined in this chapter. The objective in each case will be to establish the (v_o versus v_i) graph in a proper quantitative manner. To preserve simplicity, the threshold voltage of any given device will be considered as constant although driver and load thresholds may, in general, be different.

1. INVERTERS WITH ENHANCEMENT LOADS

The first category can be subdivided according to the load device type as follows:

(a) Resistance load R_L
(b) Saturated MOS load
(c) Non-saturated MOS load

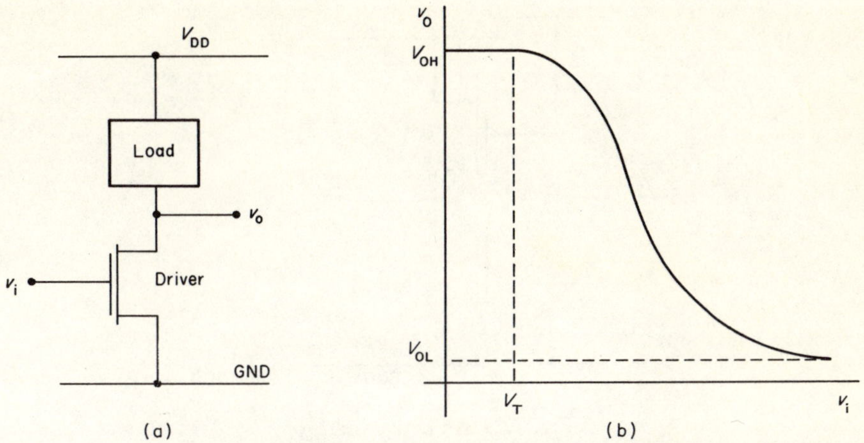

Fig. 7.1. An inverter and its static transfer curve.

The treatment is theoretical and is no more than a direct application of the elementary current equations:

Non-saturated:

$$I_{DS} = \beta[(V_{GS} - V_T)V_{DS} - 0.5V_{DS}^2] \qquad \text{with} \qquad (V_{GS} - V_T) > V_{DS} \quad (7.1)$$

Saturated:

$$I_{DS} = \frac{\beta}{2}(V_{GS} - V_T)^2 \qquad \text{with} \qquad (V_{GS} - V_T) < V_{DS} \quad (7.2)$$

In general, the driver (gain β_D, threshold V_T) goes from saturated to non-saturated operation as v_i goes high. (The transition occurs after v_i has risen and v_o fallen until v_i exceeds v_o by an amount V_T. A full-length channel is then possible.) Active loads, on the other hand, will be either saturated or non-saturated as determined by supply connections:

$$(V_{GG} - V_{TL}) > V_{DD} \qquad \text{(non-saturated)}$$
$$(V_{GG} - V_{TL}) < V_{DD} \qquad \text{(saturated)}$$

where V_{TL} is the threshold for the load device (see Fig. 7.2). For non-saturated loads, v_o can climb to an output high level of $V_{OH} = V_{DD}$. For saturated loads, v_o has an upper limit of $V_{OH} = (V_{GG} - V_{TL})$. The saturated load condition is generally achieved with only one supply $V_{GG} = V_{DD}$. For resistive loads, $V_{OH} = V_{DD}$.

The analysis procedure simply uses Eqns (7.1) and (7.2) to find expressions for i_L and i_D. Then, equating these current expressions, $i_L = i_D$, gives an identity with just two variables v_i and v_O. In this way their interrelationship (the transfer characteristic) is implicitly established.

As it happens, we can write the load current expression in a manner which makes it valid for all three load types. This reduces the total algebra and

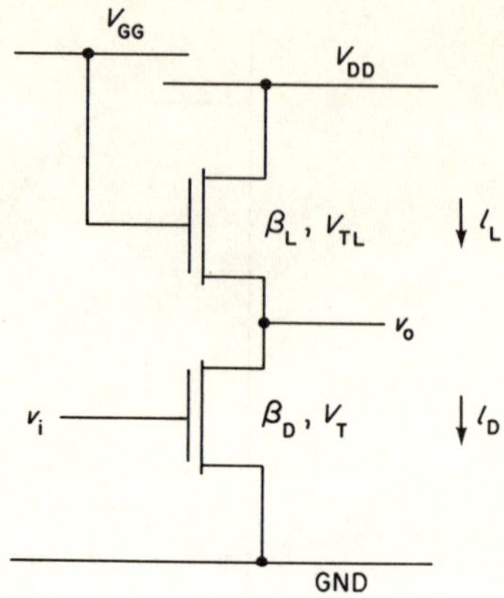

Fig. 7.2. An active load.

requires that we introduce a dimensionless factor m and that we define:

Short-circuit current: $I_{sc} = i_L]_{v_o=0}$ (7.3)

Maximum output voltage: $V_{OH} = v_o]_{i_L=0}$ (7.4)

The generalized load current expression is:

$$i_L = I_{sc}\left(1 - \frac{v_o}{V_{OH}}\right)\left(1 - m\frac{v_o}{V_{OH}}\right) \tag{7.5}$$

subject to the interpretation:

	Resistive load, R_L	Saturated MOS load	Non-saturated MOS load	
$V_{OH} =$	V_{DD}	$V_{GG} - V_{TL}$	V_{DD}	
$m =$	0.0	1.0	$\dfrac{V_{DD}}{2(V_{GG} - V_{TL}) - V_{DD}}$	(7.6)
$I_{SC} =$	$\dfrac{V_{DD}}{R_L}$	$\dfrac{\beta_L}{2}(V_{GG} - V_{TL})^2$	$\dfrac{\beta_L}{2m} V_{DD}^2$	
$T_{SC} =$	$R_L C_L$	$\dfrac{2C_L}{\beta_L(V_{GG} - V_{TL})}$	$\dfrac{2mC_L}{\beta_L V_{DD}}$	

Notice that all voltage terms in (7.5) are normalized to the ouput high level and this practice will be maintained throughout the discussion. The parameter T_{SC} is a time-constant which will be used in Chapter 8.

The proof of Eqn. (7.5) based on Eqns (7.1) and (7.2) is straightforward. To avoid clouding the main issues, it is given separately in Detail 1 at the end of the chapter. Equation (7.5) permits the use of just two general equations for all load types. Setting $i_D = i_L$ for a saturated driver:

Saturated driver, $(v_i - V_T) < v_o$:

$$\frac{\beta_D}{2}(v_i - V_T)^2 = I_{sc}\left(1 - \frac{v_o}{V_{OH}}\right)\left(1 - m\frac{v_o}{V_{OH}}\right) \tag{7.7}$$

Similarly:

Non-saturated driver, $(v_i - V_T) > v_o$:

$$\beta_D[(v_i - V_T)v_o - 0.5v_o^2] = I_{sc}\left(1 - \frac{v_o}{V_{OH}}\right)\left(1 - m\frac{v_o}{V_{OH}}\right) \tag{7.8}$$

The transfer curve (v_o versus v_i) is contained in Eqns (7.7) and (7.8). The factor m deserves comment for the non-saturated case where, by definition, $(V_{GG} - V_{TL}) > V_{DD}$. At the lower non-saturated limit, $(V_{GG} - V_{TL}) = V_{DD}$ and then, from Eqn. (7.6), $m = 1.0$. Higher V_{GG} reduces m below 1.0. Since lower V_{GG} involves saturation, values of $m > 1.0$ are meaningless. In summary:

$$(0 < m \leqslant 1.0) \quad \text{and} \quad m \to 0 \quad \text{as} \quad V_{GG} \to \infty \tag{7.9}$$

Variations of V_{GG} within a non-saturated operation cause m to change but $V_{OH} = V_{DD}$ is fixed; variations of V_{GG} within a saturated operation have m constant at 1.0 but V_{OH} is changing.

It may be helpful to write Eqns (7.7) and (7.8) in terms more specific to each of the three load conditions by naming I_{sc} and V_{OH}. This yields six equations which are justified in Detail 2 and stated here:

Resistive load:

Saturated driver, $(v_i - V_T) < v_o$:

$$1 - \frac{v_o}{V_{DD}} = \frac{\beta_D}{2} V_{DD} R_L \left(\frac{v_i - V_T}{V_{DD}}\right)^2 \tag{7.10}$$

Non-saturated driver, $(v_i - V_T) > v_o$:

$$1 - \frac{v_o}{V_{DD}} = \beta_D V_{DD} R_L \left[\left(\frac{v_i - v_T}{V_{DD}}\right)\left(\frac{v_o}{V_{DD}}\right) - 0.5\left(\frac{v_o}{V_{DD}}\right)^2\right] \tag{7.11}$$

MOS saturated load:

Saturated driver, $(v_i - V_T) < v_o$

$$\left(1 - \frac{v_o}{V_{GG} - V_{TL}}\right)^2 = \frac{\beta_D}{\beta_L}\left(\frac{v_i - V_T}{V_{GG} - V_{TL}}\right)^2 \tag{7.12}$$

Non-saturated driver, $(v_i - V_T) > v_o$

$$\left(1 - \frac{v_0}{V_{GG} - V_{TL}}\right)^2 = 2\frac{\beta_D}{\beta_L}\left[\left(\frac{v_i - V_T}{V_{GG} - V_{TL}}\right)\left(\frac{v_o}{V_{GG} - V_{TL}}\right) - 0.5\left(\frac{v_o}{V_{GG} - V_{TL}}\right)^2\right] \tag{7.13}$$

valid for $(V_{GG} - V_{TL}) < V_{DD}$.

MOS non-saturated load:

Saturated driver, $(v_i - V_T) < v_0$:

$$\left(1-\frac{v_o}{V_{DD}}\right)\left(1-m\frac{v_o}{V_{DD}}\right)=m\frac{\beta_D}{\beta_L}\left(\frac{v_i-V_T}{V_{DD}}\right)^2 \tag{7.14}$$

Non-saturated driver, $(v_i - V_T) > v_o$:

$$\left(1-\frac{v_o}{V_{DD}}\right)\left(1-m\frac{v_o}{V_{DD}}\right)=2m\frac{\beta_D}{\beta_L}\left[\left(\frac{v_i-V_T}{V_{DD}}\right)\left(\frac{v_0}{V_{DD}}\right)-0.5\left(\frac{v_o}{V_{DD}}\right)^2\right] \tag{7.15}$$

Equations (7.10) and (7.11) can be used directly to generate a family of curves (v_o/V_{DD}) versus $(v_i - V_T)/V_{DD}$ for various values of $\beta_D R_L$. In practice diffused resistors utilize space very poorly and are seldom used. In certain modern processes, however, a polysilicon medium can be used to advantage to form resistive loads.

The equations (7.12) to (7.15) for MOS loads make it clear that the circuit action is of a *ratioed* nature. It is not the size of a driver or load transistor that matters but rather the ratio of their sizes as implied by the term β_D/β_L. This is often called the 'beta ratio', β_R.

Equations (7.12) and (7.13) can be viewed as special cases of Eqns (7.14) and (7.15) respectively with $m = 1.0$ if both sets are normalized to V_{OH} as defined in (7.6). Plots of (7.14) and (7.15) are provided as design aids in Appendix C for $m = 0.3$, 0.5, 0.8, and 1.0. For each m value, ratios β_D/β_L ranging from 4 to 64 are treated. These static transfer curves are given as Figure C.1. Curves with $m < 1$ are for non-saturated loads and the $m = 1.0$ curves are for saturated loads. To allow comparison with resistive loads, the dotted curves in Fig. C.1(a) and (d) represent resistors having equal I_{SC} with loads for which $\beta_R = 9.0$.

The trigger level

The trigger level (V_{tr}) is the level for which $v_o = v_i$. It has the property that for a sequence of cascaded identical inverters, all stages pass through V_{tr} simultaneously (provided switching speed is not a limitation). Figure 7.3 indicates two important lines on a static transfer curve. The line $v_o = v_i - V_T$ divides the graph into the two regions described by Eqns (7.14) and (7.15). The line $v_o = v_i$ contains all trigger levels.

An important use of static transfer curves is in determining the LOW value of v_o which is principally a function of β_D/β_L. That is, v_o is set as the ratio of the ON resistances of the driver and load device and the v_o LOW specification generally determines β_D/β_L which is, in effect, W_D/W_L, the ratio of device widths if both have minimum length (typically about 0.2 mils).

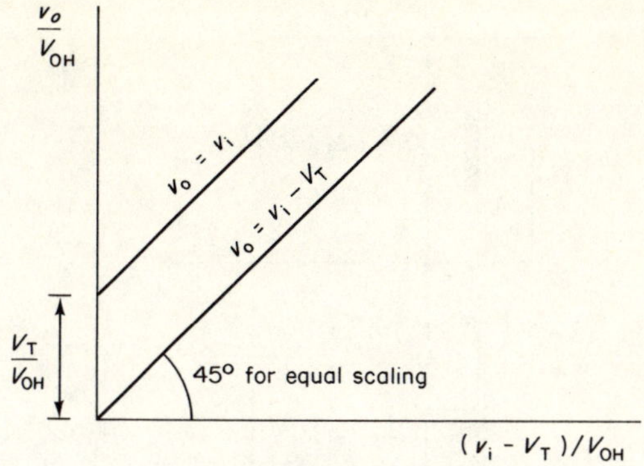

Fig. 7.3. Trigger level.

To illustrate the use of the static transfer graphs in Appendix C, consider an inverter with $V_{DD}=10$ V, $V_{GG}=20$ V, $V_T=2$ V, $V_{TL}=2$ V, and $(W/L)_D=16(W/L)_L$. This describes a non-saturated load with:

$$m=\frac{V_{DD}}{2(V_{GG}-V_{TL})-V_{DD}}=\frac{10}{2(20-2)-10}=\frac{10}{26}=0.38$$

and with a beta ratio $\beta_D/\beta_L=16$. Interpolating between the $m=0.3$ and $m=0.5$ graphs in Fig. C.1, we see that the output low level is obtained as:

$$v_o/V_{DD}=0.11 \quad \text{or} \quad v_0=1.1 \text{ V}$$

The trigger level can be found by the method already described using $V_T/V_{DD}=0.2$ as an offset for the 45° straight line. This line intersects the $(\beta_R=16)$ contour at $v_o/V_{DD}=0.46$ giving:

$$V_{tr}=0.46V_{DD}=4.6 \text{ V}$$

Thus, when $v_i=4.6$ V, $v_o=4.6$ V also.

2. INVERTERS WITH DEPLETION LOADS

Ion implantation techniques can now be used to selectively shift the thresholds of load devices so that depletion mode operation is realized. The inverter is connected as in Fig. 7.4. Current in T_L is now given by:

$$I_L=\frac{\beta'}{2}\left(\frac{W}{L}\right)_{eff}(V_{GS}-V_{TL})^2$$

$$=\frac{\beta}{2}(V_{GS}+V_I)^2 \qquad (7.16)$$

where V_{GS} and V_I are *additive.* For the n-channel case which we will discuss, the threshold V_T of a depletion device is *negative* and it is a convenient distinction to use the *inversion voltage*, V_I, such that $V_I=-V_T$ is

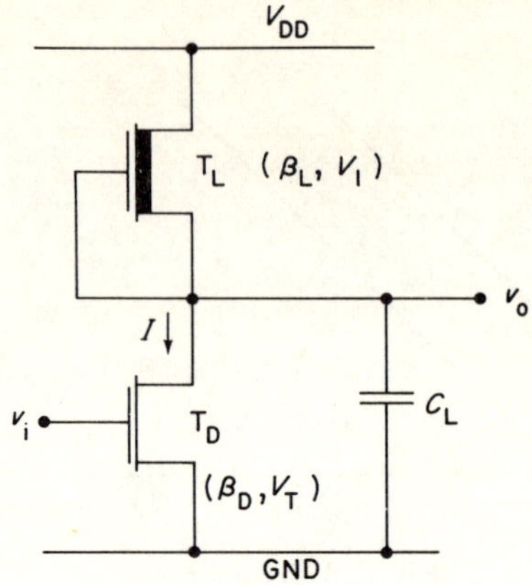

Fig. 7.4. Depletion load.

positive. The inversion voltage is a 'built-in' channel support voltage. Equation (7.16) is valid for saturated operation and, since $V_{GS}=0$, it reduces to:

$$I_L = \frac{\beta(V_I)^2}{2} \tag{7.17}$$

as long as saturated conditions prevail.

According to Eqn. (7.17), the saturated device serves as a constant current source. It will remain saturated as long as $V_{DS}=(V_{DD}-v_o)$ exceeds the 'built-in' support V_I. Thus, the load T_L delivers a constant current while v_o is less than $(V_{DD}-V_I)$ and in the subsequent non-saturated condition it behaves more like a simple resistor. This ideal-type characteristic is shown as curve 1 in Fig. 7.5. Different load types are compared here on the basis of equal short-circuit currents for all types.

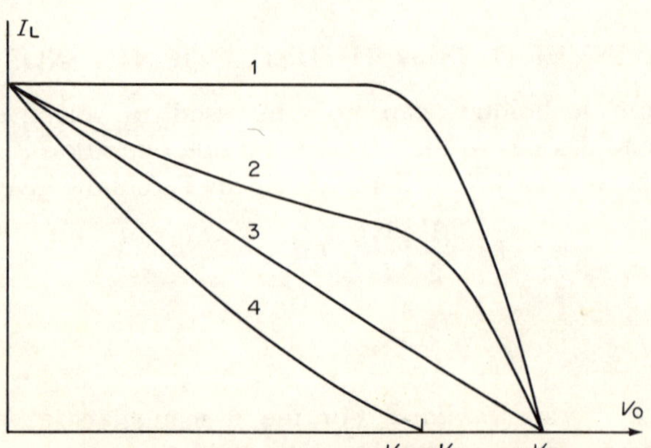

Fig. 7.5. Depletion mode load characteristics.

Comparison with curve 3 for a linear resistor load and with curve 4 for a saturated MOS load shows that, as v_o swings from 0 towards V_{DD}, the depletion load makes more current available for charging C_L at any instant and thus leads to a faster transient. The true picture (curve 2) is not quite so good since *body-effect* at the source end of T_L reduces current flow as v_o increases. In spite of this, practical depletion mode loads are superior to saturated enhancement mode devices and even to resistor loads. Note, too, that v_o can reach as high as V_{DD} with depletion load elements.

The ideal static transfer characteristics can be computed from four equations describing the conducting states of the driver (T_D) and load (T_L) as follows:

T_D saturated, $(v_i - V_T) < v_o$:

$$I_{DS} = \frac{\beta_D}{2}(v_i - V_T)^2$$

T_D non-saturated, $(v_i - V_T) > v_o$:

$$I_{DN} = \beta_D[(v_i - V_T)v_o - 0.5v_o^2]$$

T_L saturated, $(V_{CC} - v_O) > V_I$:

$$I_{LS} = \frac{\beta_L}{2} V_I^2$$

T_L non-saturated, $(V_{CC} - v_o) < V_I$:

$$I_{LN} = \beta_L[V_I(V_{CC} - v_o) - 0.5(V_{CC} - v_o)^2]$$

Consider the sequence of conducting states as v_i rises from zero. The driver if OFF until $v_i > V_T$. As v_i goes beyond V_T, the driver goes SAT while the load is initially NON-SAT, ($I_{DS} = I_{LN}$). Later, the driver becomes NON-SAT when v_o drops below $(v_i - V_T)$, ($I_{DN} = I_{LN}$). Also, the load goes SAT when v_o drops below $(V_{CC} - V_I)$, ($I_{DS} = I_{LS}$). Either of these two conditions can occur first so that, for a while, both devices are either saturated or non-saturated. In the final condition, when v_i is high, the driver is NON-SAT and the load is SAT, ($I_{DN} = I_{LS}$). Static transfer curves based on this model are readily generated with the aid of a computer. Some representative results are included in Appendix C. One graph (Fig. C.2a) shows the results with different beta ratios, $\beta_R = 1$, 4, and 16 while V_I is fixed at $0.8V_{DD}$. Another (Fig. C.2b) shows the effect of changing V_I for a fixed β_R of 4. In all cases, the driver threshold is taken as $V_T = 0.2V_{DD}$. Even though body-effect is neglected in these calculations, they can be of considerable assistance in the initial choice of device geometries and implant levels.

The method of selective implantation for threshold adjustment applies to both n-channel and p-channel devices and to various technologies, and is widely used in modern design procedures.

3. THE CMOS INVERTER

This circuit uses an n-channel and a p-channel device (a complementary pair) in series driven from the same input line and its basic operation was described in Chapter 2. It is re-drawn in Fig. 7.6 where the devices are characterized by β_N and V_{TN} for the n-channel transistor T_N and by β_P and V_{TP} for the p-channel transistor T_P. In the 'normal' condition, one or other

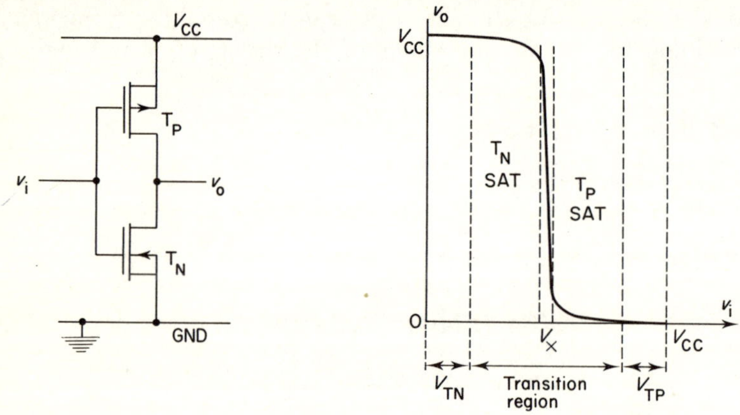

Fig. 7.6. The CMOS inverter.

of the transistors is OFF and v_o is at either V_{CC} or GND. There is no current flow and no dissipation. The main interest in this chapter is in the transition region between the two OFF states. This region is bounded by:

$$V_{TN} < v_i < (V_{CC} + V_{TP}) \qquad (7.18)$$

bearing in mind that V_{TP} is a negative number. As v_i rises beyond V_{TN}, T_N enters SAT conduction while T_P is still 'ON', i.e. in a NON-SAT condition. Then:

$$\frac{\beta_N}{2}(v_i - V_{TN})^2 = \beta_P[(V_{CC} - v_i + V_{TP})(V_{CC} - v_o) - 0.5(V_{CC} - v_o)^2] \qquad (7.19)$$

This causes v_0 to fall (towards ground) so that T_P also enters the SAT condition which satisfies:

$$\frac{\beta_N}{2}(v_i - V_{TN})^2 = \frac{\beta_P}{2}(V_{CC} - v_i + V_{TP})^2 \qquad (7.20)$$

Note that v_o does not appear in this equation! It is valid for only one value of v_i which is obtained by rearrangement:

$$v_i = V_\times = \frac{V_{CC} + V_{TP} + V_{TN}\sqrt{(\beta_N/\beta_P)}}{1 + \sqrt{(\beta_N/\beta_P)}} \qquad (7.21)$$

This unique value $V_\times$ is an input 'cross-over' voltage because the transistors switch roles as v_i passes through the $V_\times$ point. The equations used here imply a step fall in output v_o at the cross-over. This is theoretically necessary so that one device can be non-saturated at all voltages except $V_\times$. (Two saturated devices look like current sources from the output node and their current demands differ everywhere except at $V_\times$).

For $v_i > V_\times$, v_o is quite low and T_N is NON-SAT while T_P is in the SAT condition. For higher v_i, above $(V_{CC} + V_{TP})$, T_P turns OFF.

By way of an example, consider that $V_{CC} = 5$ V, $V_{TP} = -2$ V, $V_{TN} = +1$ V, and $\beta_N/\beta_P = 1.0$. Then, Eqn. (7.21) gives:

$$V_\times = \frac{5 - 2 + 1(1.0)}{1 + 1} = 2.0 \text{ V}$$

From a noise-immunity viewpoint, we may prefer to have $V_\times$ occur at half the supply level, $V_\times = 2.5$ V. This can be done, if necessary, by using unequal device sizes. In this example, $\beta_N/\beta_P = 1/9$ gives the desired result but it could hardly be described as a good trade-off because of the large resulting size of the transistor T_P. It is better to use close to minimum-size devices and to undertake appropriate threshold adjustment by ion implantation.

The slope of the transfer characteristic in the region of $V_\times$ is not an infinite one as the theory suggests. In practice, each device, when saturated, has a *finite* resistance r_{oP} or r_{oN} when viewed from the output node. In consequence of this, both devices can exist in a saturated state for a narrow band of input voltages around $V_\times$. The finite output resistance is a result of channel length modulation and will be discussed further in Chapter 10.

The transfer characteristic of the CMOS inverter is close to ideal. The noise immunity can be made close to the maximum possible figure of $V_{DD}/2$. The high slope in the region of $V_\times$ can also be put to good use in analog applications.

EXAMPLE 7.1

The p-channel inverter shown is specified as follows;

$\phi_F = 0.3$ V; $V_{TO} = -2.5$ V; $M = 0.4$; $\Delta L = 0.1$ mil; $N_B = 10^{15}$

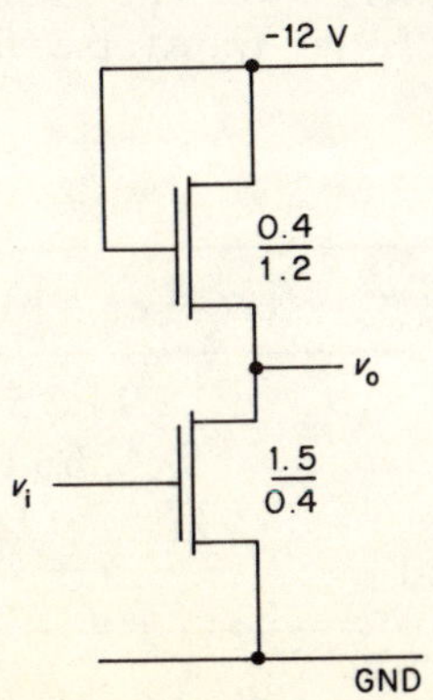

Use β' curves from Appendix K and the static transfer curves from Appendix C to determine HIGH and LOW levels for v_o if $v_{i(LOW)} = 0$ V and $v_{i(HIGH)} = -9$ V. Transistors are marked with W/L in mils (mask dimension).

Solution

$v_{o(HIGH)}$ is just V_{TCEL} volts down from the supply:

$$V_{TCEL} = -2.5 - 0.4[12 - V_{TCEL} + 0.6)^{1/2} - (0.6)^{1/2}]$$

Try $3.5 \rightarrow 3.39$

Try $3.4 \rightarrow 3.4$

$$\underline{V_{TCEL} = -3.4\ \text{V}}$$

Hence:

$$\underline{v_{o(HIGH)} = -(12 - 3.4) = -8.6\ \text{V}}$$

A first estimate of the beta ratio $\beta_R = \beta_D/\beta_L$ is $(W/L)_D/(W/L)_L$, which is:

$$\frac{1.5}{0.4} \times \frac{1.2}{0.4} = 11.25$$

Also, $m = 1.0$ for the saturated load. The $m = 1.0$ curve in Fig. C.1(d) suggests that (v_o/V_{OH}) bottoms out at 0.07 approximately, i.e. $v_o = -0.07 \times 8.6 = -0.6$ V. We use this first estimate to determine β_D and β_L more exactly.

It will also be helpful to know the relevant threshold levels. At the source end of the load,

$$V_{TSL} = -2.5 - 0.4[(0.6 + 0.6)^{1/2} - (0.6)^{1/2}] = -2.6\ \text{V}$$

The same figure applies at the drain end of the driver:

$$V_{TD} = V_{TSL} = -2.6\ \text{V}$$

and, at the source, $V_{TS} = V_{TO} = -2.5$ V.

To get an impression of the transverse field conditions, tabulate as follows:

	$V_{GS} - V_T$	V_{CB}
Driver, source	$-9 + 2.5$	0.0
Driver, drain	$-9 + 2.6 + 0.6$	-0.6
Load, source	$-12 + 2.6 + 0.6$	-0.6
Load, channel end	0.0	$-12 + 3.4$

Then, from page 250, on inspection:

$$\beta'_D = 5.5 \times 10^{-6}\ \text{A/V}^2 \text{ approx.}$$
$$\beta'_L = 6.0 \times 10^{-6}\ \text{A/V}^2 \text{ approx.}$$

The load has a shortened channel:

$$L' = \left[\frac{2k_{si}e_0 V_{TCEL}}{qN_B}\right]^{1/2} = \left[\frac{2\times 11.7\times 8.85\times 10^{-14}\times 3.4}{1.6\times 10^{-19}\times 10^{15}}\right]^{1/2}$$

$$L' = 2.1\mu = 0.08 \text{ mils.}$$

$$\beta_L = 6.0\times 10^{-6}\left(\frac{0.4}{1.2-2(0.1)-0.08}\right) = 2.61\times 10^{-6}$$

$$\beta_D = 5.5\times 10^{-6}\left(\frac{1.5}{0.4-0.2}\right) = 4.1\times 10^{-5}$$

$$\beta_R = \beta_D/\beta_L = \frac{4.1\times 10^{-5}}{2.61\times 10^{-6}} = 15.7$$

Normalized input is:

$$\frac{v_i - V_{TDD}}{V_{DD} - V_{TCEL}} = \frac{9-2.6}{12-3.4} = 0.74$$

With these β_R and input values, the $m = 1.0$ graph gives $v_o/V_{OH} = 0.05$ approx. Then:

$$\underline{v_{o(LOW)} = 0.05 \times 8.6 = 0.43\ \text{V}}$$

Note that the result is quite insensitive to small changes in β_R and, consequently, a crude estimate of β' will do very well here. Channel length modulation, likewise, has very little effect.

EXAMPLE 7.2

The inverter shown uses an *e*nhancement driver (T_1) and a *d*epletion load (T_2). It is an n-channel ED-type inverter for which:

$$V_{TO1} = 1.1\ \text{V}, \qquad \left(\frac{W}{L}\right)_1 = 4 \qquad M = 0.75,$$

$$\beta' = 25\ \mu\text{A/V}^2, \qquad 2\phi_F = 0.64, \qquad V_{TO2} = -2.4\ \text{V}$$

Find the output high level V_{OH} and determine a value for $(W/L)_2$ to give a noise margin of 0.3 V for the LOW level. Use current control equations neglecting the L' effect.

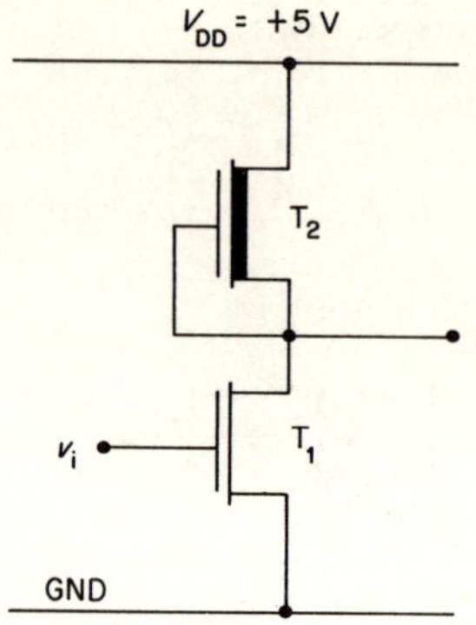

Solution

As v_o approaches 5 V, T_2 will be NON-SAT provided V_{T2} remains negative in spite of body-effect. When $v_o = 5$ V approx.:

$$V_{T2} = -2.4 + 0.75[(5+0.64)^{1/2} - (0.64)^{1/2}] = -1.2 \text{ V}$$

This confirms that v_o does reach 5 V, i.e.

$$\underline{V_{OH} = 5 \text{ V}.}$$

With v_o at its LOW value, we can take:

$$V_{TCE1} = V_{TO1} = 1.1 \text{ V}$$

since body-effect will be very small. For a noise margin of 0.3 V, $V_{OL} = 1.1 - 0.3 = 0.8$ V is required. Assume $V_{OL} = 0.8$ V. We then have T_1 in the NON-SAT condition and can calculate the driver current:

$$I_{DS1} = \beta'\left(\frac{W}{L}\right)_D [(V_{DD} - V_{TM1})V_{OL} - 0.5V_{OL}^2]$$

$$= 25 \times 10^{-6} \times 4 \times [(5-1.1)0.8 - 0.32] \text{A}$$

$$= \underline{280 \ \mu\text{A}}$$

The load T_2 is clearly saturated since the drain is $(5-0.8) = 4.2$ V *above* the gate and this exceeds the highest V_T magnitude. Then, the channel ends where $(V_{GS} - V_{TCE2}) = V_{CES}$. Since $V_{GS} = 0$ and $V_{CES} = (V_{CE} - V_{OL})$:

$$V_{TCE2} = -(V_{CE} - V_{OL})$$

Also:

$$V_{TCE2} = -2.4 + 0.75[(V_{CE} + 0.64)^{1/2} - (0.64)^{1/2}]$$

$$V_{TCE2} = -2.4 + 0.75[(V_{OL} - V_{TCE2} + 0.64)^{1/2} - (0.64)^{1/2}]$$

$$V_{TCE2} = -2.4 + 0.75[(1.44 - V_{TCE2})^{1/2} - (0.64)^{1/2}]$$

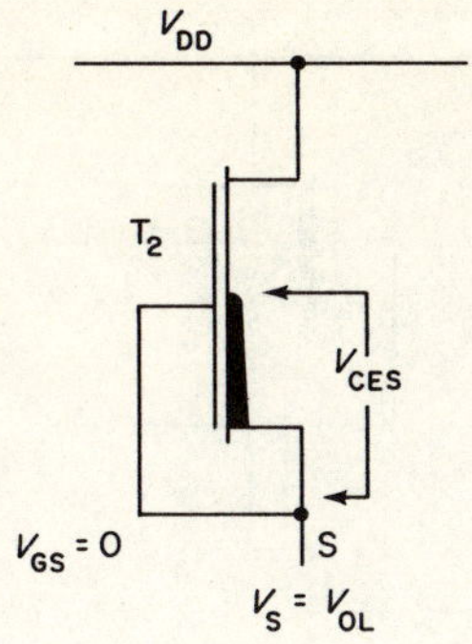

By iteration:

$$\underline{V_{TCE2} = -1.67}$$

Note that this gives $V_{CE} = V_{OL} - V_{TCE2} = 2.47$ V = channel end voltage to GND.

The mean load threshold is $V_{TM2} = -(2.4 + 1.67)/2 = -2.0$ V
Then

$$I_{DS2} = \frac{\beta'}{2}\left(\frac{W}{L}\right)_L (V_{GS} - V_{TM})^2$$

$$= \frac{\beta'}{2}\left(\frac{W}{L}\right)_L (V_{TM})^2 = I_{DS1} = 280\ \mu\text{A}$$

from which:

$$\left(\frac{W}{L}\right)_L = \frac{280 \times 2}{\beta'(V_{TM})^2} = \frac{280 \times 2}{25(2.00)^2}$$

$$= \underline{5.6}$$

Note that a simpler but overoptimistic view would use the threshold voltage at the source end of the depletion load, i.e. $V_{TSL} = -2.4$ V approx.

EXAMPLE 7.3

The inverter shown uses a depletion load device with a negative threshold of -4 V. Stated otherwise, the load has a 'build-in' inversion voltage $V_I = 4$ V. Assume that v_i is gradually increased from zero towards $+5$ V. Sketch the transitions between states (SAT or NON-SAT) for T_L and T_D and find (using iteration if necessary) the values of v_o at which state transitions occur. Find the current I at each transition. Ignore body-effect.

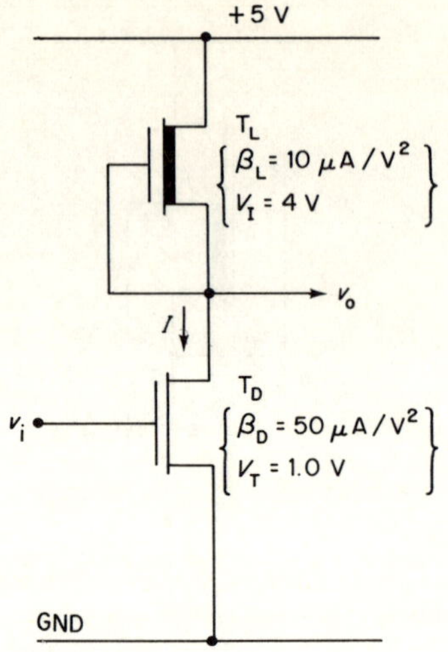

Solution

T_L goes NON-SAT for $v_o > (5 - V_I)$, i.e. for $v_o > 1$ V. T_D changes state when $v_o = v_i - V_T = v_i - 1.0$. We will assume, initially, that this occurs while $v_o > 1$ V, i.e. with T_L NON-SAT. Consider $v_o = 1.5$ V:

$$I = \beta_L[V_I(5 - v_o) - 0.5(5 - v_o)^2]$$
$$= 10[4(3.5) - 0.5(3.5)^2]\ \mu A$$
$$= \underline{78.7\ \mu A}$$

Assuming T_D to be SAT:

$$I = 0.5\beta_D(v_i - V_T)^2$$

so that:

$$v_i = V_T + \left(\frac{2I}{\beta_D}\right)^{1/2}$$

$$v_i = 1.0 + \left(\frac{2 \times 78.7 \times 10^{-6}}{50 \times 10^{-6}}\right)^{1/2} = 2.77\ V$$

This would give $v_i - v_o = 2.77 - 1.5 = 1.27$ V which $> V_T$:

Try $v_o = 1.6$: $\quad I = 78.2\ \mu A, \quad v_i = 2.76$

Try $v_o = 1.73$ $\quad I = 77.3\ \mu A, \quad v_i = 2.76$

Try $v_o = 1.76$: $\quad I = 77.1\ \mu A, \quad v_i = 2.756$

This is close enough. At $v_i = 2.76$ V:

$$v_o = 1.76\ V, \qquad v_i - v_o = V_T = 1.0$$

and

$$I = 77.1\ \mu A$$

T_D goes NON-SAT for rising v_i. When v_o reaches 1.0 V, T_L goes SAT also. Then:

$$I = \tfrac{1}{2}\beta_L V_I^2 = \frac{10 \times 10^{-6}}{2}(4)^2 = 80\ \mu\text{A}$$

and

$$I = \beta_D[(v_i - V_T)v_o - \tfrac{1}{2}v_o^2]$$

$$80 = 50[(v_i - 1) - 0.5] \qquad \text{for } v_o = 1.0\ \text{V}$$

Then:

$$v_i = 1.5 + \tfrac{80}{50} = 3.1\ \text{V}$$

When:

$$v_i = 3.1\ \text{V},\ v_o = 1.0\ \text{V}, \quad \text{and} \quad I = 80\ \mu\text{A}$$

T_L goes SAT while v_i rises further.

v_i	
5.0	
	T_L SAT, T_D NON-SAT
3.1	
	T_L NON-SAT, T_D NON-SAT
2.7	
	T_L NON-SAT, T_D SAT
0	

DETAIL 1

For a resistive load

$$i_L = \frac{V_{DD} - v_o}{R_L} = \frac{V_{DD}}{R_L}\left(1 - \frac{v_o}{V_{DD}}\right)$$

$$= I_{sc}\left(1 - \frac{v_o}{V_{OH}}\right)\left(1 - m\frac{v_o}{V_{OH}}\right)$$

for $I_{sc} = V_{DD}/R_L$ and $V_{OH} = V_{DD}$ with $m = 0$. QED

For a MOS SAT load

$$i_L = \frac{\beta_L}{2}(V_{GG} - V_{TL} - v_o)^2$$

$$= \frac{\beta_L}{2}[(V_{GG} - V_{TL})^2 - 2v_o(V_{GG} - V_{TL}) + v_o^2]$$

$$= \frac{\beta_L}{2}(V_{GG} - V_{TL})^2\left[1 - \frac{2v_o}{V_{GG} - V_{TL}} + \frac{v_o^2}{(V_{GG} - V_{TL})^2}\right]$$

$$= \frac{\beta_L}{2}(V_{GG} - V_{TL})^2\left(1 - \frac{v_o}{V_{GG} - V_{TL}}\right)\left(1 - \frac{v_o}{V_{GG} - V_{TL}}\right)$$

$$= I_{sc}\left(1 - \frac{v_o}{V_{OH}}\right)\left(1 - m\frac{v_o}{V_{OH}}\right)$$

with I_{sc}, V_{OH}, and m as given in Eqn. (7.6). QED

For a MOS NON-SAT load

$$i_L = \beta_L[(V_{GG} - V_{TL} - v_o)(V_{DD} - v_o) - 0.5(V_{DD} - v_o)^2]$$

$$= \beta_L\left[V_{DD}(V_{GG} - V_{TL}) - V_{DD}v_o - (V_{GG} - V_{TL})v_o + v_o^2 - \frac{V_{DD}^2}{2} + V_{DD}v_o - \frac{v_o^2}{2}\right]$$

$$= \beta_L\left[V_{DD}\left(V_{GG} - V_{TL} - \frac{V_{DD}}{2}\right) + \frac{v_o^2}{2} - v_o(V_{GG} - V_{TL})\right]$$

$$= \beta_L V_{DD}\left(V_{GG} - V_{TL} - \frac{V_{DD}}{2}\right)\left[1 + \frac{v_o^2/2 - v_o(V_{GG} - V_{TL})}{V_{DD}(V_{GG} - V_{TL} - V_{DD}/2)}\right]$$

$$= \beta_L V_{DD}\left(V_{GG} - V_{TL} - \frac{V_{DD}}{2}\right)$$

$$\times\left[1 + \frac{v_o^2/2 - v_o(V_{GG} - V_{TL} - V_{DD}/2) - v_o V_{DD}/2}{V_{DD}(V_{GG} - V_{TL} - V_{DD}/2)}\right]$$

$$= \beta_L V_{DD}\left(V_{GG} - V_{TL} - \frac{V_{DD}}{2}\right)\left(1 - \frac{v_o}{V_{DD}}\right)\left(1 - \frac{v_o/2}{V_{GG} - V_{TL} - V_{DD}/2}\right)$$

Note that:

$$I_{sc} = \frac{\beta_L}{2m} V_{DD}^2 \qquad \text{from Eqn. (7.6)}$$

$$= \frac{\beta_L V_{DD}^2}{2} \frac{2(V_{GG} - V_{TL} - V_{DD}/2)}{V_{DD}} = \beta_L V_{DD}\left(V_{GG} - V_{TL} - \frac{V_{DD}}{2}\right)$$

and with $V_{OH} = V_{DD}$:

$$i_L = I_{sc}\left(1 - \frac{v_o}{V_{OH}}\right)\left(1 - m\frac{v_o}{V_{OH}}\right) \qquad \text{QED}$$

Thus, Eqn. (7.5) with parameters given in Eqn. (7.6) describes all three load conditions.

DETAIL 2

For a resistive load

Equation (7.7) becomes:

$$\frac{\beta_D}{2}(v_i - V_T)^2 = \frac{V_{DD}}{R_L}\left(1 - \frac{v_o}{V_{DD}}\right)$$

Rearranging yields Eqn. (7.10). QED

Equation (7.8) becomes:

$$\beta_D\left[(v_i - V_T)v_o - \frac{v_o^2}{2}\right] = \frac{V_{DD}}{R_L}\left(1 - \frac{v_o}{V_{DD}}\right)$$

Rearranging yields Eqn. (7.11). QED

For a MOS SAT load

Equation (7.7) becomes:

$$\frac{\beta_D}{2}(v_i - V_T)^2 = \frac{\beta_L}{2}(V_{GG} - V_{TL})^2\left(1 - \frac{v_o}{V_{GG} - V_{TL}}\right)\left(1 - \frac{v_o}{V_{GG} - V_{TL}}\right)$$

Rearranging yields Eqn. (7.12).

Equation (7.8) becomes:

$$\beta_D\left[(v_i - V_T)v_o - \frac{v_o^2}{2}\right] = \frac{\beta_L}{2}(V_{GG} - V_{TL})^2\left(1 - \frac{v_o}{V_{GG} - V_{TL}}\right)^2$$

Rearranging yields Eqn. (7.13).

For a MOS NON-SAT load

Equation (7.7) becomes:

$$\frac{\beta_D}{2}(v_i - V_T)^2 = \frac{\beta_L V_{DD}^2}{2m}\left(1 - \frac{v_o}{V_{DD}}\right)\left(1 - m\frac{v_o}{V_{DD}}\right)$$

which yields Eqn. (7.14), and from Eqn. (7.8):

$$\beta_D\left[(v_i - V_T)v_o - \frac{v_o^2}{2}\right] = \frac{\beta_L V_{DD}^2}{2m}\left(1 - \frac{v_o}{V_{DD}}\right)\left(1 - m\frac{v_o}{V_{DD}}\right)$$

giving Eqn. (7.15).

CHAPTER 8

MOS Inverters: Transient Responses

This chapter is concerned with the transient responses of MOS inverters. Response times determine the highest speed at which digital logic circuits can work satisfactorily. It is therefore of importance to understand how speed of operation is influenced by device size, by the type of load transistor, by voltage levels including bulk bias, and by inherent features of the process in use.

In a simple inverter model, such parameters as V_T, V_{TL}, β', channel length, and output node capacitance C_L are regarded as constants. In practice, thresholds vary due to body-effect (especially for n-channel circuits), β' varies due to transverse fields E_G and E_B, while channel length and part of capacitance C_L are both subject to change because of the voltage dependence of depletion layer widths. The approach adopted here will be to produce analytical solutions from simple models which treat these parameters as constants. The values proposed for the parameters can be made to reflect the actual conditions of use in some approximate way. This is usually a good compromise position between the inadequacies of a very simple model and the complexity of a more accurate one. It is, of course, possible to compute parameter values at all stages of a switching transient and, consequently, a computer-aided design effort could produce very precise results using iterative methods.

The transient analysis of MOS devices given here offers a basis on which the maximum operating speed for MOS digital circuits can be predicted. More often, the analysis permits a target speed to be achieved by a suitable choice of transistor geometries.

The treatment is simpler than similar efforts on bipolar transistors in that the MOS device exhibits no storage or transit time effects; ordinary lumped circuit concepts suffice. Also, all output node capacitances are lumped together into a simple equivalent value C_L.

The analysis amounts to determining $v_o(t)$ for step changes in v_i (see Fig. 8.1). When v_i is high, v_o is near the ground. A negative step in v_i with a final value $v_{iF} < V_T$ turns off the driver and then C_L *charges* through T_L to V_{OH}. Subsequently, a positive step in v_i permits C_L to *discharge* through T_D and v_o goes low. During this transient we can usually neglect the current in T_L since the effective ON resistance of T_D is far smaller than that of T_L, as design examples will show. The transient begins with T_D in saturation and moves to non-saturated operation when v_o goes below $(v_i - V_T)$.

Solutions are obtained using the current control equations treating W/L,

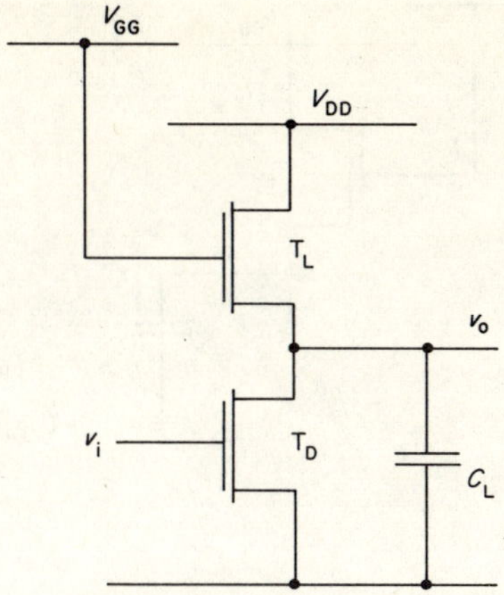

Fig. 8.1. The inverter.

β', V_T, and C_L as constants. The expressions for transistor currents are just as before but now these currents are used to charge and discharge the output node capacitance C_L. The resulting transients are described in terms of simple differential equations which are easily solved.

CHARGING TRANSIENT: ENHANCEMENT LOADS

The generalized expression for load current introduced in the previous chapter, and valid for all enhancement mode load devices, can be used here to permit simultaneous treatment of their transient responses.

A negative-going step in v_i turns T_D OFF at $t=0$. Then C_L charges through T_L. Equating i_L to i_c for the general case (Fig. 8.2):

$$i_L = I_{sc}\left(1-\frac{v_o}{V_{OH}}\right)\left(1-m\frac{v_o}{V_{OH}}\right) = C_L\frac{dv_o}{dt}$$

Rearranging terms with:

Normalized output: $$v_{on} = \frac{v_o}{V_{OH}} \tag{8.1}$$

Time constant: $$T_{sc} = \frac{V_{OH}C_L}{I_{sc}} = R_{sc}C_L \tag{8.2}$$

and applying integral signs:

$$\int dt = \int T_{sc}\frac{dv_{on}}{(1-v_{on})(1-mv_{on})}$$

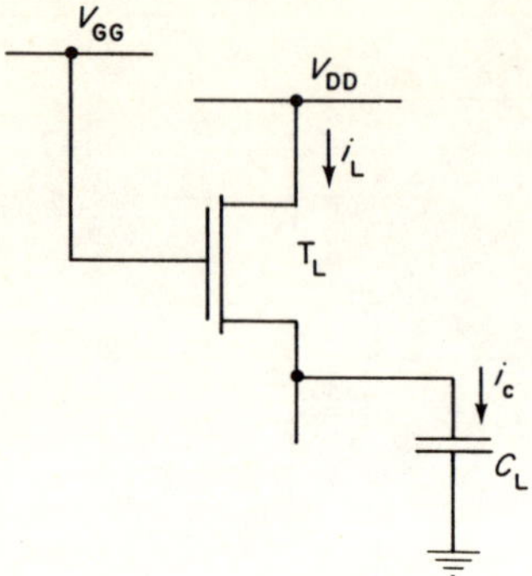

Fig. 8.2.

Then:

$$t = \frac{T_{sc}}{1-m} \text{Ln} \frac{1-mv_{on}}{1-v_{on}} + \text{const.} \tag{8.3}$$

Equation (8.3) is indeterminate for $m=1$. In that case:

$$\int dt = \int T_{sc} \frac{dv_{on}}{(1-v_{on})^2}$$

giving:

$$t = \frac{T_{sc}}{1-v_{on}} + \text{const.} \tag{8.4}$$

Differentiation of Eqns (8.3) and (8.4) will quickly verify the integration formulae used. If the constants in (8.3) and (8.4) are evaluated using $v_o = 0$ at $t=0$, the equations become:

$$t_r = \frac{T_{sc}}{(1-m)} \text{Ln}\left(\frac{1-mv_{on}}{1-v_{on}}\right), \qquad m < 1.0, \tag{8.5}$$

and

$$t_r = T_{sc}\left(\frac{v_{on}}{1-v_{on}}\right), \qquad m = 1.0 \tag{8.6}$$

Note that T_{sc} is specified for each load type in the set of equations (7.6).

Appendix C provides plots of v_{on} versus time for various m values (Fig. c.3). The time constant used for the rise is $T_r = T_{sc}/2m = C_L/\beta_L V_{OH}$. This is chosen in preference to T_{sc} because it is not affected by changes in the m value. The response shown dotted is for a resistor having the same short-circuit current I_{sc} as a saturated load. From (7.6), $R_L = 2/\beta_L V_{OH}$ so that $R_L C_L = 2C_L/\beta_L V_{OH} = 2T_r$. As expected, resistor performance is superior to that of the saturated load and is roughly comparable to an m value of 0.5.

It may be required to know the rise time $(t_2 - t_1)$ from v_{o1} to v_{o2} (Fig. 8.3).

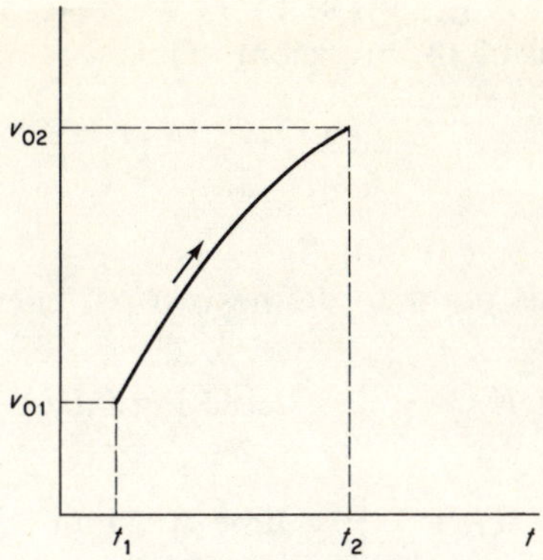

Fig. 8.3. Rise time.

Appropriate expressions are readily deduced from Eqns (8.3) and (8.4) using definite limits of integration. The saturated and non-saturated cases are thereby taken separately.

CHARGING TRANSIENT: DEPLETION LOADS

This transient begins with T_L in saturation and switches to the non-saturated condition when $V_{DS} = (V_{DD} - v_o) < V_I$, i.e. when v_o goes above $(V_{DD} - V_I)$ (see Fig. 8.4).

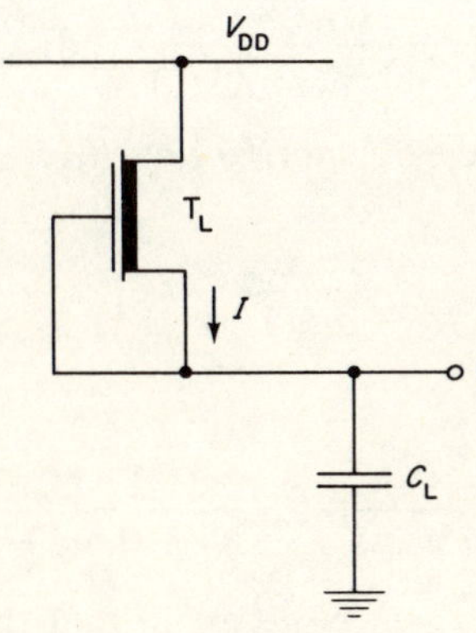

Fig. 8.4.

The saturated operation is described by:

$$I = \frac{\beta_L}{2} V_I^2 = \text{const.}$$

so that $v_o = It/C_L$ becomes:

$$v_o = \frac{\beta_L}{2C_L} V_I^2 t$$

to a maximum of $v_{os}=(V_{DD}-V_I)$ at $t=t_s$, where t_s represents the duration of the saturated portion of the transient. Thus:

$$t_s=\frac{2C_L}{\beta_L V_I^2}(V_{DD}-V_I) \tag{8.7}$$

on the assumption that $v_o=0$ initially.

It is convenient at this point to define two parameters as follows:

$$K_D=\frac{V_I}{V_{DD}}=\text{depletion factor} \tag{8.8}$$

$$T_D=\frac{C_L}{\beta_L V_I}=\text{time constant} \tag{8.9}$$

K_D is a normalized measure of built-in channel support for the device and T_D is a time constant of the system. Then:

$$t_s=2T_D\left(\frac{1}{K_D}-1\right) \tag{8.10}$$

The non-saturated portion of the transient is more complex and is governed by:

$$I=C_L\frac{dv_o}{dt}=\beta_L[V_I-0.5(V_{DD}-v_o)](V_{DD}-v_o)$$

Using normalized voltage $v_{on}=v_o/V_{DD}$ and rearranging:

$$\frac{dt}{T_D}=\frac{K_D}{[K_D-0.5(1-v_{on})](1-v_{on})}dv_{on}$$

The standard integral $\int dx/[x(ax+b)]=(1/b)\,\text{Ln}[x/(ax+b)]$ provides a solution resulting in:

$$\frac{t_2-t_s}{T_D}=-\text{Ln}\frac{1-v_{on}}{K_D-0.5(1-v_{on})}\Bigg|_{v_{os}/V_{DD}}^{v_{o2}/V_{DD}} \tag{8.11}$$

which, on expansion, becomes:

$$-\text{Ln}\frac{1-v_{o2}/V_{DD}}{K_D-0.5(1-v_{o2}/V_{DD})}+\text{Ln}\frac{1-v_{os}/V_{DD}}{K_D-0.5(1-v_{os}/V_{DD})}$$

Equation (8.11) is valid for all $t_2>t_s$. Note again that the voltage v_{os} at t_s is simply $(V_{DD}-V_I)$. With this substitution, the right-hand term reduces to Ln(2) and, since $\text{Ln}(A)-\text{Ln}(B)=\text{Ln}(A/B)$, we get:

$$\frac{t_2-t_s}{T_D}=-\text{Ln}\frac{0.5(1-v_{o2}/V_{DD})}{K_D-0.5(1-v_{o2}/V_{DD})}$$

which becomes:

$$t_2=t_s+T_D\,\text{Ln}\frac{2K_D-(1-v_{o2}/V_{DD})}{1-v_{o2}/V_{DD}} \tag{8.12}$$

The two equations, (8.10) and (8.12), describe the charging transient for the depletion load. Results are shown graphically in Appendix C (Figure C.4) for $K_D = 0.2$, 0.3, 0.4, 0.6, and 0.8. Evidently, large values of K_D mean a low-resistance path through T_L and rapid charging of C_L. Note the time-scale normalization in terms of time constant $T_r = T_D K_D$ which is simply $C_L/\beta_L V_{DD}$, independent of V_I.

It is readily shown that a resistor of the same short-circuit current value I_{sc} satisfies the condition $R_L C_L = 2T_r/K_D^2$. The resistor curve shown dotted is for comparison with a depletion load having $K_D = 0.4$. The performance of the depletion load is very much better than that of the resistor. This is due to it's constant-current characteristic mentioned earlier.

THE DISCHARGE TRANSIENT

A positive step change in input to a high value v_i at $t=0$ turns ON the driver and permits C_L to discharge (see Fig. 8.5). v_i is fixed at this value during the transient. The requirement that $v_{o(LOW)}$ be near to ground requires that $\beta_D \gg \beta_L$. This circumstance allows us to ignore the load device during discharge with small error.

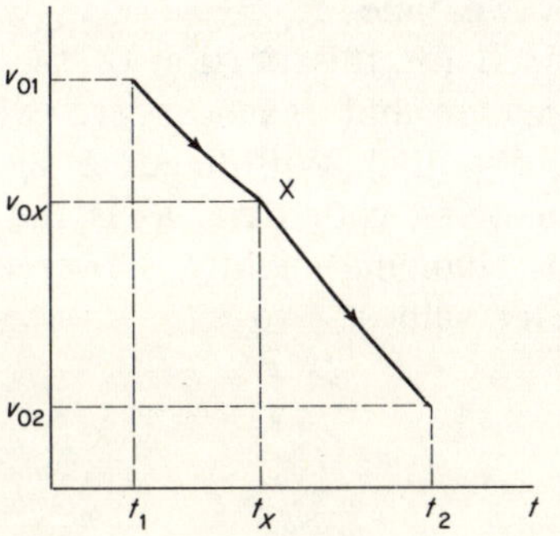

Fig. 8.5. Discharge transient.

Initially, the driver is saturated until, at time t_x,

$$v_{ox} = v_i - V_{TCED}$$

As v_o falls further, the driver is non-saturated. These intervals have to be treated separately.

Discharge to saturated driver

Since $i_D = -i_C$ (Fig. 8.6):

$$\frac{\beta_D}{2}(v_i - V_T)^2 = -C_L \frac{dv_o}{dt}$$

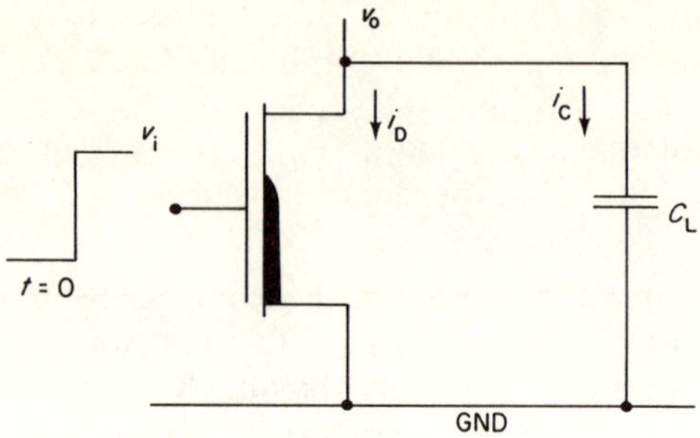

Fig. 8.6.

Rearranging and integrating:

$$\int dt = -\int \frac{2C_L}{\beta_D(v_i - V_T)^2}\, dv_o = -\frac{2C_L}{\beta_D(v_i - V_T)^2}\int dv_o$$

remembering that v_i is fixed. Thus:

$$t_x - t_1 = \frac{-2C_L}{\beta_D}\frac{v_o}{(v_i - V_T)^2}\Bigg|_1^x \tag{8.13}$$

provided $v_{ox} \geqslant (v_i - V_{TCED})$; normally we take $v_{ox} = (v_i - V_{TCED})$. Channel end voltage, $V_{CE} = (v_i - V_{TCED})$, is fixed for this portion of the transient. Since v_o is falling, L' is getting smaller and reaches zero when $v_{ox} = (v_i - V_{TCED})$. The source is at ground potential. With fixed voltage levels at both ends of the channel, β' is unchanging with time, as is the threshold voltage V_T, but the effective W/L is changing slightly. Observations like these are helpful in choosing parameter values.

Discharge to non-saturated driver

As before (see Fig. 8.7), $i_D = -i_C$:

$$\beta_D[(v_i - V_T)v_o - 0.5v_o^2] = -C_L\frac{dv_o}{dt}$$

$$\frac{dv_o}{v_o[2(v_i - V_T) - v_o]} = -\frac{\beta_D}{2C_L}dt$$

We again use the standard integral:

$$\int \frac{dx}{x(ax+b)} = \frac{1}{b}\text{Ln}\frac{x}{ax+b}$$

With $a = -1$, $b = 2(v_i - V_T)$, we get:

$$\frac{1}{2(v_i - V_T)}\text{Ln}\frac{2(v_i - V_T) - v_o}{v_o} = \frac{\beta_D t}{2C_L} + \text{const.}$$

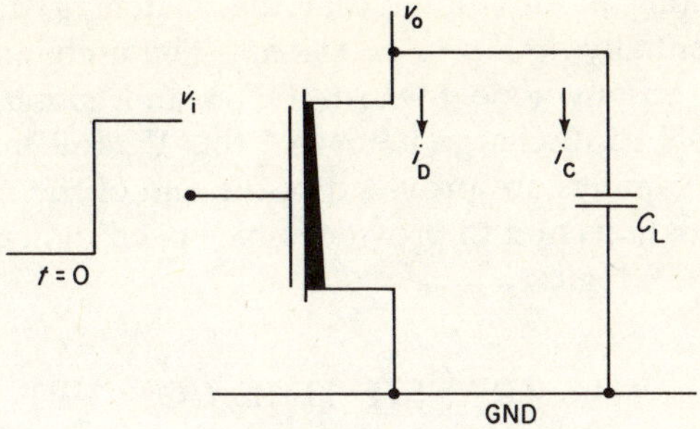

Fig. 8.7.

or, between limits:

$$t_2 - t_x = \frac{C_L}{\beta_D(v_i - V_T)} \operatorname{Ln} \frac{2(v_i - V_T) - v_o}{v_o} \Bigg|_x^2 \qquad (8.14)$$

This part of the transient starts from the point $v_{ox} = (v_i - V_T)$. Evaluated at this point, we get:

$$\operatorname{Ln} \frac{2v_{ox} - v_{ox}}{v_{ox}} = \operatorname{Ln}(1) = 0$$

Thus the *total* non-saturated time to a final value v_{o2} is:

$$t_{\text{non-sat}} = \frac{C_L}{\beta_D(v_i - V_T)} \operatorname{Ln} \frac{2(v_i - V_T) - v_{o2}}{v_{o2}} \qquad (8.15)$$

During this interval the source conditions are fixed. The drain-end threshold is V_{TD} and is varying and the associated transverse field conditions are changing. In fact, C_L, β_D, v_o, and V_T are all changing.

Total discharge time

Total discharge time is obtained by adding (8.13) and (8.15). If the saturated and non-saturated portions are designated t_s (from v_{o1} to v_{ox}) and t_{ns} (from v_{ox} to v_{o2}) respectively, the equations can be re-arranged as:

$$\frac{t_2 - t_1}{T_f} = \frac{t_s}{T_f} + \frac{t_{ns}}{T_f} = 2\left[\frac{v_{o1}}{v_i - V_T} - 1\right] + \operatorname{Ln}\left[\frac{2(v_i - V_T)}{v_{o2}} - 1\right] \qquad (8.16)$$

with time constant

$$T_f = C_L / \beta_D(v_i - V_T). \qquad (8.17)$$

The saturated interval involves a linear fall in output voltage because discharge current is constant during this time. In the subsequent non-saturated transient, the output falls to 10% of $(v_i - V_T)$ in about $3T_f$ seconds. The full response is plotted in Appendix C (Fig. C.5).

It is worth noting at this point that the switching transients in CMOS inverters are essentially *discharge* transients. The n-channel device goes ON to discharge the output node towards GND and, similarly, the p-channel transistor goes ON to discharge it toward the V_{cc} rail instead.

A number of examples are provided at the end of the chapter to illustrate the use of the equations and to provide some speed indications for switching operations on MOS logic.

THE TRANSIT TIME CONCEPT

When a MOS transistor is in the non-saturated condition with low V_{DS}, the channel is like a simple resistor of length L, width W, and very small depth x_c. Channel charge has mobility $\bar{\mu}$ and is transported under the influence of the electric field $E = V_{DS}/L$. The *transit time* is the time taken for a conducting particle (electron or hole) to pass through the channel:

$$\text{Transit time } T = \frac{\text{length}}{\text{velocity}} = \frac{\text{L}}{\bar{\mu}E} = \frac{L^2}{\bar{\mu}V_{DS}} \tag{8.18}$$

It follows that the total charge in the channel at a given instant can be moved out in T seconds. This idea has some interesting implications when we examine the case of an inverter driving another inverter which we will assume, for now, to have the same device sizes. The situation is illustrated in Fig. 8.8.

The transient amounts to charging C_1 so that C_2 may then discharge through the driver. If the stages are alike, $C_1 = C_2$ approx. and the charge Q that must pass through the driver equals the gate charge on C_1. It is now necessary to recall that:

$$\text{Gate charge} = (\text{channel charge}) + (\text{threshold-inducing charge})$$

If the gate voltate $\gg V_T$, the channel charge at any instant is only slightly less

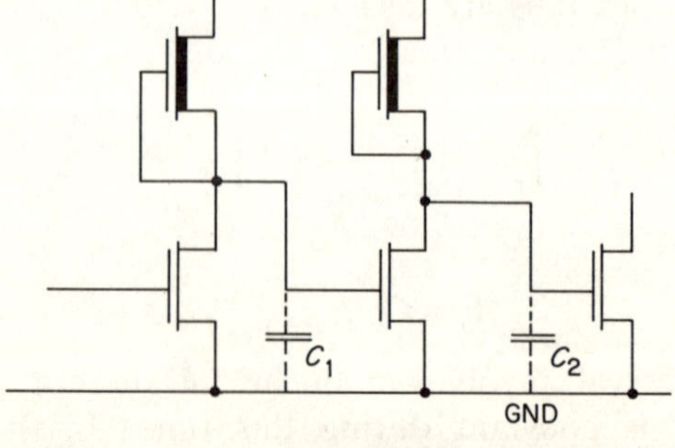

Fig. 8.8. Charge transfer between inverters.

than the charge on the gate. We saw that this quantity can be moved through in T seconds. It follows that the charge Q on C_2 can be 'dumped' in little more than one transit time. In evaluating transit time, there is some concern over a suitable choice of V_{DS}. It should represent 'average' conditions during the transient.

Some assistance is available from Eqn. (8.15) if we make a few reasonable assumptions. Suppose we choose $v_i = V_{DD}$, $V_T = 0.2V_{DD}$, and $v_{o2} = 0.2V_{DD}$ also. We are assuming that C_L is a gate capacitance, i.e. $C_L = C_{ox}WL$. We will also ignore the brief saturated interval when v_o exceeds $0.8V_{DD}$. With these provisions and $\beta' = C_{ox}\bar{\mu}$:

$$\begin{aligned} t_2 - t_1 &= \frac{C_{ox}WL}{C_{ox}\bar{\mu}(W/L)(V_{DD} - 0.2V_{DD})} \mathrm{Ln} \frac{1.6V_{DD} - 0.2V_{DD}}{0.2V_{DD}} \\ &= \frac{L^2}{\bar{\mu}(0.8V_{DD})} \mathrm{Ln}(7.0) = \frac{L^2(1.94)}{\bar{\mu}(0.8V_{DD})} \\ &= \frac{L^2}{\bar{\mu}(0.41V_{DD})} \end{aligned} \tag{8.19}$$

A comparison of Eqn. (8.19) with the transit time expression in Eqn. (8.18) shows that transit time T is a surprisingly good measure of switching speed. It is a good yardstick of inverter performance and a useful first estimate of discharge time.

It is also useful, as a first estimate, to consider that the charging transient takes $\beta_R T$ seconds, where β_R is the beta ratio, i.e. the ratio of driver size to load size. As a general design approach, one might start by selecting minimum dimensions for the load device which will provide a specified rise time. The next step would probably be to select a driver large enough to make $v_{o(LOW)}$ acceptably small, i.e. less than V_T by a reasonable noise margin. One could then check whether the resulting fall time was acceptable. All such estimates should be made under worst-case conditions. This calls for a good physical insight into the interplay of the various parameters. A few guidelines are given in the next section.

WORST-CASE DESIGN CONDITIONS

Worst-case design is essential for reliable circuit operation. It requires a full understanding of the MOS device itself. To illustrate, some worst-case conditions for design of an inverter circuit are as follows:

(a) Conditions which restrict current flow: high V_{TL}, low V_{DD}, low V_{GG}, low β'.

(b) Conditions which raise V_{OL}: high V_{DD}, high V_{GG}, low V_{TL}.

It is necessary to base decisions on the worst combination of circumstances affecting each performance requirement.

EXAMPLE 8.1

Assume that the inverter of Example 7.1 is driven by a square-wave input with fast-rising edges between levels of -1.0 and -8.5 V and drives an output-node load capacitance of 2 pF. Determine the output rise and fall times between -0.5 and -6.0 V. Use β' data from Appendix K.

Solution

Charging from SAT-load (rise time): $m = 1.0$. Also $V_{OH} = V_{DD} - V_{TCEL} = -8.6$ V. Then, using Eqns. (7.6) and (8.2)

$$T_{SC} = \frac{V_{OH}C_L}{I_{SC}} = \frac{2V_{OH}C_L}{\beta_L(V_{DD} - V_{TL})^2} = \frac{2C_L}{\beta_L V_{OH}}$$

A little thought suggests that the 'mean' level of (overhead) channel support voltage during the transient is no more than a few volts. Accordingly, we select a fairly high value of β' from page 000. We take $\beta' = 6.5\ \mu A/V^2$. We also ignore L' effect. Then

$$T_{SC} = \frac{2C_L}{\beta'(W/L)V_{OH}} = \frac{2\times 2\times 10^{-12}}{6.5\times 10^{-6}\times[0.4/(1.2-0.2)]\times 8.6}$$

$$T_{SC} = 179 \text{ nsec} \quad \text{and} \quad T_r = T_{SC}/2 = 90 \text{ nsec}$$

The normalized output moves from $(0.5/8.6) = 0.06$ to $(6.0/8.6) = 0.70$. Reading from the $m = 1$ graph (Fig. C.3)

$$\frac{t_2 - t_1}{T_r} = 4.6 - 0.2 = 4.4$$

Thus, $(t_2 - t_1) = 4.4(90 \text{ nsec}) = 394$ nsec

$$\underline{\text{Rise time} = 0.394\ \mu\text{sec}}$$

A more precise result might be obtained by direct application of Eqn. (8.4). The graphical approach is preferred because it permits the user to see at a glance the portion of the transient over which his solution lies. This, in turn, makes for sensible choice of upper and lower limits.

The discharge transient

Here the transition level is $v_{ox} = v_i - V_T$. With $v_i = -8.5$ V, v_{ox} will be close to -5.5 V, allowing for some body-effect on threshold. For this v_{ox}:

$$V_T = -2.5 - 0.4[(5.5 + 0.6)^{1/2} - (0.6)^{1/2}]$$

$$V_T = -3.2$$

We conclude that $v_{ox} = -(8.5 - 3.2) = \underline{-5.3\text{ V}}$

Referring to Fig. C.5:

$$\frac{t_s}{T_f}=2\left[\frac{v_{o1}}{v_i-V_T}-1\right]=2\left[\frac{6}{8.5-3.2}-1\right]$$

$$t_s=0.26T_f.$$

Also $v_{o2}/(v_i-V_T)=v_{o2}/v_{ox}=0.5/5.3=0.094$. Reading from Fig. C.5:

$$t_{ns}=3.0T_f$$

In total, the fall time is $t_s+t_{ns}=3.26T_f$.

To find T_f, the previously used value of β' seems appropriate here too. Then

$$T_f=\frac{C_L}{\beta'(W/L)(V_i-V_T)}=\frac{2\times10^{-12}}{6.5\times10^{-6}\times(1.5/0.2)\times5.3}$$

$$=7.7 \text{ nsec}$$

It follows that the fall time is $3.26T_f$ or

$$\underline{\text{Fall time}=25.2\ \mu\text{sec}}$$

The large beta ratio required for ratioed operation (11.5 in this example) is primarily responsible for rise and fall times which differ greatly, by a factor of 16 approximately.

EXAMPLE 8.2

An n-channel process uses depletion loads for which the following data is applicable:

$$\beta'=20\ \mu\text{A/V}^2, \qquad V_{TO}=-4\text{ V}$$

$$M=1.5, \qquad 2\phi_F=0.6 \text{ approx.}$$

Calculate the time required for a square device to charge a 2 pF load to 80% of supply level if a 5 V supply is used.

Solution

If the influence of body-effect is neglected, the solution proceeds very simply as follows:

$$K_D=\frac{V_I}{V_{DD}}=\frac{-V_{TO}}{V_{DD}}=\frac{4}{5}=0.8$$

$$T_D=\frac{C_L}{\beta_L V_I}=\frac{2\times10^{-12}}{20\times10^{-6}\times4}=2.5\times10^{-8}=25 \text{ nsec}$$

$$T_DK_D=(0.8)(25)=20 \text{ nsec.}$$

From Fig. C.4 with $K_D=0.8$, the normalized output reaches 80% of V_{DD}

when $t = 3.1 T_D K_D$. Then

$$\text{Rise time} = (3.1)(20) = \underline{62\ \text{nsec}}$$

This very rapid rise is overoptimistic. Body-effect will reduce the effective V_I. An attempt will be made to quantify this influence.

Much of the charging transient is in the non-saturated region due to the high value of V_I. The *actual* V_I at the drain for a full-length channel is:

$$V_{ID} = 4 - 1.5[\sqrt{(5+0.6)} - \sqrt{(0.6)}] = 1.61\ \text{V}$$

The large value of M has greatly reduced the channel support at the drain. Body-effect at the source increases as v_o rises. The mean V_{IS} over the interval is about $(4+1.6)/2 = 2.8$ V. Then, for purposes of current estimates, a reasonable overall figure is $(V_{IS} + V_{ID})/2 = (2.8+1.6)/2 = 2.2$ V

$$V_{IM} = 2.2\ \text{V}$$

Then:

$$K_D = 2.2/5 = 0.44$$

$$T_D = (25\ \text{nsec}) \times 4/2.2 = 45\ \text{nsec}$$

$$T_D K_D = (0.44)(45) = 20\ \text{nsec} \quad \text{(as before!)}$$

Referring to Fig. C.4, the time scale is unaffected by changes in V_I. Only K_D changes. This time we find by interpolating:

$$\text{Rise time} = 9.0 T_D K_D = \underline{180\ \text{nsec}}$$

This is three times longer than was predicted when body-effect was ignored. It is an important consideration when M values are large.

CHAPTER 9

MOS Processing Variations

Since the early sixties, the performance capabilities of MOS circuits have been steadily improving. This improvement has come through a succession of processing variations aimed at stable thresholds, high breakdown voltages, improved gain, and, especially, higher speed. This chapter looks at the performance capabilities of some established technologies and also examines some of the processing methods that are responsible for the high performance that is currently achieved.

SOME p-CHANNEL TECHNOLOGIES

Standard p-channel MOS is long established. It uses (111) crystal orientation with a typical Q_{ss}/q of about 4×10^{11} (often less nowadays). Gate oxide thickness is typically 1000 Å. If $N_B=10^{15}$, we get $V_T=-3.2$ V approx. with field inversion voltage $V_{TF}=-35$ V for a 15,000 Å field oxide. These are satisfactory and highly reproducible but a lower threshold is desirable especially in the interests of compatibility with bipolar circuits. Some low-threshold processes are described below.

The (100) *silicon p-channel process* uses the (100) crystal orientation because this leads to lower $Q_{ss}/q=1.5\times10^{11}$ approx. Then V_T reduces to about -2 V; unfortunately V_{TF} falls too, to about -17 V, which is uncomfortably low. An increase of field oxide thickness to a maximum practical limit of about 20,000 Å brings V_{TF} to:

$$V_{TF}=-0.6-\phi_F-\frac{Q_B}{C_{ox}}-\frac{Q_{ss}}{C_{ox}}$$

with:

$$\phi_F=0.026\,\mathrm{Ln}\frac{N_B}{n_i}=0.026\,\mathrm{Ln}\frac{10^{15}}{1.43\times10^{10}}=0.29\text{ V}$$

$$C_{ox}=\frac{k_{ox}e_0}{t_{ox}}=\frac{3.9\times8.85\times10^{-14}}{20{,}000\times10^{-8}}$$

$$=1.72\times10^{-9}\text{ F/cm}^2$$

$$Q_B=[2k_{si}e_0qN_B(2\phi_F)]^{1/2}=1.38\times10^{-8}\text{ C/cm}^2$$

giving:

$$V_{TF}=-0.6-0.29-8.02-13.95=-22.9\text{ V}$$

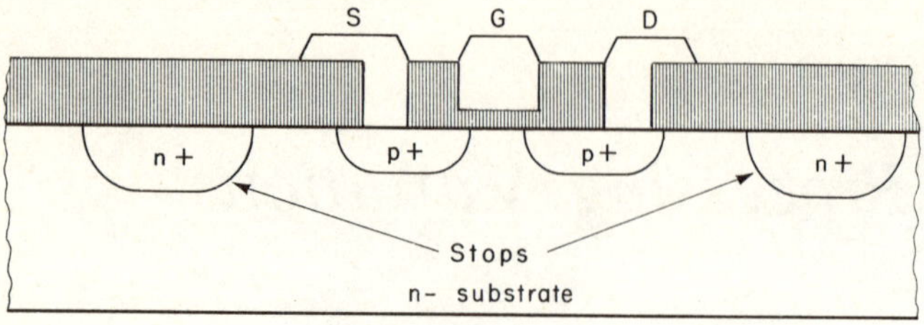

Fig. 9.1. Channel stops.

which is a small improvement. Another method is to use *field doping* to increase the surface N_B in field regions only; this increase must be quite small to avoid breakdown voltage problems. A second field-doping strategy is the use of *channel stops* (Fig. 9.1). The channel-stop bars may be more heavily doped but must be sufficiently far from the p regions to avoid restricting the spread of their depletion layers. Easier processing is obtained at a cost of extra chip area.

The MNOS technology uses a metal–nitride–oxide–silicon gate structure to increase C_{ox} to a higher value, call it C_{nox}, and thereby reduce V_T while still using (111) silicon for its high field thresholds. A silicon nitride (Si_3N_4) dielectric offers a much higher dielectric constant (k_{ni}) than oxide but process difficulties do not permit a direct substitution. A commonly used stable arrangement is a sandwich made up of a thin oxide layer covered by 400 Å of nitride. Then:

$$C_{nox} = \frac{C_{ox}C_n}{C_{ox}+C_n}$$

for the series combination of dielectrics. This method can yield thresholds around -2 V if Q_{ss} is reduced somewhat from the 4×10^{11} figure by improved processing. It can also give some advantage in the case of plastic encapsulation since it offers higher protection against sodium ions.

The *silicon gate technology* and the use of *ion implantation* are two other means by which low-threshold p-channel devices can be fabricated. These methods have much broader implications and are treated at a later point.

n-CHANNEL TECHNOLOGY

In the pursuit of high-speed operation, n-channel devices have the obvious attraction of higher electron mobility (×2 approx.) compared with hole conduction. Despite this, the n-channel technology arrived late due to more critical processing requirements. The threshold voltage for an n-channel device is:

$$V_{TO} = -0.6 + \phi_F + \frac{Q_B}{C_{ox}} - \frac{Q_{ss}}{C_{ox}}$$

and the graphical implementation of this equation in Appendix J shows at a glance the difficulties of ensuring enhancement mode operation (i.e. positive V_{TO}). The use of (100) orientation to give a minimum Q_{ss}/q around 1×10^{11} with $N_B=10^{16}$ and $t_{ox}=1000$ Å gives a small positive threshold of about 0.5 V. There are several difficulties with this approach:

(a) Field inversion voltage is low (around 15 V) and gate modulated avalanche breakdown is a further danger due to high N_B. Problems with V_{TF} are inherently more difficult than with the p-channel device. Thick oxide regions tend to have higher ionic charge (Q_{ss}, Q_{Na^+}) than gate regions. While this *improved* the situation for the p-channel device by increasing V_{TF}/V_{TO}, it *disimproves* the ratio V_{TF}/V_{TO} for n-channel devices. As a further complication, the positive voltages used for n-channel devices cause Na^+ ions in field (as well as gate) regions to drift slowly towards the silicon dioxide–silicon interface where their effect is maximized and V_{TF} falls due to long-term drift. The result can be unexpected failure in service.

(b) The high value of N_B increases scattering effects and reduces surface mobility as already mentioned. In fact, the n-channel device at $N_B=10^{16}$ offers no mobility advantage over the p-channel one at $N_B=10^{15}$.

(c) Higher N_B also increases the junction capacitances from S and D diffusions to bulk (C is proportional to $\sqrt{N_B}$ by the step junction approximation) and thus reduces speed further.

(d) Higher N_B increases the body-effect factor M (which is proportional to $\sqrt{N_B}$) substantially and leads to much higher threshold variations with the n-channel device. Note that a high threshold further degrades the speed by reducing channel support voltage.

To get around these difficulties one can select a relatively low N_B to give depletion mode devices and use substrate bias (negative V_{BS}) to secure enhancement operation. At the cost of an extra supply, much higher mobility is possible and V_{BS} also reduces junction capacitances substantially by reverse bias. An increased gate oxide thickness t_{ox} can aid in producing a suitable V_{TO} with back bias. This approach can give considerably faster operation than the p-channel device. As an example, consider a 10 Ω · cm wafer using $t_{ox}=1500$ Å and $t_{oxF}=15{,}000$ Å with $Q_{ss}/q=1.25\times10^{11}$. The corresponding N_B (Appendix D) is 1.3×10^{15} atoms/cm³ which gives a mobility around twice the typical p-channel value. To determine thresholds:

$$Q_B=[2k_{si}e_0qN_B(2\phi_F)]^{1/2}$$

with:

$$\phi_F=0.294 \qquad \text{giving} \qquad Q_B=1.59\times10^{-8}\ \text{C/cm}^2$$

Also:

$$C_{ox}=\frac{k_{ox}e_0}{t_{ox}}=2.3\times10^{-8}\ \text{F/cm}^2.$$

Then:

$$V_{TO} = -0.6 + 0.294 + \frac{1.59 \times 10^{-8} - 2.0 \times 10^{-8}}{2.3 \times 10^{-8}}$$

$$= -0.6 + 0.294 - 0.18$$

$$= \underline{-0.48 \text{ V}}$$

The field threshold is:

$$V_{TOF} = -0.6 + 0.29 - 1.8 = -2.1 \text{ V}$$

Now calculate:

$$M = \frac{t_{ox}}{k_{ox}} \left(\frac{2k_{si} q N_B}{e_0} \right)^{1/2}$$

for gate and field regions:

$$M_G = 0.90, \qquad M_F = 9.0$$

Calculate thresholds with $V_{BS} = -5$ V:

$$V_{TO}^B = -0.48 + 0.9(\sqrt{5.59} - \sqrt{0.59}) = \underline{0.95 \text{ V}}$$

$$V_{TOF}^B = -2.1 + 9(\sqrt{5.59} - \sqrt{0.59}) = \underline{12.2 \text{ V}}$$

Also, since $C = [qk_{si}e_0 N_B / 2(\psi_0 - V)]^{1/2}$, the back bias reduces capacitance by the factor $(5.59/0.59)^{1/2} = 3.1$. Gate overlap capacitances are not affected.

The merits of the above scheme are obvious and it has wide usage. It does, however, suffer from low field thresholds. An important point is that the V_{TF} of 12.2 produced in this way is dominated by N_B and is not much affected by ionic contamination; in a word, it is farily stable.

To permit higher voltage and more reliable operation, *field doping* is used. N_B is increased in field areas only and the resulting high M_F can easily give very high V_{TF}; e.g. $V_{TF} = 70$ V with $V_T = 1.5$ V is quite feasible. A number of techniques exist to increase V_{TF} in this way.

CAPACITANCE REDUCTION

For a *depletion mode* MOSFET it is not essential that the gate metal extend as far as the drain since a channel exists even without gate influence. Thus, an *offset gate* MOSFET (Fig. 9.2) greatly reduces C_{GD} and improves speed.

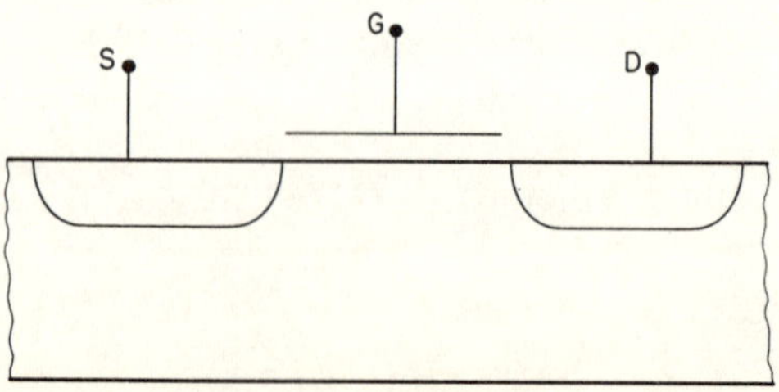

Fig. 9.2. Offset gate.

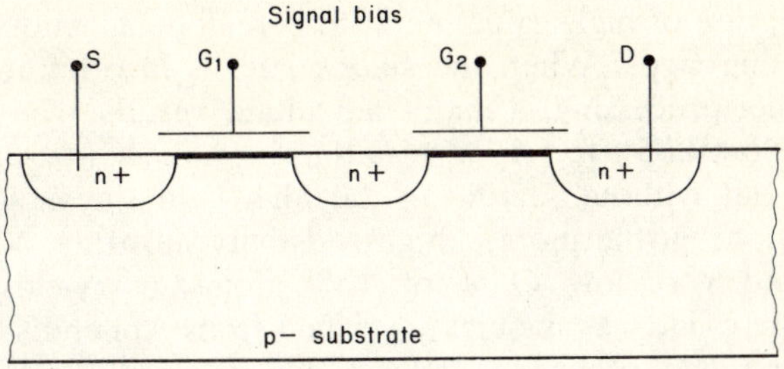

Fig. 9.3. Dual-gate MOS transistor.

The cost is a slightly higher channel resistance which is easily kept within acceptable limits. Offset gate devices are usually n-channel ones, consistent with high-speed requirements and the ease of obtaining n-channel depletion mode operation. Note that the lack of gate metal over the drain end of the thin oxide invites ionic contamination and countermeasures are required.

C_{GD} is almost completely eliminated in the *dual-gate device* (Fig. 9.3) which also has other very useful features. Its operation is similar to that of the *cascode* amplifier using bipolar transistors. Variation of the d.c. bias on gate 2 provides *gain control.* Applications that stem from this property include mixers, modulators, and demodulators. Very high gain bandwidth is possible.

SILICON GATE TECHNOLOGY

Gate metal overlap capacitances are minimized in technologies which first provide a gate electrode and subsequently *use the gate electrode as a mask* to define the source and drain regions. With such *self-aligned* structures, the overlap is reduced to that caused by lateral diffusion (about $0.75x_j$). Aluminium is not suitable as a gate metal since it will melt at temperatures used for diffusion of dopants. *Molybdenum*, with a melting point of 2600 °C, is sometimes used and is an established self-aligned technology. Of greater importance, as the next section shows, is self-aligned silicon gate technology. This scheme employs a polycrystalline silicon gate metal which, like molybdenum, is able to serve as a gate diffusion mask. The diffusion step which defines the source and drain regions also heavily dopes the polysilicon gate in the same type of impurity. A low-resistivity silicon layer results and, as well as serving as a gate electrode, it provides a useful additional interconnect medium running under the aluminium interconnect pattern. Polysilicon interconnects are restricted by not being able to run over diffused regions (except in the role of the gate metal). They nevertheless have a considerable advantage. Silicon gate processing also permits very small x_j, around 1 μm, which further reduces overlap capacitance. One disadvantage, compared with the standard aluminium gate process is that the silicon gate does not provide the option of high-value metal-to-p-region capacitors without additional processing steps. When the silicon gate is incorporated in standard (111) p-channel processing, a major advantage results. The gate threshold reduces by (0.6+0.55) or 1.15 V resulting in *another low-threshold* (*about* 2 V) *process* but without sacrificing the high field thresholds afforded by (111) orientation. Furthermore, a high-resistivity substrate, $N_B = 3 \times 10^{14}$ (or 15 Ω · cm), and very low Q_{ss}/q of 10^{10} atoms/cm^2 yields $V_T = 0$ V for 1000 Å of gate oxide—as is quickly verified from Appendix J. It is evident that *depletion mode* p-channel devices are marginally possible by these methods. The Silicon gate is equally applicable to n-channel processes. Compared with the aluminium gate, it gives a threshold voltage which is 0.05 V more positive and has a minimal effect. However, there is an obvious advantage in combining high n-channel mobility with low silicon gate overlap capacitance to give a high-speed process. As with conventional n-channel devices, field doping is necessary to raise V_{TF}; if bulk bias is also used, V_{TF} can become very high indeed. Finally, silicon-gate yields improved packing density by virtue of its self-aligned structure.

DEVICE ISOLATION METHODS

In most cases, isolation between devices is achieved via the non-conducting property of reverse biased p–n junctions (junction isolation). While this method is the most widely used, it suffers from leakage effects which may in some cases be objectionable. A more serious limitation is the inherent junction capacitance which seriously limits switching speed.

To reduce or eliminate these problems, various forms of *dielectric* isolation have evolved. One example of this arose in Chapter 3 in the description of a silicon gate process (Fig. 3.7). In that process, a method of component isolation called LOCOS (LOCal Oxidation of Silicon) was used. Thick oxide was grown selectively in the regions separating the devices using a nitride mask. This resulted in a smaller capacitance of the diffused layers. The method also permits close spacing of (unconnected) devices because depletion layer spreading is not a consideration. This, and the self-aligned gate structure, lead to a high packing density and higher speed. It is also worth mentioning that the oxide isolation method avoids the large steps (from thick oxide to thin oxide) that were characteristic of earlier processes. Such steps caused reliability problems relating to the aluminium interconnects which had to maintain continuity in crossing over them. That problem is aggravated by *electromigration*, i.e. the tendency of the aluminium to 'flow' when subjected to high current densities.

While the LOCOS process produced partial dielectric isolation, complete isolation is also possible. The technique of preferential etching is important in this context and will now be described.

An etchant is used which etches much faster along the (111) plane than along the (100) plane. Because of this, if a (100)-oriented material is

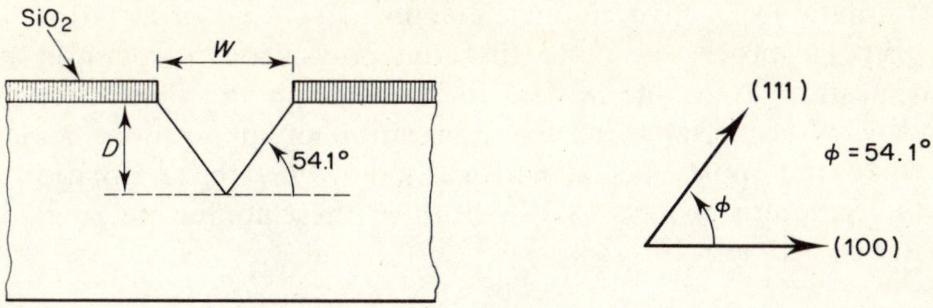

Fig. 9.4. Preferential etching.

oxidized and patterned in the usual way, the subsequent preferential etch forms V-grooves in the silicon under the windows, as shown in Fig. 9.4. Since the side-wall angle is fixed by the crystal structure, the depth D of the groove is determined by the window width W according to:

$$\tan(54.1°) = \frac{D}{W/2}$$

so that:

$$D = 0.69W \tag{9.1}$$

These V-grooves or 'moats' are commonly used to provide isolation in the horizontal direction by simply removing the silicon as described and then oxidizing the wafer so that the walls of the grooves are covered by an insulating silicon dioxide layer. Figure 9.5 shows a silicon island separated laterally by V-grooves and below by a layer of silicon dioxide. It has full dielectric isolation. The process starts with a thin monocrystalline wafer which has a layer of silicon dioxide grown on its surface. A layer of polycrystalline silicon is then deposited over the oxide. The wafer is then inverted and the other surface (the single-crystal silicon) is etched until the remaining layer is no more than several micrometres in thickness. The V-grooves are then etched in this surface and re-oxidation forms an insulating film. The isolation islands are subsequently used in the usual way

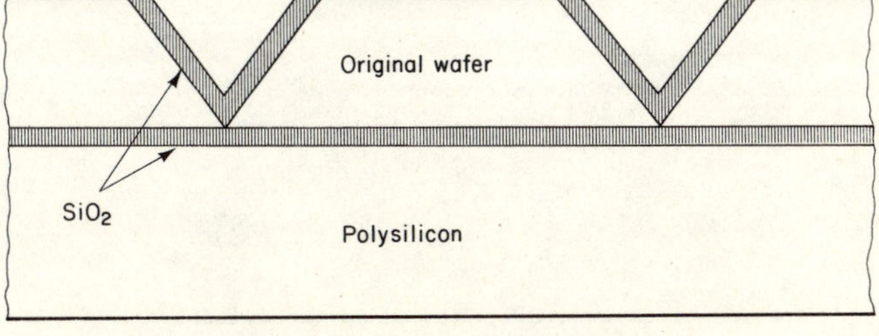

Fig. 9.5. V-groove isolation.

for the formation of active circuit elements. This kind of isolation is often used in bipolar processes. Note that the polysilicon is required to lend mechanical strength to the wafer. In the context of MOS circuitry, the V-groove is of importance to the fabrication of high-power devices (V-MOS) while the most spectacular method of dielectric isolation is the Silicon-On-Sapphire process (SOS). Both of these topics are treated separately later.

ION IMPLANTATION TECHNIQUES

Ion implantation can be used as illustrated in Fig. 9.6 to produce extremely small values of gate overlap capacitance—far better even than Silicon gate self-aligned methods. For p-channel devices, implantation is with boron. Thick oxide and gate metal (aluminium or polysilicon) are equally effective as implantation masks. High ion doses around 10^{13} ions/cm^2 are necessary to produce sufficiently low sheet resistance in the p-region extensions. Considerable crystal damage results and, with aluminium gates, there are severe leakage current side effects. With silicon gates, post-implantation high-temperature anneal can be performed.

The high sheet resistivity of the implanted p-regions can be used as a basis for implanting high-value resistors. Such resistors are implanted at the same time as source and drain extensions. Resistor contacts are p+ diffusions made at the same time as source and drain diffusions. Such resistors are suitable as inverter loads and also in forming precision resistor ladder networks for MOS digital-to-analog converters. Typical sheet resistance is about 3000 Ω/□.

The n-channel ion-implanted device gives very good high-frequency performance. Also, annealing at only 500 °C will activate implanted phosphorus atoms giving sheet resistance as low as a few hundred ohms per square.

Ion implantation is equally useful in shifting threshold voltages. Wide-ranging accurately controlled shifts are easily achieved. Implantation is through the thin gate oxide (Fig. 9.7) to reduce contamination. A good

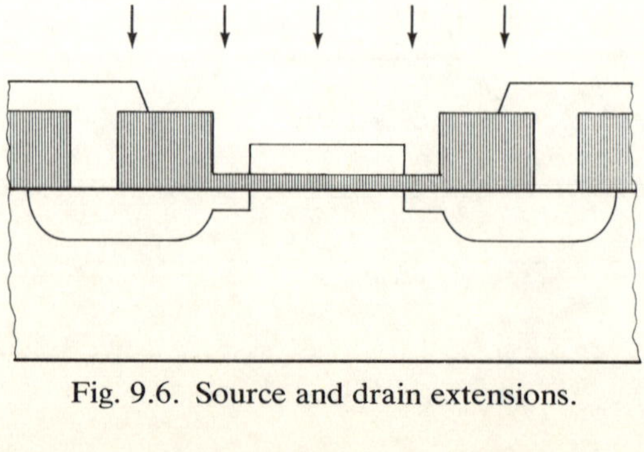

Fig. 9.6. Source and drain extensions.

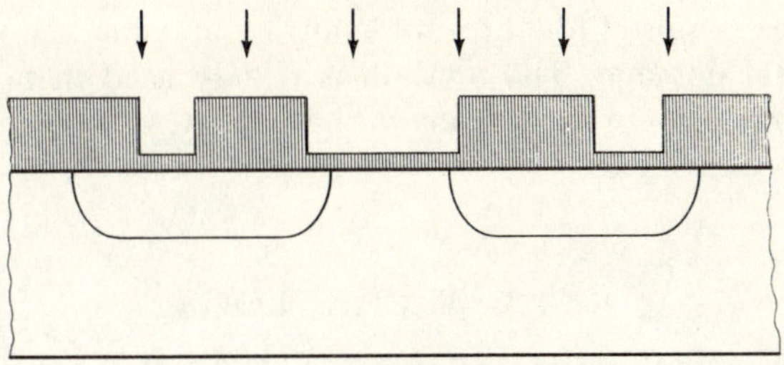

Fig. 9.7. Threshold shifting.

example is implantation of regular high-threshold (111) p-channel devices to produce a lower V_{TO}.

Consider a gate region with 1000 Å oxide dielectric and $Q_{ss}/q = 4\times10^{11}$. Assume $\phi_F = 0.3$ V and $N_B = 10^{15}$. *Then*:

$$Q_B = [2k_{si}e_0qN_B(2\phi_F)]^{1/2} = 1.4\times10^{-8}\ \text{C/cm}^2$$

and

$$V_{TO} = -0.6 - \phi_F - \frac{Q_B}{C_{ox}} - \frac{Q_{ss}}{C_{ox}}$$

$$= -0.6 - 0.3 - 0.41 - 1.9$$

$$= -3.21\ \text{V}$$

where
$$C_{ox} = \frac{k_{ox}e_0}{t_{ox}} = 3.4\times10^{-8}$$

Ion implantation can reduce V_{TO} to the desired level. If a threshold of -1.5 V is required, it is necessary to implant ions giving an effective Q'_B/C_{ox} of $(3.21-1.5)$ or 1.71 V. Thus:

$$Q'_B = 1.71C_{ox} = 5.8\times10^{-8}\ \text{C/cm}^2$$

This Q'_B is of *opposite* type of Q_B. This means that *acceptor* ions (boron) are implanted at the surface of the n-type wafer under the gate. The effective dose is:

$$\frac{Q'_B}{q} = 3.6\times10^{11}\ \text{ions/cm}^2$$

The implanted dose would be somewhat higher because not all ions are electrically active. Unlike diffused layers, maximum ion concentration occurs somewhat below the surface and concentration falls in a Gaussian manner above and below that level.

Selective implantation can be used to produce enhancement-type drivers and depletion-type loads on the same chip—a major advantage. Photoresist protects selected devices from the ion beam at a cost of one extra masking step.

The ion doses required for threshold shifting are small enough to produce negligible crystal damage. The method is widely used in p- and n-channel technologies and also in *complementary* MOS (CMOS) which is the next topic to be discussed.

CMOS TECHNOLOGY

From the foregoing discussion of threshold difficulties with p-channel and n-channel devices, it is clear that the fabrication of well-matched complementary devices with low consistent thresholds calls for very accurate doping concentrations around 10^{16} for the p- tub and for the n+ diffusions (Fig. 9.8). Note, too, that Q_{ss}/q tends to raise p-channel thresholds while

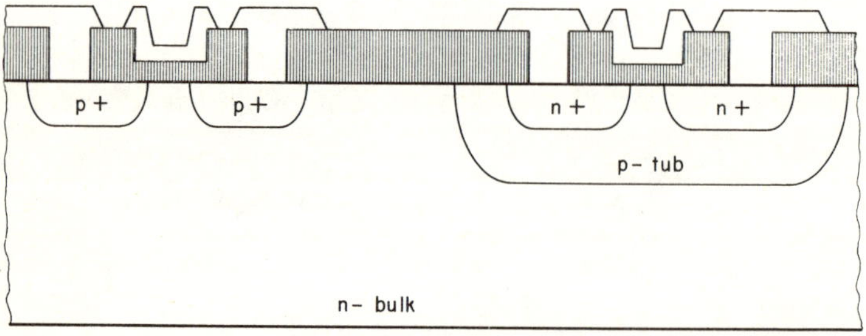

Fig. 9.8. Complementary MOS.

lowering n-channel thresholds, which further complicates matters. Producing a satisfactory CMOS was at first very difficult but eventually came to pass with the aid of low-level diffusion and epitaxial refill techniques. More recently, ion implantation methods allowed surface doping concentrations to be accurately controlled, resulting in good control of threshold voltages also. In fact, CMOS circuits can now operate off 1.5 V batteries (wrist-watch applications) with thresholds around 0.5 V readily attainable.

Apart from the obvious power advantages at low to moderate frequencies, CMOS offers *higher switching speeds* and *high noise immunity*. The higher speed is explained by the fact that output node capacitance is always charged or discharged through a transistor which is driven 'hard ON'—in contrast, for example, to charging through a satuated or even a non-saturated load. Such a load would have to have a high impedance to reduce power dissipation and to minimize size.

The disadvantages associated with CMOS are principally poorer packing density (p-tubs are very space-consuming) and increased process complexity.

SILICON-ON-SAPPHIRE (SOS) PROCESS

The SOS process uses a polished sapphire substrate. Because this is an *insulating* substrate, it offers dielectric isolation and its special property is that it permits the growth of thin single-crystal silicon films on its surface (even though the crystal structure is different) with good carrier mobilities. These silicon films are only 1 to 2 μm thick and are epitaxially deposited. They consist of p-type islands where n-channel devices are to be fabricated and of n-type islands where p-channel devices are to exist. There is no silicon on the sapphire surface outside the actual device regions.

The islands are appropriately doped at either side to form the source and drain regions and, between these, there remains a lightly doped portion of the island where channelling occurs under the control of the overhead gate. Figure 9.9 illustrates the construction of complementary devices on a sapphire substrate—CMOS on sapphire. In the CMOS process, all the

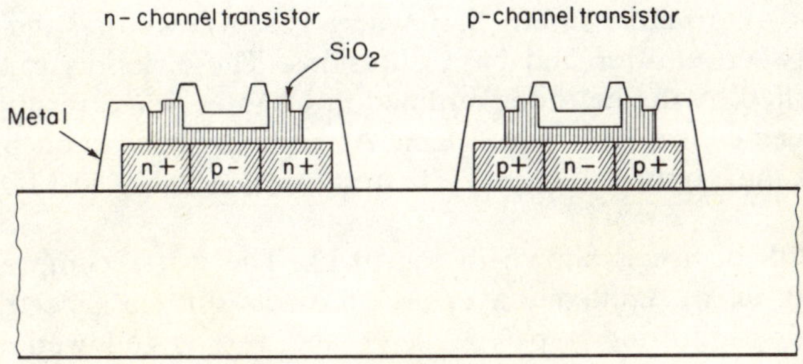

Fig. 9.9. CMOS on sapphire.

islands are originally lightly doped n-type epitaxial layers and the second masking operation serves to convert some of the islands to p-type layers, wherever n-channel transistors are to be formed. The next two masking steps define regions which are to be heavily doped, p-type and n-type regions in turn. Subsequent masking operations are concerned with gate oxide growth and with metallization over the gate and the provision of source and drain contacts.

In the SOS process, the metal interconnects pass directly over the sapphire substrate and in contact with it while the transistor diffusions have almost complete dielectric isolation. Both of these factors result in very low parasitic capacitances and, consequently, in very high speed.

POWER FET STRUCTURES

A low value of 'ON' resistance, R_{on}, is an important attribute of any power switch. In the case of a FET structure, this requires large $(W/L)_{eff}$ and, to avoid excessively large device sizes, the channel length L should be as small as possible.

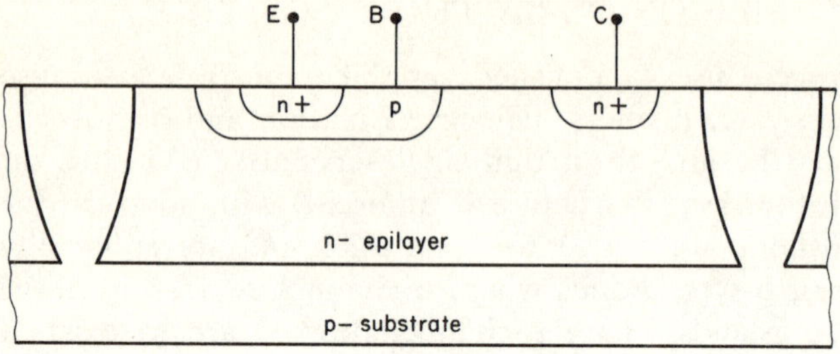

Fig. 9.10. A bipolar n–p–n transistor.

Channel length is normally limited by lithographic considerations to about 5 μm. This is because of the accuracy limits on photomasking operations. Very short channels can be achieved in essentially the same way that narrow base regions are possible in bipolar transistors. The method can be seen from the bipolar structure in Fig. 9.10 where the base width is the difference in depth between emitter and base diffusions. These depths can be accurately controlled by the methods outlined in Chapter 4 and resulting typical base width values are less than 1 μm. Accordingly, the same principle is utilized in some popular power FET structures, notably D-MOS and V-MOS.

The V-MOS device is shown in Fig. 9.11. The wafer comprises an n+ substrate with an n− epitaxial layer grown on its surface. The transistor is formed by first diffusing a p-type layer and this is followed by an n+ diffusion into the p-type layer. Then a V-groove is anisotropically etched and a layer of silicon dioxide is grown. The source and gate contacts are then deposited. The n+ substrate is the drain terminal. Current flows in the vertical direction between two n+ layers which are separated by a thin p region. No conduction is possible until the gate potential causes inversion of

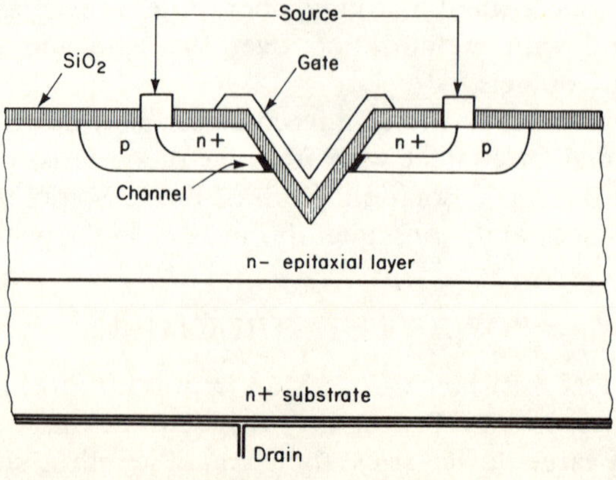

Fig. 9.11. A V-MOS power FET.

the p layer where it touches the side walls of the groove. The resulting channel is semi-vertical and its length is the difference in junction depth of the two diffusions. Below this very narrow channel, the epitaxial layer behaves as a drift region. Electrons flow to the n+ substrate under electric field influence. (The length L' of a conventional saturated MOSFET transports charge in the same manner.)

The epitaxial layer is an essential feature of high-voltage operation. It must have high resistivity so that the avalanche breakdown voltage is high. Its thickness must be sufficient to ensure that the depletion layer of the reverse biased junction does not extend all the way to the n+ substrate. Note that these requirements result in higher resistance in the epilayer as the required breakdown voltage figure is increased. This resistance is part of the total ON resistance and the result is that R_{on} increases very rapidly with breakdown voltage rating. To keep R_{on} low, a large die area is required. This, in turn, reduces the yield per wafer and increases the cost.

The advantages of the V-groove structure are not limited to the small channel length. Note that the structure is also economical in its use of wafer surface area because the source-to-drain spacing is vertical rather than horizontal and because the drain is absent from the upper surface. A further advantage is very low values of gate–drain capacitance and, consequently, rapid switching. The shape of the groove does have one disadvantage: its sharp bottom edge can cause high field gradients between the gate and drain. This can be remedied by halting the anisotropic etch before it has reached full depth. The resulting shape is shown in Fig. 9.12 for the completed device. This structure is effective for withstanding voltages up to about 150 V. At higher voltages, the double-diffused (D-MOS) structure is better.

In the D-MOS device (Fig. 9.13), short channels are achieved in the same way as before with one notable difference, namely that the channel is bounded by diffusion side walls rather than vertically. It is made possible by conducting two diffusion operations in succession through the same window.

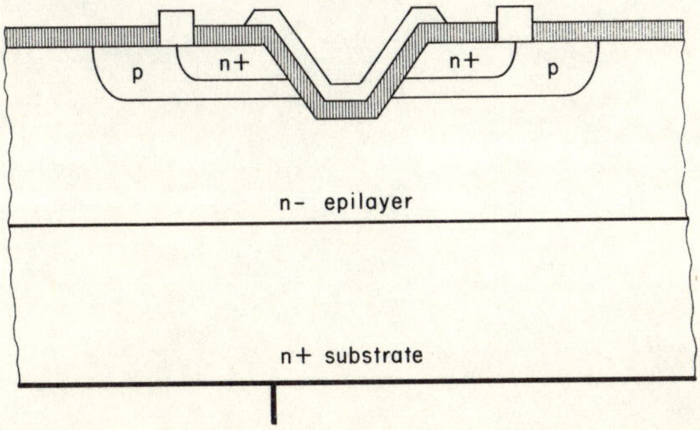

Fig. 9.12. Modified V-MOS.

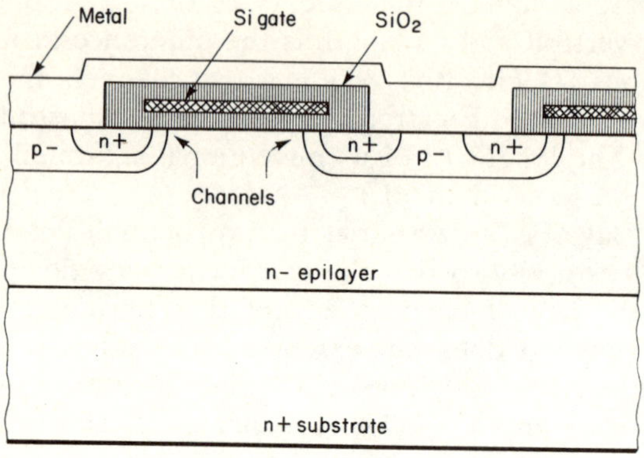

Fig. 9.13. D-MOS.

Current flow is vertical in the epitaxial layer and horizontal in the channels. A polysilicon gate can be used to produce channelling. This device is fabricated in various shapes. One example, the HEX FET, has hexagonal p regions resulting in a long channel perimeter for a given chip area.

At the present stage of development the power MOSFET can claim considerable performance advantages over the bipolar product. It is gradually gaining preference where speed, cost, and high input impedance are decisive factors.

CHAPTER 10

Analog MOS: Parameters and Circuits

The action of MOS devices and circuits in performing analog functions is explored in this chapter. The MOSFET will emerge as a poor second to the bipolar transistor in most analog roles. Nevertheless, there is much pressure to perform analog tasks using MOSFETs. This stems from the convenience and cost-saving of single-chip circuits and relatively low-cost processing. The most important examples of mixed digital and analog roles on one chip arise in the case of analog-to-digital and digital-to-analog conversion products.

SMALL-SIGNAL PARAMETERS

Among the better features of the MOSFET is its high input resistance, appropriate to both digital and analog roles. One analog use to which the MOS transistor is well suited is as an *electronic switch*. Unlike the bipolar transistor, there is no inherent offset. Since V_{DS} is essentially very small for the 'ON' device (ensured by a high circuit-impedance level), the operation is non-saturated (Fig. 10.1) with *ON resistance* (R_{on}) calculated as follows:

$$I_{DS} = \beta[(V_{GS} - V_T)V_{DS} - 0.5V_{DS}^2]$$

$$\frac{dI_{DS}}{dV_{DS}} = \beta(V_{GS} - V_T - V_{DS}) = \frac{1}{R_{on}} \qquad (10.1)$$

$$R_{on} = \frac{1}{\beta(V_{GS} - V_T - V_{DS})}$$

and, for negligible V_{DS},

$$R_{on} = \frac{1}{\beta(V_{GS} - V_T)} \qquad (10.2)$$

The operating point Q is at the origin. R_{on} varies from less than a hundred ohms to a few kilo-ohms for a large W/L transistor. This may cause significant error even in a high-impedance situation. In CMOS circuitry, the R_{on} variation is largely avoidable, as seen later.

Transconductance is an important 'gain' quantity:

$$g_m = \left.\frac{\delta I_D}{\delta V_{GS}}\right|_{V_{DS}=\text{const.}}$$

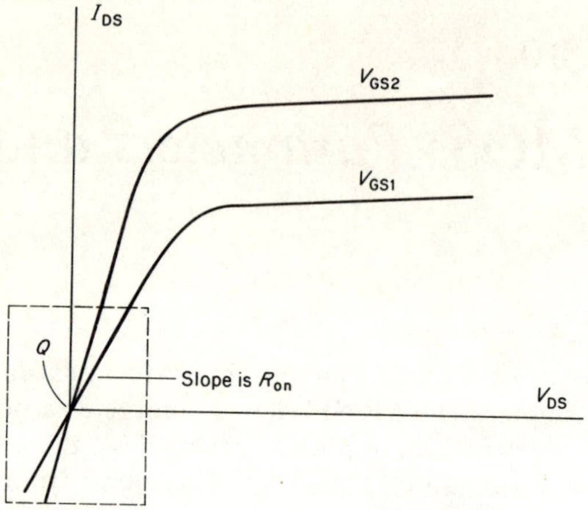

Fig. 10.1. MOSFET analog switch.

(see Fig. 10.2) and for saturated operation, which gives the best gate control:

$$I_D = 0.5\beta(V_{GS} - V_T)^2$$

so that:

$$g_m = \beta(V_{GS} - V_T) = \sqrt{(2\beta I_{DS})} \tag{10.3}$$

Compare with the bipolar case:

$$g_m = \frac{I_c}{kT/q} = 40 I_c \qquad \text{(approx.)} \tag{10.4}$$

Using $\beta' = 5 \times 10^{-6}$ and $(W/L)_{eff} = 10$ with $I_D = 100\ \mu A$, Eqn. (10.3) yields a

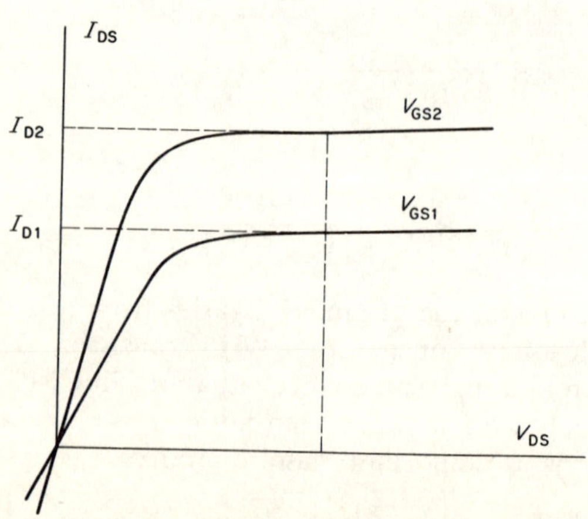

Fig. 10.2. Defining transconductance.

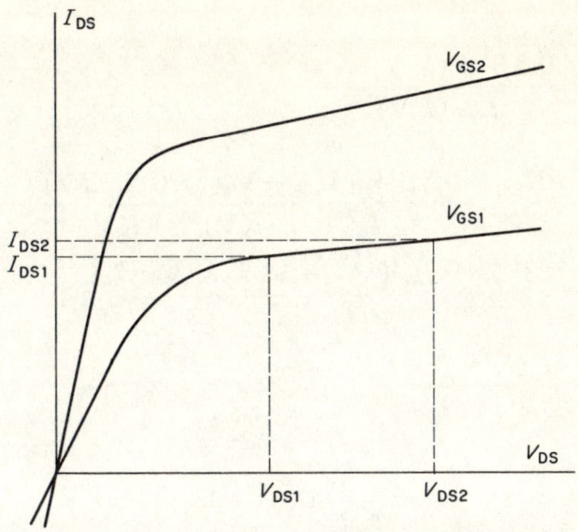

Fig. 10.3. Defining output resistance.

g_m of 0.1 m℧ while Eqn. (10.4) gives $g_m = 4.0$ m℧ for the bipolar case. A better type of comparison would be for devices *of similar areas* at similar current levels and would show the bipolar case to be about an order of magnitude better. The low g_m of the IGFET is one of its main drawbacks as an analog circuit element.

Output resistance r_{o_2} is theoretically infinite in the saturated mode with $I_{DS} = 0.5\beta(V_{GS} - V_T)^2$ *independent* of V_{DS}:

$$r_o = \left.\frac{\delta V_{DS}}{\delta I_{DS}}\right|_{V_{GS}=\text{const.}} = \frac{V_{DS2} - V_{DS1}}{I_{DS2} - I_{DS1}}$$

as seen in Fig. 10.3. In practice, r_o varies from the low megohm region for small V_{GS} to several kilo-ohms for large V_{GS} for typical devices. The variation is caused by *channel-length modulation* which effectively increases β for large V_{DS} by increasing $W/(L - L')$ as L' increases. The comparable phenomenon for bipolar transistors is *base-width modulation.*

By taking account of L' in the saturated current equation, a useful expression for r_o can be obtained:

$$I_{DS} = \frac{\beta'}{2}\frac{W}{L - L'}(V_{GS} - V_T)^2$$

where

$$L' = \left\{\frac{2k_{si}e_o}{qN_B}[V_{DS} - (V_{GS} - V_T)]\right\}^{1/2} = (KV)^{1/2}$$

using:

$$K = \frac{2k_{si}e_o}{qN_B}$$

$$V = [V_{DS} - (V_{GS} - V_T)]$$

Thus:

$$I_{DS}=\frac{0.5\beta' W(V_{GS}-V_T)^2}{L-(KV)^{1/2}}$$

and

$$\frac{dI_{DS}}{dV_{DS}}=\frac{dI_{DS}}{dV}=\frac{0.5\beta' W(V_{GS}-V_T)^2\times 0.5(K/V)^{1/2}}{L^2-2L(KV)^{1/2}+KV}=\frac{1}{r_o}$$

$$\frac{1}{r_o}=\frac{(\beta'/2)(W/L)(V_{GS}-V_T)^2(K/V)^{1/2}.L}{2(L-L')^2}$$

$$=\frac{I_o(K/V)^{1/2}.L}{2(L-L')^2}$$

where:

$$I_o=\frac{\beta'}{2}\frac{W}{L}(V_{GS}-V_T)^2$$

is the current value at the onset of saturation when $L'=0$. Thus,

$$r_o=\frac{2L}{I_0}\left(1-\frac{L'}{L}\right)^2\frac{[V_{DS}-(V_{GS}-V_T)]^{1/2}}{[2k_{si}e_0/q\,.\,N_B]^{1/2}} \tag{10.5}$$

Two points are worth noting. Firstly, r_o falls rapidly in the short-channel case where L' is a large part of L. The second point is that r_o increases with V_{DS} to give a more nearly horizontal drain characteristic provided L' is small compared with L.

As an example, consider a device with $W=L=8\ \mu$m, $(V_{GS}-V_T)=2$ V, and $V_{DS}=6$ V. Then, using $N_B=10^{15}$ atoms/cm^3, L' becomes 2.27 μm and, for an assumed β' of 5 μA/V^2, $I_0=10^{-5}$ A from which;

$$r_o=\frac{2\times 8\times 10^{-4}(1-0.28)^2(6-2)^{1/2}}{10^{-5}\times 1.14\times 10^{-4}}$$

$$=1.4\ \mathrm{M\Omega}\ \text{(approx.)}$$

Large values of r_o are useful in signal amplification roles. Examination of Eqn. (10.5) shows that r_o increases in proportion to L^2 rather than L because I_o falls with increasing length.

MOSFET SWITCHES AND MULTIPLEXERS

The action of the MOSFET as a switch is evident from Eqn. (10.2):

$$R_{on}=\frac{1}{\beta(V_{GS}-V_T)} \tag{10.6}$$

For a square device ($W/L=1$) with $(V_{GS}-V_T)=10$ V and $\beta'=5\ \mu$A/V^2, the ON resistance is $R_{on}=20$ kΩ. Thus, quite large devices are necessary for low ON resistance. For $W/L=100$, R_{on} becomes 200 Ω and this can be considered a 'reasonably low' value.

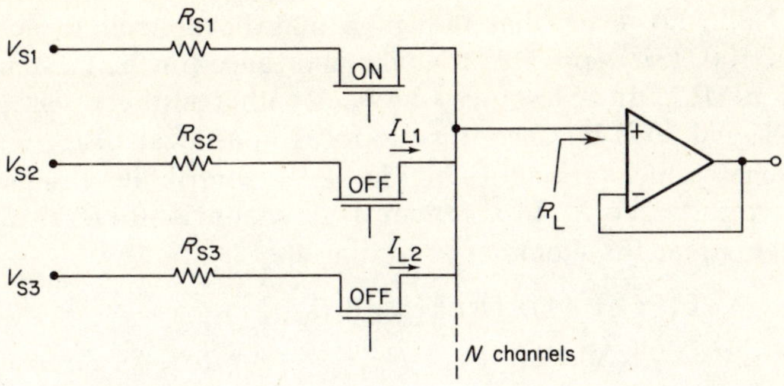

Fig. 10.4. Analog multiplexing.

The question of what R_{on} value is acceptable depends very much on the application. If the switch is a pass transistor in digital logic circuitry, quite high R_{on} is probably acceptable because the only load is a small capacitance and the only concern is the charging time constant. If the switch is part of an analog signal multiplexer, the requirements are much more stringent.

Figure 10.4 shows three input channels of a multiplexer which uses MOS switches. Each channel has a source resistance R_{Sj} (for channel j) and only one switch may be ON, at any given time, to connect its channel to the input of a unity-gain buffer. The buffer serves mainly to present a high impedance to the multiplexer and this impedance is modelled as a high-value resistor R_L.

The circuit shows channel 1 ON so that:

$$\frac{V_{in}}{V_{S1}} = \frac{R_L}{R_{S1} + R_{on} + R_L}$$

(see Fig. 10.5). To ensure that $V_{in} = V_{S1}$ to good accuracy, it is necessary that $(R_{S1} + R_{on}) \ll R_L$. It is easy to achieve high R_L, in the region of 1 MΩ. Then, if $(R_{S1} + R_{on}) = 1$ kΩ, for example, V_{in} is within 0.1 per cent of V_{S1}. This describes 10-bit accuracy (since $2^{10} = 1000$, approx.). Higher bit resolution calls for lower $(R_{S1} + R_{on})$, by a factor of 2 for each additional bit. Thus, high-resolution digital systems impose tough requirements on the analog

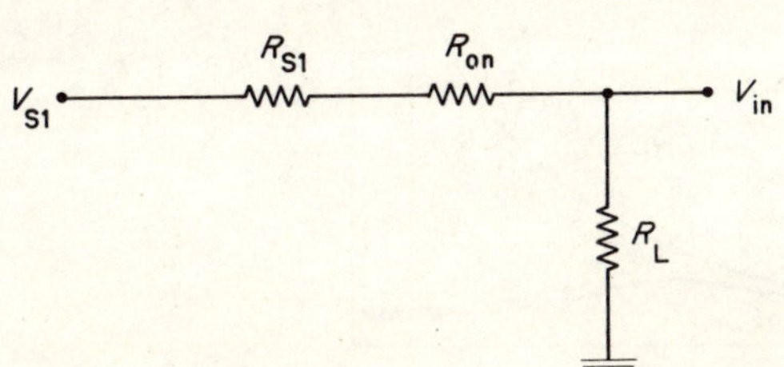

Fig. 10.5.

interfaces. Note, however, that though it may be difficult to achieve R_{on} below about 100 Ω, it is possible to make allowance for the existence of any *fixed* value of R_{on}. It is less easy to deal with resistance variation as a function of signal level because this produces non-linear distortion.

The switches which are OFF in Fig. 10.4 contribute a small leakage current of typical value I_L. In a system of N channels, these produce a net offset voltage E_L at the amplifier according to:

$$E_L = (N-1)I_L[R_L \| (R_{S1} + R_{on})]$$
$$= (N-1)I_L(R_{S1} + R_{on}) \qquad \text{(approx., for large } R_L\text{)}$$

This error may not be important unless the channel has a high source impedance or in the case of high bit-accuracy requirements. It does advise against combining a large number of channels in one group. For example, if $R_{S1} = 10\ \text{k}\Omega$, $I_L = 5\ \text{nA}$, and $N = 8$, the error is approximately:

$$E_L = 7 \times 5 \times 10^{-9} \times 10^4 = 0.35\ \text{mV}$$

assuming $R_{on} \ll 10\ \text{k}\Omega$. This error is not small where 12-bit resolution or better is expected on normal voltage ranges.

Wherever analog signals are passed through MOSFET switches, the allowable analog signal range is limited by the requirement that $V_{GS} > V_T$ at all times. Thus, for n-channel switches with gates at $+V_{cc}$ during the ON state, the analog signal must not exceed $(V_{cc} - V_T)$. In practice, it may be desirable to further limit the analog signal swing so as to avoid excessive variations in R_{on}.

Figure 10.6 shows typical variations in R_{on} versus signal level $V_{in} = V_D = V_S$ (assuming $V_{DS} = 0$) for the case of a single 15 V supply. If resistance variations must be minimized, the CMOS switch is a very good answer (Fig. 10.7). It combines a p-channel and an n-channel device in parallel so that when one has high ON resistance the other is low. This is evident from the

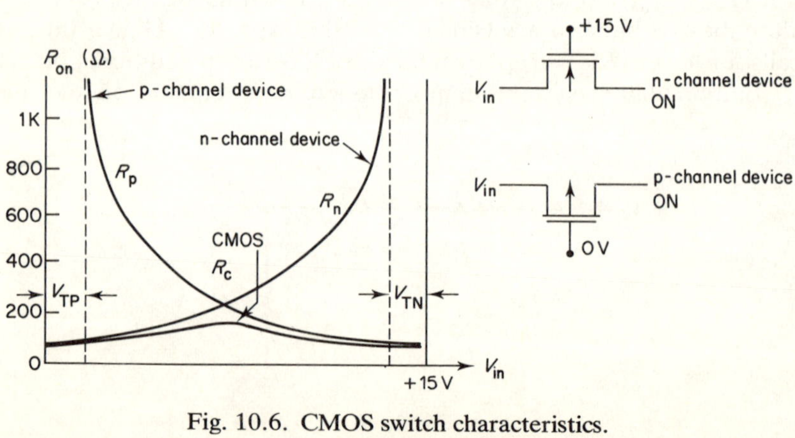

Fig. 10.6. CMOS switch characteristics.

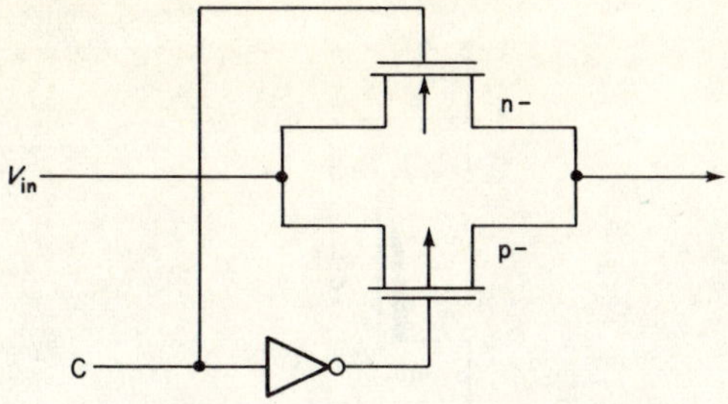

Fig. 10.7. A CMOS switch.

'CMOS' graph in Fig. 10.6 where:

$$R_c = R_p \parallel R_n \qquad \text{for all } V_{in}$$

The result is low R_{on} with little variation over the analog signal range. The price of this performance is two complementary switch transistors plus a CMOS inverter for switch control. The control line C (Fig. 10.7) goes HIGH to turn the switch ON. It takes the n-channel device gate to +15 V and the p-channel device gate to 0 V. The signal V_{in} lies between these two levels and can swing as far as either supply rail.

SMALL-SIGNAL AMPLIFICATION

A number of basic amplifying modes are possible and some of these modes will now be examined. Unless d.c. bias conditions are specifically mentioned, they will be assumed to be appropriate to the situation in hand and the discussion will be in terms of small-signal quantities only.

The 'load device' shown in Fig. 10.8 is often used in analog roles and the small-signal impedance r_L at the source node is of interest:

$$r_L = \frac{1}{di/dv} = \frac{1}{dI_{DS}/dV_{GS}} = \frac{1}{g_m}$$

Changes in V_{GS} control current flow. The condition that $V_{DS} = \text{const.}$ is

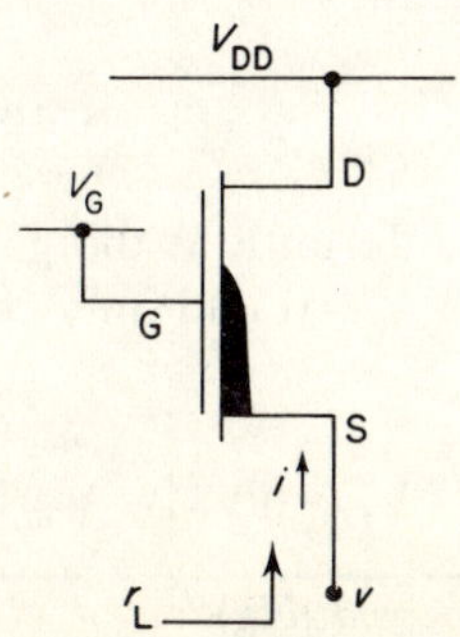

Fig. 10.8.

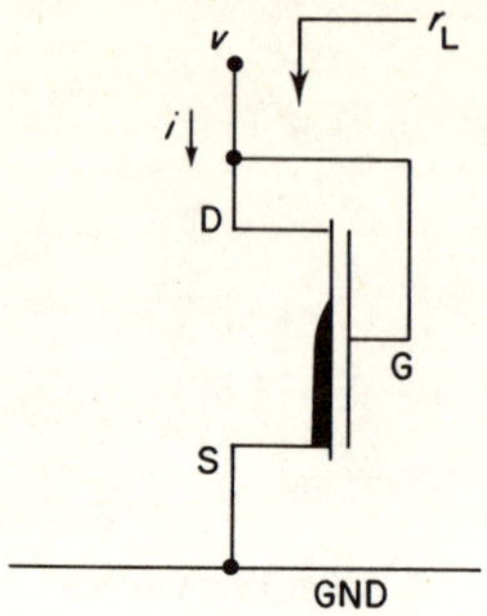

Fig. 10.9.

not met but note that the channel end voltage is fixed at $(V_G - V_{TCE})$ and L' does not vary. Hence, the drain has a negligible influence on r_L. For a square device using $\beta' = 10\ \mu A/V^2$ and $(V_{GS} - V_T) = 5$ V:

$$r_L = \frac{1}{g_m} = \frac{1}{\beta(V_{GS} - V_T)} = \frac{1}{10^{-5} \times 5} = 20\ \text{k}\Omega$$

This is generally much smaller than the resistance r_o associated with channel-length modulation. The same loading effect arises in the case illustrated by Fig. 10.9. Again V_{GS} modulates current flow and:

$$r_L = \frac{1}{g_m} = \frac{1}{(2\beta I_{DS})^{1/2}}$$

where I_{DS} is the bias current level. Thus, low bias current is conducive to high r_L values.

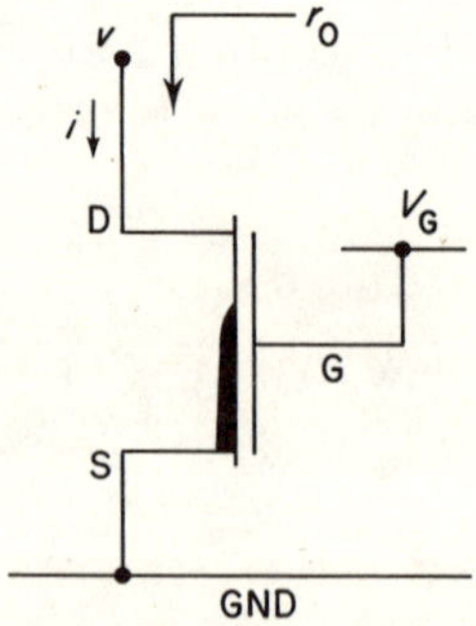

Fig. 10.10.

The same device behaves quite differently if the gate is tied to a suitable bias level V_G (Fig. 10.10). Then $\Delta V_{GS} = 0$ and only channel-length modulation can influence current flow:

$$r_o = \frac{\dfrac{2L}{I_o}\left(1 - \dfrac{L'}{L}\right)^2 [V_{DS} - (V_{GS} - V_T)]^{1/2}}{(2k_{si}e_0/qN_B)^{1/2}}$$

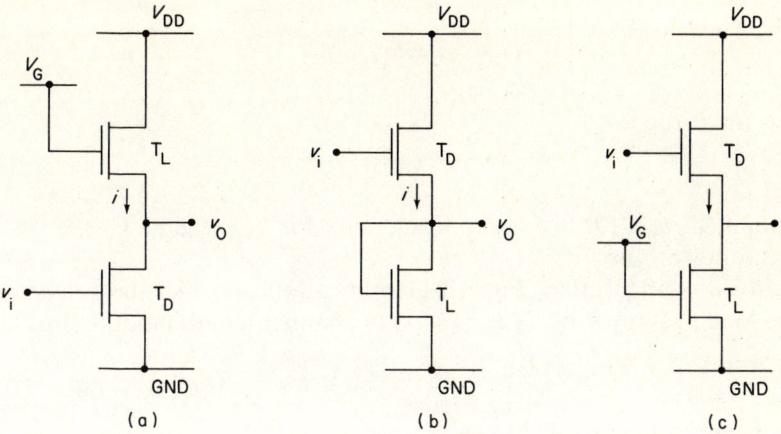

Fig. 10.11. Some single-stage amplifiers in MOS technology.

where $I_o = (\beta'/2)(W/L)(V_{GS} - V_T)^2$, as already established. For a 'long' device ($W/L \ll 1$), r_o can be quite high and favours high voltage gain values.

It is now possible to examine some complete gain stages and to rapidly assess their performance. Note that *saturated* device operation is used throughout.

In Fig. 10.11(a), the loading due to T_L is $r_L = 1/g_{mL}$. The driver T_D controls current flow:

$$i = g_{mD} v_i \qquad \text{and} \qquad v_o = -i r_L = \frac{-i}{g_{mL}}$$

from which:

$$\frac{v_o}{v_i} = \frac{-g_{mD}}{g_{mL}} \tag{10.7}$$

This resembles the 'common emitter' configuration in bipolar circuitry. The expression ignores the loading of T_D itself due to L' variations. Since this r_{oD} appears in parallel with $1/g_{mL}$ and is generally much larger, Eqn. (10.7) is good enough for most situations. Since $g_m = (2\beta I_{DS})^{1/2}$ for each device, it follows from Eqn. (10.7) that:

$$\frac{v_o}{v_i} = -\left(\frac{\beta_D}{\beta_L}\right)^{1/2} = -\left[\frac{(W/L)_D}{(W/L)_L}\right]^{1/2} \tag{10.8}$$

This implies that a large device size (a wide driver or a long load) is necessary for a modest gain. For example, to achieve a gain of 20 requires that $\beta_D/\beta_L = 400$!

The circuit in Fig. 10.11(b) uses a load device T_L with $r_L = 1/g_{mL}$. Thus:

$$v_o = i r_L = \frac{i}{g_{mL}} \qquad \text{while } i = g_{mD}(v_i - v_o)$$

Then:

$$v_o = \frac{g_{mD}}{g_{mL}}(v_i - v_o)$$

Solving for v_o/v_i:

$$\frac{v_o}{v_i} = \frac{g_{mD}}{g_{mD} + g_{mL}} \tag{10.9}$$

In general, $v_o/v_i < 1.0$ but approaches unity for $g_{mD} \gg g_{mL}$. This is clearly a 'source follower' circuit.

The third configuration, Fig. 10.11(c), is a follower like the previous one but the load presented by T_L is r_{oL} due to channel-length modulation. Then:

$$\frac{v_o}{v_i} = \frac{g_{mD} r_{oL}}{1 + g_{mD} r_{oL}} \approx 1.0 \tag{10.10}$$

since $r_{oL} \gg 1/g_{mD}$ in most cases.

The use of a depletion load offers the possibility of a fairly high inverting gain that does not rely on a large device size. The circuit is shown in Fig. 10.12. To a first approximation, T_L (assumed saturated) behaves like a

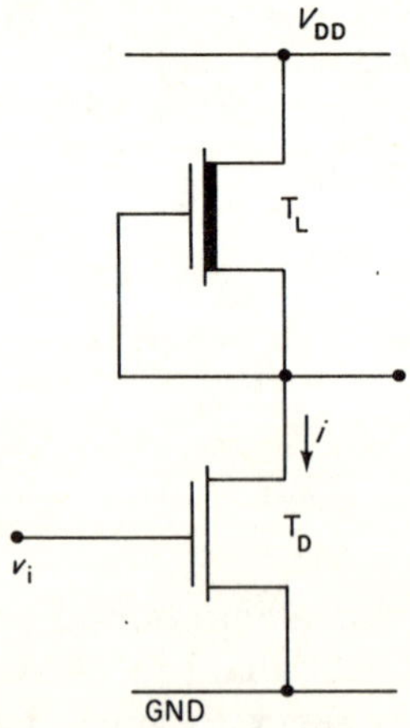

Fig. 10.12.

constant current source. However, for a 'built-in' support voltage V_I, the channel pinches off at a voltage $(V_I + v_o)$ using the ground reference. Thus, the voltage across L' is $[V_{DD} - (V_I + v_o)]$. This means that the small-signal output v_o modulates L' and the loading effect is r_{oL}. The driver has a similar loading effect which cannot be ignored in this case. The output node impedance is $(r_{oL} \| r_{oD})$ and the voltage gain is determined from:

$$i = g_{mD} v_i \qquad \text{while } v_o = -i(r_{oL} \| r_{oD})$$

giving:

$$\frac{v_o}{v_i} = -g_{mD}(r_{oL} \| r_{oD}) \tag{10.11}$$

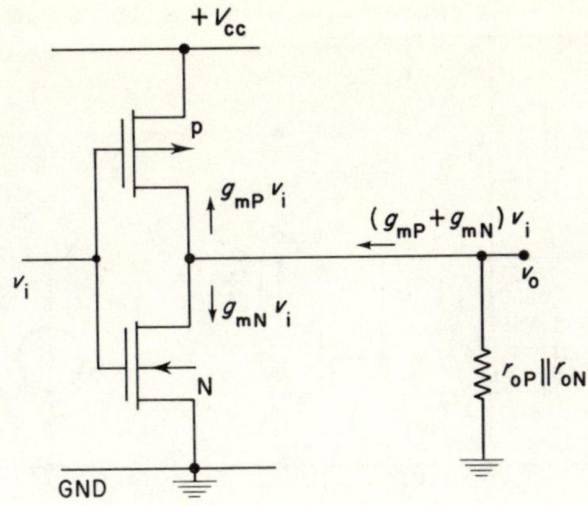

Fig. 10.13.

Equation (10.11) merely expresses the slope of the static transfer characteristic at the operating point. In the idealized presentation given in Chapter 7, the slope was infinite (a vertical line segment) with both devices saturated because L' variation was not taken into account.

The CMOS inverter also exhibits a very steep slope in the transition region and is suitable for delivering high gain. In this case, however, both transistors are driven by v_i. Figure 10.13 shows the small-signal current responses to a small-signal input v_i, assuming that both devices are saturated. (For a fully symmetrical circuit this will occur with d.c. levels $v_i = v_o = v_{cc}/2$). Then:

$$v_o = -(g_{mP} + g_{mN})v_i(r_{oP} \parallel r_{oN})$$
$$\frac{v_o}{v_i} = -(g_{mP} + g_{mN})(r_{oP} \parallel r_{oN}) \qquad (10.12)$$

This is a further improvement over the gain realizable from previous inverter stages.

BIASING THE AMPLIFIERS

A number of methods are available for providing suitable d.c. bias. A frequent option is the use of a *current repeater.* This is illustrated in Fig. 10.14 where the current I_3 in T_3 is set by the current I_2 in T_2. In fact, for identical devices, $I_3 = I_2$ because V_{GS} is the same for both. The statement holds as long as T_3 is saturated. Devices sizes can be varied to alter the ratio I_3/I_2. The action of the circuit is summarized in terms of the *constant current sink,* I_3, shown on the right. Sometimes the action is inverted to provide a *constant current source* as suggested in Fig. 10.15.

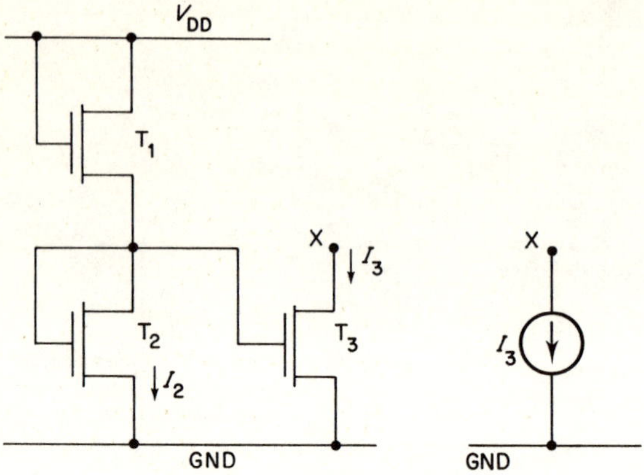

Fig. 10.14. A current repeater.

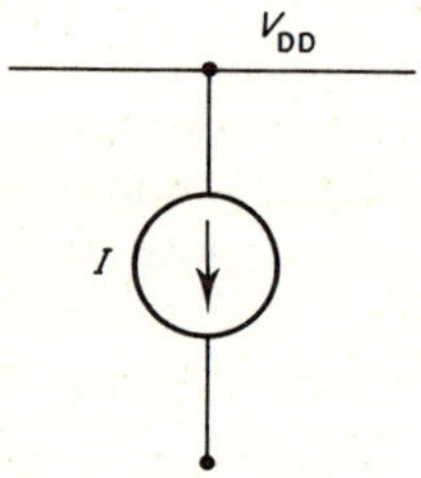

Fig. 10.15. Constant current source.

The CMOS amplifier in Fig. 10.16 is driven in this way. The amplifier formed by T_3 and T_4 has its bias current I_3 set by the current I_1 in T_1. A signal which is coupled into the gate of T_4 can utilize the full available gain of the CMOS amplifier stage.

CMOS inverters can be made to self-bias into the transition region for purposes of signal amplification and then used in a manner similar to operational amplifiers.

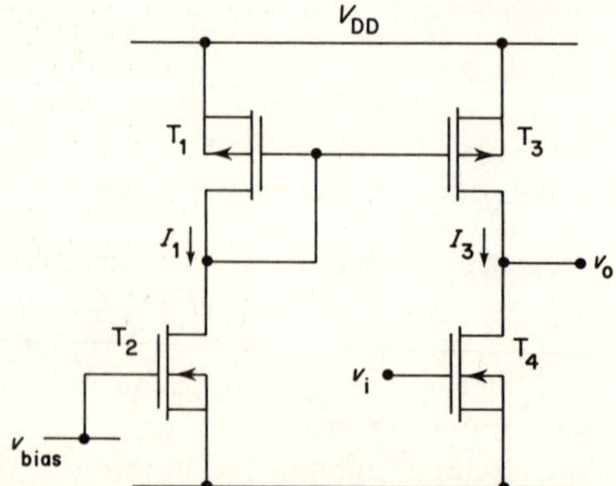

Fig. 10.16. A CMOS amplifier.

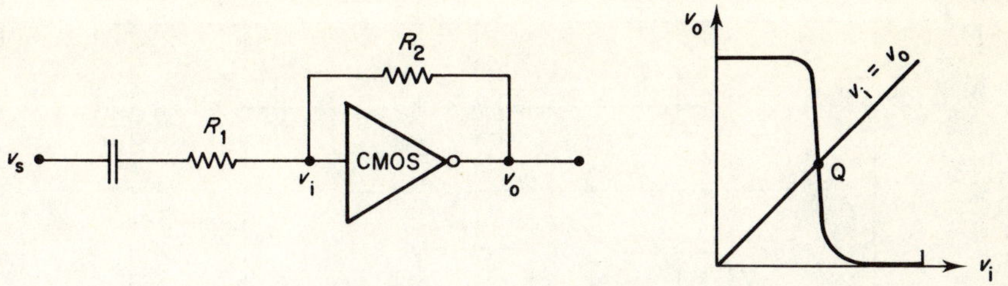

Fig. 10.17. CMOS 'op-amp'.

In Fig. 10.17, the d.c. condition $v_i = v_o$ is valid and the operating point Q is as shown. R_2 also closes the loop for small-signal variations and, if the incremental gain is high, the closed-loop gain becomes approximately:

$$\frac{v_o}{v_s} = -\frac{R_2}{R_1}$$

This is analogous to the action of the inverting op-amp configuration. High incremental gain can be achieved by using three inverters rather than one. This is shown in Fig. 10.18. It carries the usual benefits of feedback, notably, stable closed-loop gain and much-improved linearity. Other op-amp attributes, such as differential input and improved output drive capability, are absent.

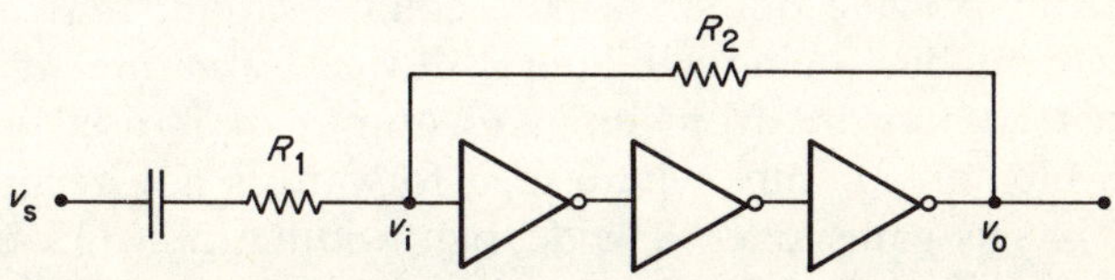

Fig. 10.18. Improved 'op-amp'.

A CMOS OP-AMP

A simplified version of a 'true' CMOS op-amp is shown in Fig. 10.19. A constant current I is used to set bias for T_2 which, in turn, biases T_1, T_4, and T_9 directly. T_3 is biased by the current from T_1 and is further used to set the bias for T_{12}.

T_4 serves as a constant-current supply $(2I_0)$ for the differential input stage (T_5, T_6). Under balanced conditions $(v_1 = v_2)$, T_5 and T_6 carry equal currents I_0. A differential input signal introduces a current imbalance $\pm\Delta I$. The current in T_7 is *repeated* in T_8 and this differs from the T_6 current by a net $2\Delta I$. This current difference is injected to the next stage comprising T_9 and T_{10}. These are complementary devices offering reasonably high inverting gain. The output from this stage drives T_{11} which is an n-channel source-follower device biased by T_{12}.

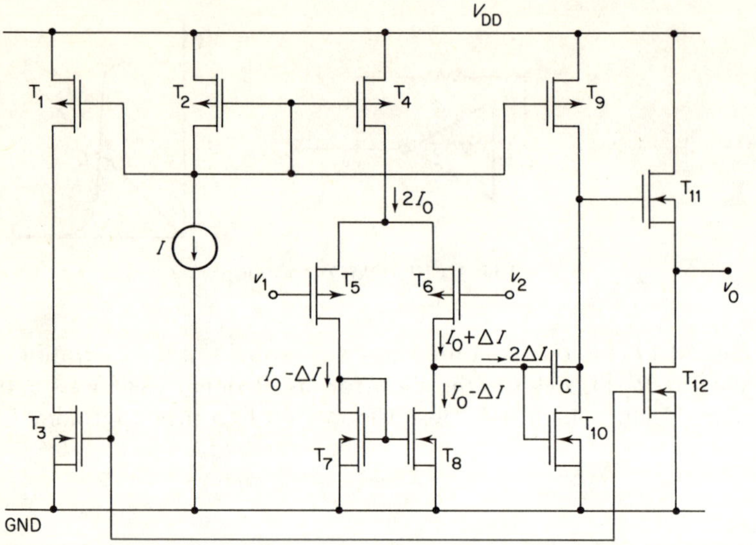

Fig. 10.19. A simplified CMOS op-amp.

The amplifier consists, therefore, of three stages. The first stage offers differential inputs and some voltage gain. The second stage provides further voltage gain and the third is, approximately, a unity-gain buffer. A capacitor C is included to provide dominant-pole compensation. It is intended as a feed*back* element but, being a bilateral device by nature, it also has the capability to feed *forward* from input to output. It is used in exactly the same way in bipolar op-amps where feed-forward is not a significant effect due to high stage gains over a wide bandwidth. In MOS op-amps, the

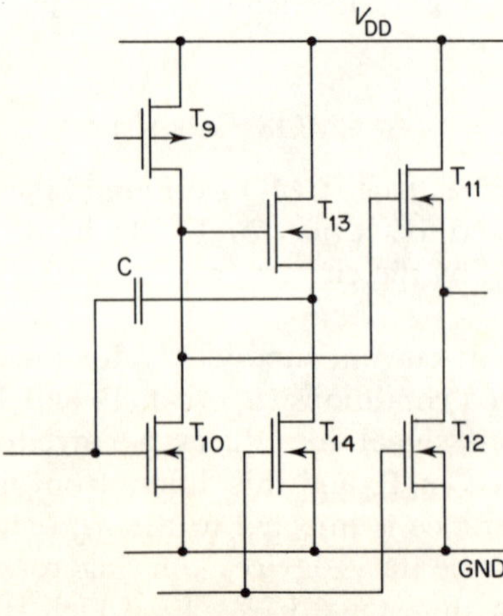

Fig. 10.20.

second-stage gain is not so high and feed-forward tends to by-pass the stage at moderate frequencies with resulting stability problems. These problems can be alleviated by connecting the capacitor in series with a unity-gain buffer which feeds back but not forward. The arrangement is shown in Fig. 10.20. The buffer is simply a follower circuit involving T_{13} and T_{14}, both n-channel devices. T_{14} is biased in the same way as T_{12}.

NMOS AMPLIFIERS

Single-channel technologies are not at all as well suited as CMOS to the realization of amplifier circuitry. Nevertheless, the inclusion of op-amp type functions on the same chip with n-channel digital circuits has many attractions. The most compelling attraction is that of very low cost compared with the alternatives.

A differential NMOS stage, suitable as an op-amp front end, is shown in Fig. 10.21. T_5 serves as a current source to provide bias and give reasonable common mode rejection while T_1 and T_2 are saturated load elements. For differential input signals ($v_1 = -v_2$), the potential at N does not change and then:

$$(v_{o2} - v_{o1}) = -\frac{g_{m4}}{g_{m2}} v_2 + \frac{g_{m3}}{g_{m1}} v_1 = -\frac{g_{m3}}{g_{m1}} (v_2 - v_1)$$

assuming symmetrical working.

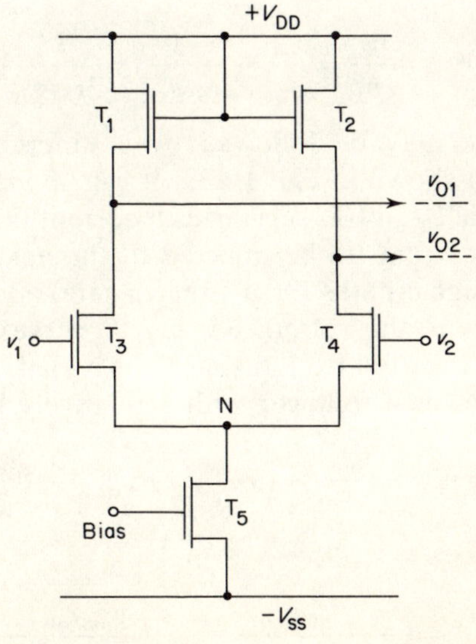

Fig. 10.21.

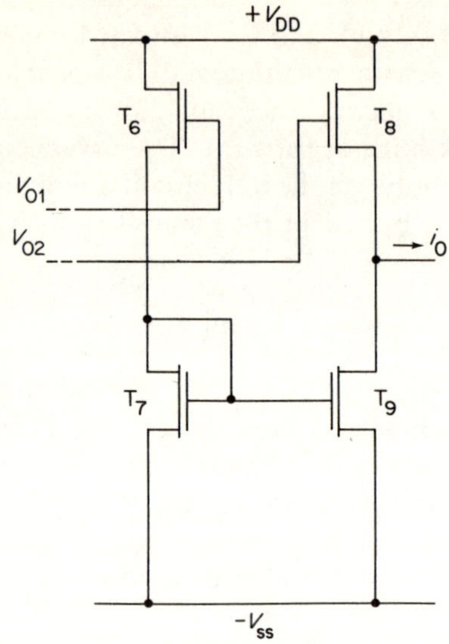

Fig. 10.22.

For common mode signals ($v_1 = v_2 = v_{cm}$), T_3 and T_4 transmit the small-signal changes to point N with a gain close to unity, i.e. $v_N = v_{cm}$ approx. due to the relatively large output resistance r_{o5} of T_5. This resistance is 'shared' between T_3 and T_4, equivalent to two parallel resistors of $2r_{o5}$ ohms each. Currents on either side of the circuit are varying by an amount $v_{cm}/2r_{o5}$ leading to a common mode gain:

$$A_{cm} = \frac{v_{o1}}{v_1} = \frac{v_{o2}}{v_2} = \frac{-1}{2r_{o5}g_{m1}} = \frac{-1}{2r_{o5}g_{m2}}$$

The differential stage may be followed by a differential to single-ended converter such as that shown in Fig. 10.22. It can be viewed as a voltage-to-current converter with T_7 and T_9 serving as a current mirror causing $i_o = 2\Delta I$, twice the unbalance current, to be injected to the next stage.

Alternatively, the signal paths for v_{o1} (at T_6 gate) and v_{o2} (at T_8 gate) can be followed separately to the output. For v_{o1}, T_6 serves as a source follower and the signal is then inverted by T_9 using T_8 as a load. For v_{o2}, no inversion occurs and T_8 behaves as a follower with a relatively high load r_{o9}. Thus:

$$v_o = v_{o1}\frac{g_{m6}}{g_{m6}+g_{m7}}\left(-\frac{g_{m9}}{g_{m8}}\right) + v_{o2}\frac{g_{m8}r_{o9}}{1+g_{m8}r_{o9}}$$

and, assuming $v_{o2} = -v_{o1} = v_D/2$:

$$\frac{v_o}{v_D} = \frac{g_{m6}g_{m9}}{2(g_{m6}+g_{m7})g_{m8}} + \frac{g_{m8}r_{o9}}{2(1+g_{m8}r_{o9})} \tag{10.13}$$

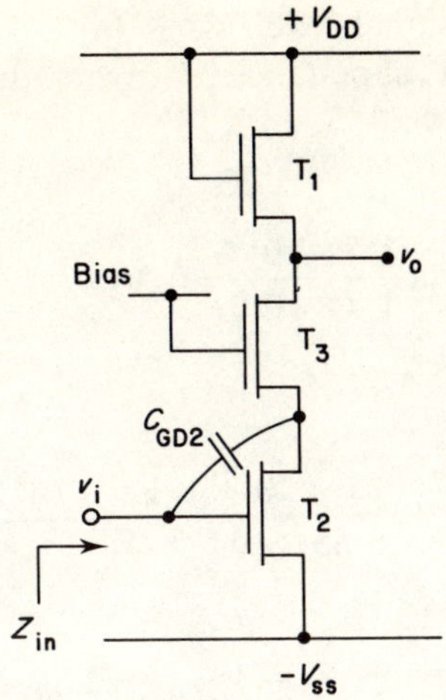

Fig. 10.23. A cascode amplifier.

If such a circuit is designed with $g_{m6} = g_{m8}$, $g_{m7} = g_{m9}$, and $g_{m7}, g_{m9} \gg g_{m6}$, g_{m8}, and if $g_{m8} r_{o9} \gg 1$, then both terms in the gain becomes approximately 0.5 and:

$$\frac{v_o}{v_D} = 0.5 + 0.5 = 1.0 \qquad \text{(approx.)}$$

For common mode signal components, the two terms in Eqn. (10.13) subtract to give a common mode gain which is very low.

One final circuit that deserves mention is the *cascode* amplifier (Fig. 10.23). It uses an extra device T_3 but does *not* offer improved gain. In fact:

$$\frac{v_o}{v_i} = -\frac{g_{m2}}{g_{m1}}$$

as though T_3 were absent. The advantage offered is that the input impedance Z_{in} does not include a large capacitance due to the Miller effect on C_{GD2} because the source of T_3 undergoes relatively small signal excursions. The net result is reduced loading on the previous stage and, consequently, wider bandwidth.

EXAMPLE 10.1

The following data applies to the MOS multiplex switch shown:

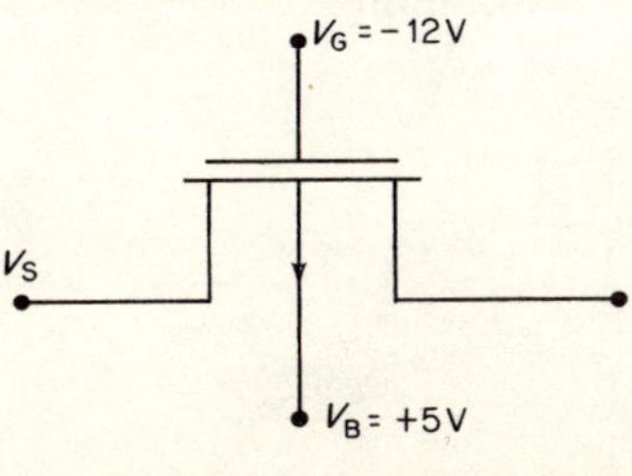

$$V_{TO} = -2.5 \text{ V}$$

$$N_B = 1.5 \times 10^{15}$$

$$t_{ox} = 1000 \text{ Å}$$

$$\left(\frac{W}{L}\right)_{eff} = 100$$

β' as given in Appendix K

Find the ON resistance R_{on} when the signal level V_S is (a) +5 V and (b) −5 V. What must be the signal circuit impedance if errors due to R_{on} variations are to be less than 0.1 per cent?

Solution

$$\phi_F = \frac{kT}{q} \text{Ln} \frac{N_B}{n_i} = 0.0258 \text{ Ln} \frac{1.5 \times 10^{15}}{1.4 \times 10^{10}} = 0.30$$

$$M = \frac{t_{ox}}{k_{ox}} \left(\frac{2k_{si} q N_B}{e_o} \right)^{1/2}$$

$$= \frac{1000 \times 10^{-8}}{3.9} \left(\frac{2 \times 11.7 \times 1.6 \times 10^{-19} \times 1.5 \times 10^{15}}{8.85 \times 10^{-14}} \right)^{1/2}$$

$$= 0.64$$

For $V_s = +5$ V: This results in $V_{BS} = 0$, giving $V_T = V_{TO} = -2.5$ V.

$$R_{on} = \frac{1}{\beta'(W/L)_{eff}(V_{GS} - V_T)}$$

$\beta'(17 - 2.5, 0) = \beta'(14.5, 0) = 4.3 \times 10^{-6}$ A/V² from appendix K

$$R_{on} = \frac{1}{4.3 \times 10^{-6} \times 100 \times 14.5} = \underline{160\ \Omega}$$

For $V_s = -5$ V:

$$V_T = -2.5 - 0.64[(10 + 0.6)^{1/2} - (0.6)^{1/2}]$$

$$= -4.1 \text{ V}$$

$\beta'(12 - 5 - 4.1, 10) = \beta'(2.9, 10) = 6.2 \times 10^{-6}$ A/V²

$$R_{on} = \frac{1}{6.2 \times 10^{-6} \times 100 \times 2.9} = \underline{556\ \Omega}$$

$$\Delta v_1 = V_S \frac{R_{on1}}{R_L + R_{on1}} = V_S \frac{R_{on1}}{R_L} \qquad \text{(approx.)}$$

$$\Delta v_2 = V_S \frac{R_{on2}}{R_L}$$

$$\frac{\Delta v_2 - \Delta v_1}{V_S} = \frac{R_{on2} - R_{on1}}{R_L} = \frac{556 - 160}{R_L} = 0.001$$

$$R_L = \frac{556 - 160}{0.001} = 396 \text{ k}\Omega \qquad \text{(minimum)}$$

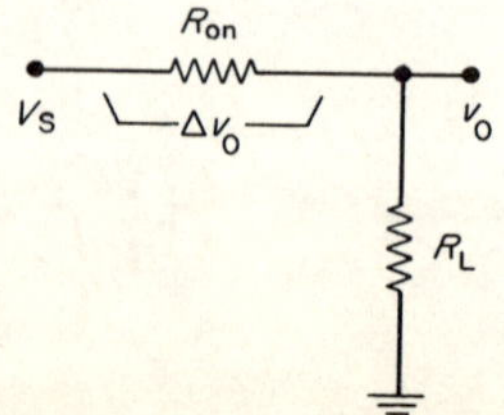

EXAMPLE 10.2

Examine the small-signal characteristics of the self-bias amplifier shown.

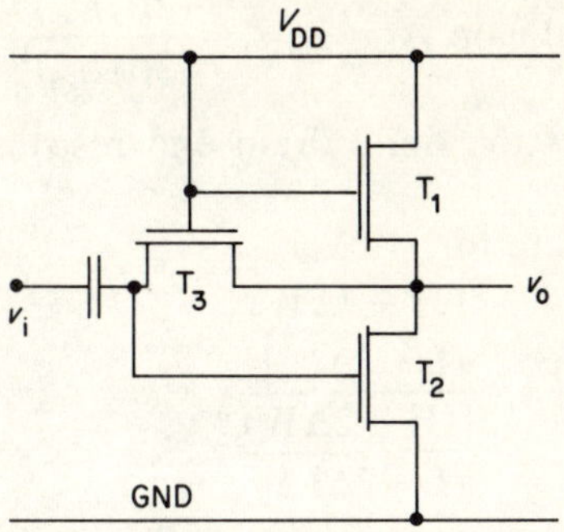

Find a suitable expression for the dynamic resistance of T_1, as seen from the output node, as a function of device current I_{DS}. Hence, get an expression for the voltage gain of the amplifier. Given the data:

	$V_{TO} = 1.0$ V,	$M = 0.6$
	$2\phi_F = 0.6$ V,	$V_{DD} = 5$ V
Driver:	$W = 10$ mil,	$L = 0.4$ mil
Load:	$W = 0.4$ mil,	$L = 1.0$ mil
	$\Delta L = 0.05$ mil,	$N_B = 10^{16}$

find the quiescent current I_{DS}, output voltage v_o, power dissipation on standby for each device, and the small-signal gain. Use β' data from Appendix K assuming n-channel operation. Neglect the influence of L' variations on signal gain.

Solution

T_1 is inherently saturated so that:

$$I_{DS} = \frac{\beta_1}{2}(V_{DD} - V_{T1} - v_O)^2$$

The small-signal resistance is r_1 where:

$$\frac{1}{r_1} = -\frac{dI_{DS}}{dv_o} = +\beta_1(V_{DD} - V_{T1} - v_o) = (2\beta_1 I_{DS})^{1/2}$$

Note that the result is not complicated by an L' variation!

T_2 is biased just like T_1 in an inherently saturated condition which obeys:

$$I_{DS} = \frac{\beta_2}{2}(v_i - V_{T2})^2$$

Then:

$$g_{m2} = \frac{dI_{DS}}{dv_i} = \beta_2(v_i - V_{T2}) = (2\beta_2 I_{DS})^{1/2}$$

$$\text{Small-signal gain } A_v = g_{m2} r_1 = \frac{(2\beta_2 I_{DS})^{1/2}}{(2\beta_1 I_{DS})^{1/2}} = \left(\frac{\beta_2}{\beta_1}\right)^{1/2}$$

(This expression ignores the finite drain end resistance of T_2 caused by a variation in L_2'.)

We take a first estimate for:

$$\beta_R = \frac{\beta_2}{\beta_1} = \frac{\dfrac{(W+2\Delta W)}{(L-2\Delta L)]_2}}{\dfrac{[(W+2\Delta W)}{(L-2\Delta L)]_1}}$$

$$= \frac{(10+0.1)/(0.4-0.1)}{(0.4+0.1)/(1-0.1)} = \frac{33.66}{0.55} = 61$$

We may now use the $m = 1.0$ curve of Appendix C with $\beta_R = 61$ and find where this curve intersects the line $v_i = v_o$. The large β_R will yield low v_o so we neglect body-effect on the threshold and take $V_{TD} = 1.0$ V for T_2. For T_1 we require the output high level $V_{OH} = V_{DD} - V_{TCE}$, where:

$$V_{TCE} = V_{TO} + M[(V_{DD} - V_{TCE} + 2\phi_F)^{1/2} - (2\phi_F)^{1/2}]$$
$$= 1.0 + 0.6[(5.6 - V_{TCE})^{1/2} - (0.6)^{1/2}]$$

By iteration:

$$\underline{V_{TCE} = 1.7 \text{ V}}$$

$V_{OH} = 5 - 1.7 = 3.3$ V and $V_{TD}/V_{OH} = 1.0/3.3 = 0.30$. Draw a 45° line through 0.30 on the vertical axis of the $m = 1$ graph. From the graph, $\underline{v_o = 0.38 V_{OH} = 1.25 \text{ V}}$ approx. With this value of v_o we can reevaluate our parameters more accurately:

$$V_{TD} = 1.0 + 0.6[(1.25 - V_{TD} + 2\phi_F)^{1/2} - (2\phi_F)^{1/2}]$$
$$= 1.0 + 0.6[(1.85 - V_{TD})^{1/2} - (0.6)^{1/2}]$$

By iteration:

$$\underline{V_{TD} = 1.06 \text{ V}}$$

For the load:

$$L_1' = \left(\frac{2k_{si}e_0 V_{TCEL}}{qN_B}\right)^{1/2} = 3.59 \times 10^{-5}(V_{TCEL})^{1/2}$$
$$= 3.59 \times 10^{-5}(1.7)^{1/2} = 4.7 \times 10^{-5} \text{ cm}$$
$$= 0.47\ \mu\text{m} = 0.018 \text{ mil}$$

It is clear that L_1' and L_2' have a very small effect. Also, while variations in β' may alter the beta ratio by perhaps 20 per cent, the effect on v_o will be fairly small and we will not reiterate.

T_2 gives:

$$I_{DS} = \frac{\beta}{2}(v_o - V_{TD})^2 = \frac{33.66}{2}(1.25 - 1.0)^2\beta'$$

$$= \underline{1.05\beta'}$$

T_1 gives:

$$I_{DS} = \frac{\beta}{2}(V_{DD} - v_o - V_{TCEL})^2 = \frac{0.55}{2}(5 - 1.25 - 1.7)^2\beta'$$

$$= \underline{1.15\beta'}$$

These are in reasonable agreement as expected!

From Appendix K, a reasonable β' is about 18 $\mu A/V^2$, taking approximate account of field influences. Then:

$$\underline{I_{DS} = 1.1\beta' = 20\ \mu A}$$

$$\text{Dissipation in } T_1\text{:}\ \ P_1 = (5 - 1.25)(20 \times 10^{-6}) = \underline{75\ \mu W}$$

$$\text{Dissipation in } T_2\text{:}\ \ P_2 = 1.25 \times 20 \times 10^{-6} = \underline{25\ \mu W}$$

$$\text{Signal gain} = (\beta_R)^{1/2} = (61)^{1/2} = \underline{7.8}$$

CHAPTER 11

Digital Logic and Memory Elements

The possibility of information storage through charge retention on (often parasitic) capacitive elements is a very special feature of MOS circuitry. It has resulted in a wide variety of options for implementing digital logic functions and memory elements. The more important 'building blocks', usually involving just a few transistors, are the subject of this chapter. They range from static combinational logic gates through clocked dynamic logic circuits to multi-phase schemes and memory circuits. Examples will be drawn, as appropriate, from single-channel technologies and from CMOS alternatives.

STATIC LOGIC CIRCUITS

The basic logic element is the inverter. More complex logic structures are realized by replacing the driver of the inverter with two or more devices in series and/or parallel combinations. The NOR gate in Fig. 11.1 uses a saturated load. The NAND gate in Fig. 11.2 is illustrated with a depletion load device. The load type is in no way related to the logic function but it is important that the beta ratio β_R of the standard inverter be preserved in all cases. Appropriate W/L values are shown bracketed for each device, assuming minimum geometry loads with $\beta_R = 10$. The series connection of input devices in Fig. 11.2 requires individual transistors of twice the normal W/L so that, when both are ON, they have the equivalent total ON

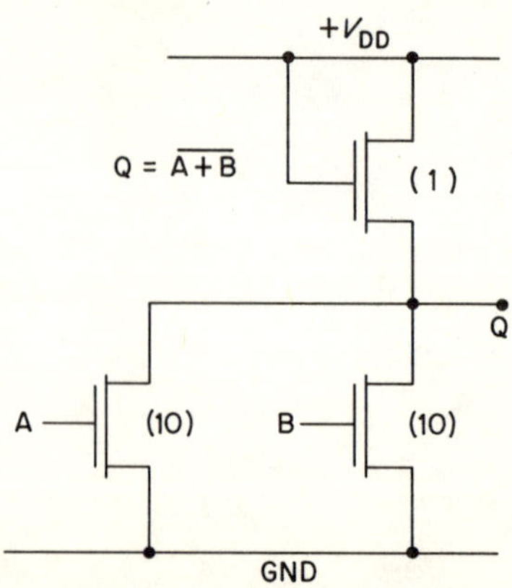

Fig. 11.1. NOR gate.

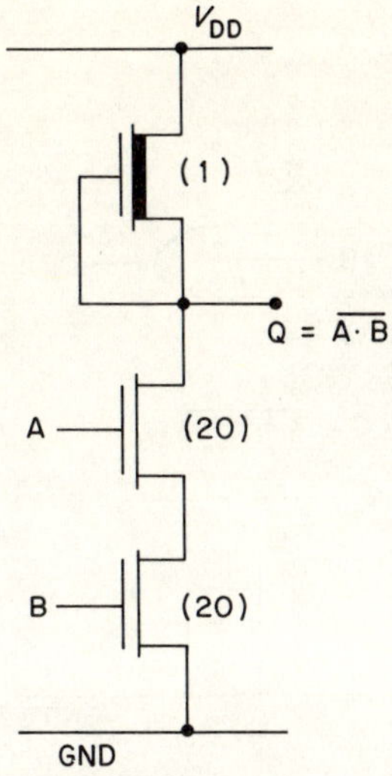

Fig. 11.2. NAND gate.

resistance of a single driver. This is wasteful of chip area and can frequently be avoided by the preferred use of NOR logic elements.

More complex logical functions can easily be implemented with a single inversion, as suggested by Fig. 11.3. Logic functions in CMOS technology require more devices, as illustrated in Fig. 11.4. Each logical input uses a p-channel and an n-channel device so that one or the other is always OFF. While space requirements on chip are relatively large, low-power dissipation is achieved through the absence of a d.c. path. Also, the concept of 'drivers' which have to be much larger than 'loads' is absent, permitting minimum-size devices. The p-channel transistors, however, are typically larger than the n-channel ones to compensate for their inherent lower mobility.

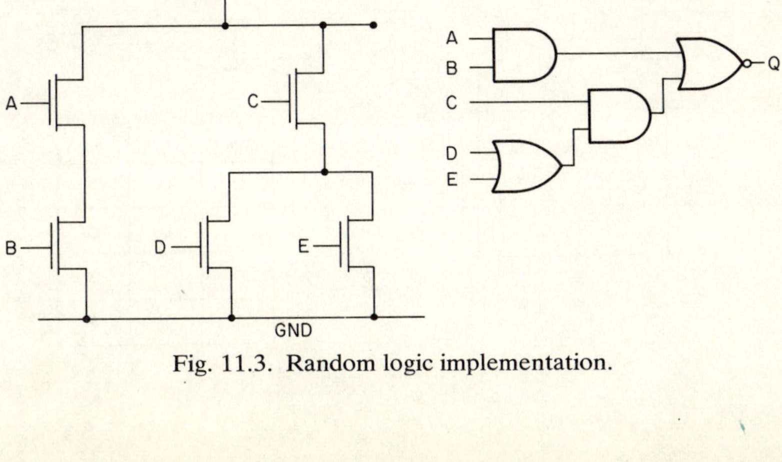

Fig. 11.3. Random logic implementation.

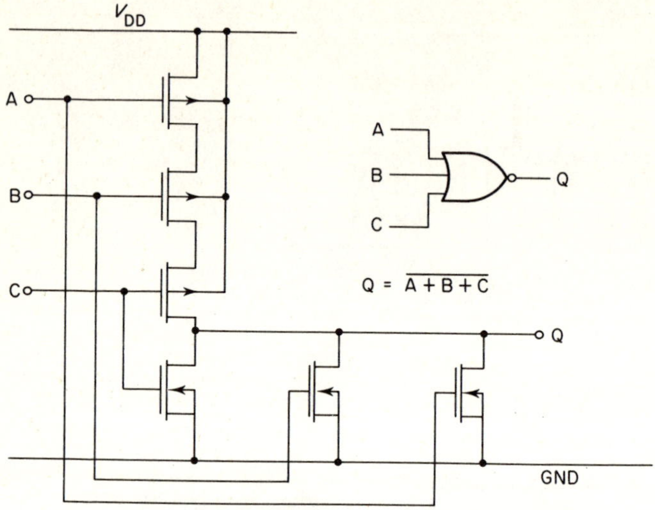

Fig. 11.4. A NOR gate in CMOS technology.

STATIC AND SEMI-STATIC MEMORY

The static memory cell fashioned from cross-coupled inverters was mentioned in Chapter 2 and is reproduced in Fig. 11.5 as an RS flip-flop. In the 'normal' state, S and R are low and the output Q can be either 0 or 1. The information is retained indefinitely by positive feedback. If S is taken high, it forces Q′ to 0 which in turn (since R = 0) forces Q to 1. This '1' is fed back to maintain the new state after S has returned to low. Thus, it is only necessary to apply a brief 'high' pulse to the set input (S) in order to write a '1' into the memory element. Likewise, a high pulse to the reset input (R) causes a '0' to be stored.

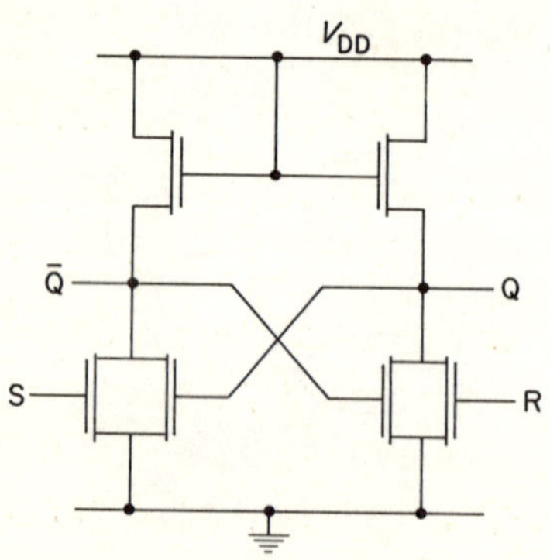

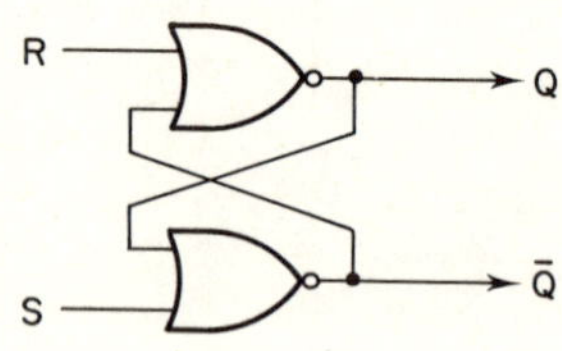

R_n	S_n	Q_{n+1}
0	0	Q_n
1	0	0
0	1	1
1	1	?

Fig. 11.5. Static RS flip-flop.

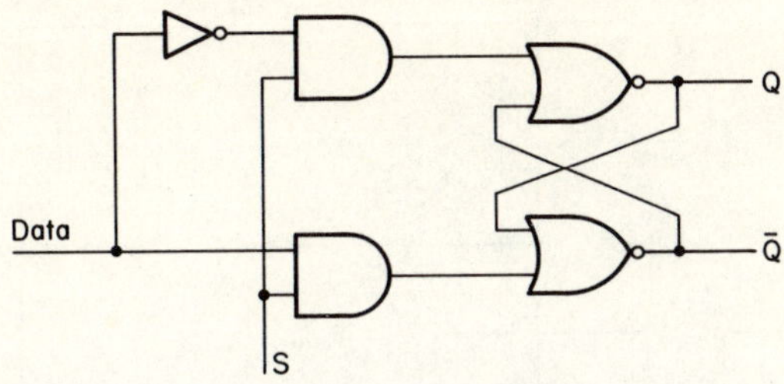

Fig. 11.6. Flip-flop with data entry facility.

Additional logic is required to load information from a data line to such a memory cell and Fig. 11.6 illustrates how this may be done. The strobe (S) is taken high to admit data to the memory. The data is retained after S returns to low. If this circuit drives another identical one which is enabled by S′ rather than S, the result is the well-known 'master-slave' flip-flop. Data are entered while S is high and become valid at the output after S returns to low. A single memory cell with data entry is implemented in the manner shown in Fig. 11.7. The extra logic is included at existing nodes except for the data inverter which is built separately.

A semi-static alternative to this circuit appears in Fig. 11.8. Information is normally stored in the fully static sense by positive feedback when S is low and T_F is ON to complete the loop. When S is high, the feedback is interrupted by turning T_F OFF and T_D goes ON to admit new data. Subsequently, T_D goes OFF and T_F goes ON to re-close the feedback loop. Even if signal timing is such as to leave T_D and T_F both OFF for a short interval, the state is maintained by charge storage on C_1; hence the term 'semi-static'. The circuit in Fig. 11.8 can be operated in a fully asynchronous

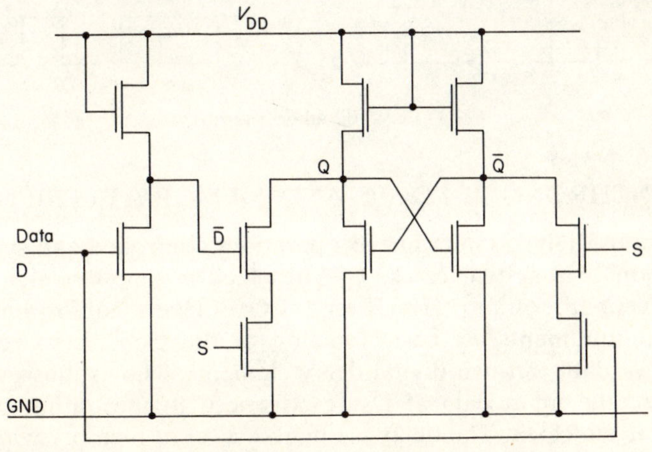

Fig. 11.7. Flip-flop circuitry.

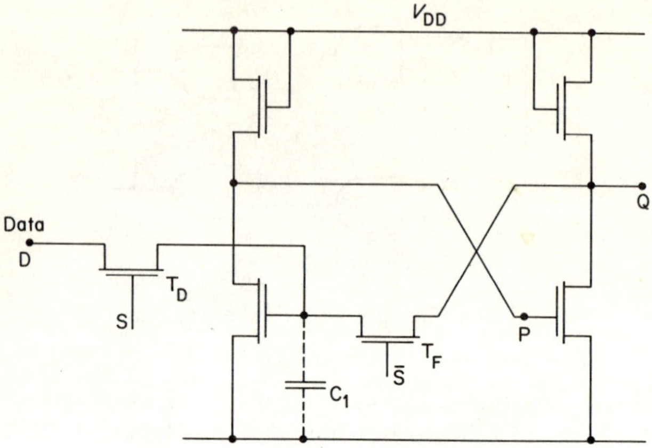

Fig. 11.8. Semi-static memory cell.

mode, i.e. data can be entered at any time via the strobe S. An alternative scheme (Fig. 11.9) has already been mentioned in chapter 2 (Fig. 2.17). It uses two-phase non-overlapping clocks ϕ_1 and ϕ_2. Data are admitted or refreshed during ϕ_1 time and are valid during ϕ_2 time. Notice that during ϕ_2 time, because ϕ_1 is OFF, T_D and T_F are OFF and the information is retained on the gate capacitance. This capactance also includes capacitance to the bulk due to the diffused layer to which it is linked.

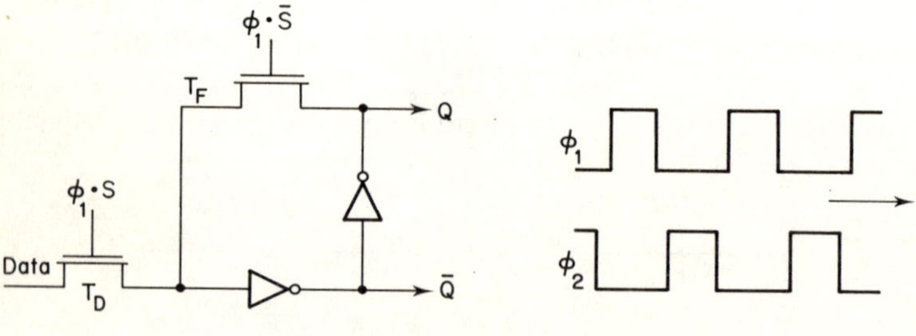

Fig. 11.9. Clocked memory.

SYNCHRONOUS LOGIC AND MEMORY ELEMENTS

A synchronous digital system has its operations controlled and synchronized by the action of a system clock, C. The clock is simply a high-frequency square wave with output HIGH and LOW levels conforming to logic switching requirements. A basic function of the clock is to regulate the movement of data through digital delay elements. This is illustrated in Fig. 11.10 where the input data at D are advanced to the output point Q by the action of the clock. The delay element is a bit of memory which may be static or dynamic. If it is a master–slave flip-flop, data are entered while C is

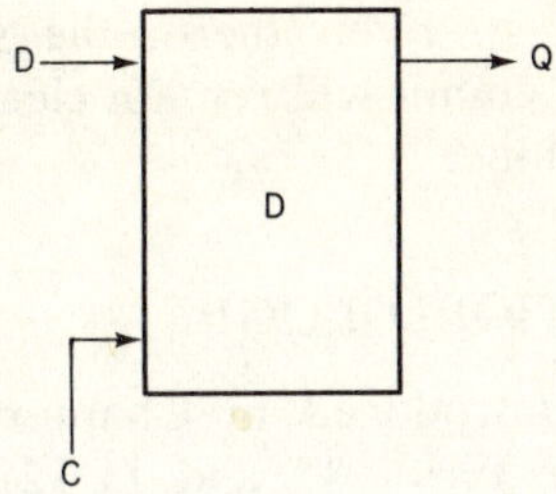

Fig. 11.10. Delay element.

high and transferred to output when C goes low. Alternatively, information may be transferred on the positive-going or on the negative-going edge of the clock signal (but not on both).

Even the most complex of digital machines can be built using simple delay elements and combinational logic. One relatively simple illustration of a digital machine is the functioning of a J–K flip-flop. Its action is summarized by the accompanying characteristic table (Fig. 11.11). When J and K are both low, G_3 outputs a '0' while G_2 transmits Q_n without inversion. Hence Q_n appears as input data and, after the clock transition, $Q_{n+1} = Q_n$. In a similar way, if J = 1 and K = 0, G_3 transmits Q_n with inversion while G_2 transmits Q_n without inversion. Therefore, at least one input to G_1 will be high ensuring that the new value of Q, i.e. Q_{n+1}, will be 1. The remaining circuit features can be similarly verified. An important requirement for D is that it has input–output isolation. This means that new data appearing at the output must not be able to immediately feed back via external circuitry so as to modify the conditions that just caused it and, thereby, cause a further change of state. The master–slave principle avoids this difficulty. Master–slave action is easily added to the semi-static cell of Fig. 11.8. All that is necessary is an additional pass transistor at point P controlling data transfer to the second (right-hand) inverter. This is held OFF while input data is being admitted and goes ON a little later, *after* T_D has been turned OFF.

Further examples of a more general nature will be seen in Chapter 12 where the physical layout of circuits will be considered side by side with their electrical design. In fact, a completely general approach to both

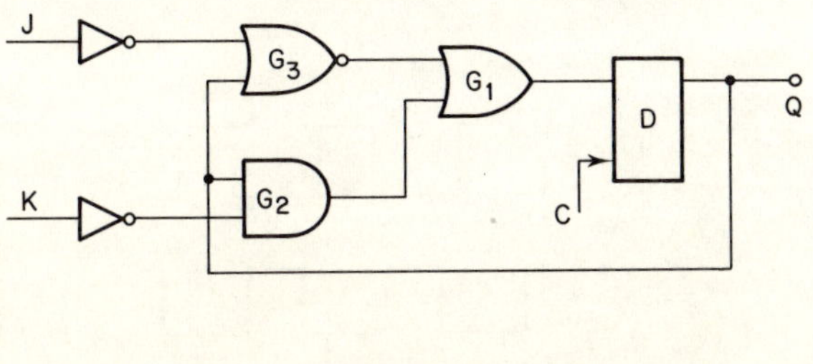

J	K	Q_{n+1}
0	0	Q_n
1	0	1
0	1	0
1	1	$\bar{Q}_n$

Fig. 11.11. A J–K flip-flop.

electrical design and layout will emerge. Even there, the simple delay mechanism and the combinational logic circuit will remain clearly visible as the component parts of the digital machine.

TWO-PHASE RATIOED LOGIC

The concept of two-phase logic was introduced in Chapter 2 with the discussion of a simple shift register (see Figs 2.15 and 11.12). It is worth reexamining the properties of this scheme. Clocks are non-overlapping with clock period T seconds, referred to as 'one bit time'. A data bit D_n reaches P_1 when ϕ_1 goes high. It reaches P_2 as D_n' when ϕ_2 goes high after $T/2$ seconds of delay. Finally, after another inversion, D_n reaches P_3 when ϕ_1 again goes high. Thus, in T seconds, D_n is transferred intact from P_1 to P_3. Also, because clocks are non-overlapping, there is true isolation at all times between P_1 and P_3. It follows that the circuitry between these points qualifies as a unit of delay with effective master–slave isolation. It can therefore be used to implement the delay elements of any digital machine. As well as this, the inverters can each incorporate a good deal of combinational logic. The result is a type of circuitry which is extremely efficient in its use of devices and which has built-in memory capability. If it offered no more than that, it would be an attractive basis for digital system implementations. In fact, a number of additional benefits are also possible. These will become evident as other two-phase options are discussed.

In the shift register circuit just considered, the inverters were shown symbolically because their circuit details did not influence the operation. Conventional ratioed inverters using enhancement- or depletion-type loads are entirely satisfactory for two-phase working. They do, however, require large-sized drivers and they do present a path for d.c. current flow with consequent high-power dissipation.

The two-phase scheme of Fig. 11.13 offers some improvement because the load elements are driven by the phase clocks and are therefore turned

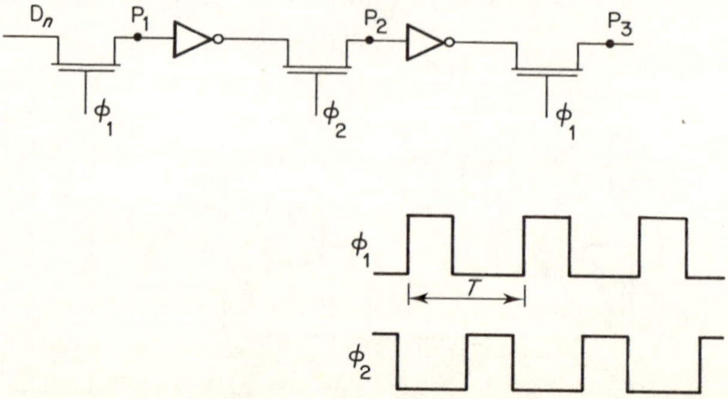

Fig. 11.12. A two-phase shift register.

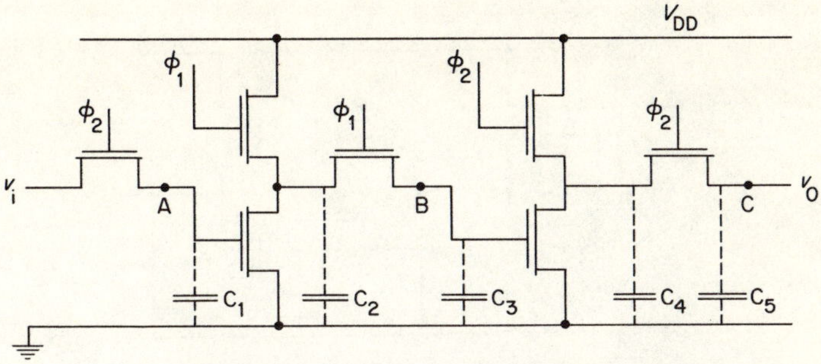

Fig. 11.13. Ratioed two-phase logic.

OFF for over 50 per cent of the time. This reduces power dissipation substantially. The operation is as follows. During ϕ_2 time, data are admitted to A from the left and stored on C_1. This may discharge C_2 but does not affect data at B. During ϕ_1 time, inverter operation sets $B = A'$. This is a ratioed operation and driver dimensions must ensure that a low state at B is below threshold. A further inversion during the next ϕ_2 time sets $C = B' = A$. Each data transfer, A to B and B to C involves $T/2$ seconds of delay. Thus, data at A reaches C after a delay of 1 bit time. Note the necessity of a series coupling device between two inverters clocked by *different* phases. The driver of any inverter can be replaced by combinational logic. Also, there is no objection to a tandem of two or more inverters (and associated logic) clocked by *the same* phase *without* intermediate coupling devices—as long as the increased data transfer time is acceptable. In other words, the maximum number of 'logic levels' per bit time is limited only by the required operating frequency. (The term 'logic level' refers to the amount of logic associated with a single inverter.) Speed is, of course, determined by the time required to charge and discharge the node and gate capacitances. In this scheme, rise time is the critical factor since nodes are charged through small-area load devices.

TWO-PHASE RATIOLESS LOGIC

This scheme is illustrated in Fig. 11.14. As previously, two non-overlapping clock phases are used. The operation is as follows. During ϕ_1 time the data at A are admitted to C_1 and by inversion $B = A'$. For a LOW at B, the voltage on C_2 is somewhere between ground and V_{DD} and is not yet a valid output. After ϕ_1 goes off the data at B become valid: for a '0' entered on A, C_2 has charged to V_{DD} during ϕ_1 time and stays there after ϕ_1 returns to zero; for a '1' entered on A, C_2 discharges to ground through the driver *after* ϕ_1 goes off so that a LOW at B is actually the ground potential. In this way, a conditional discharge of C_2 replaces the previous ratioed scheme and allows both driver and load to be minimum-geometry devices. During ϕ_2 time, data

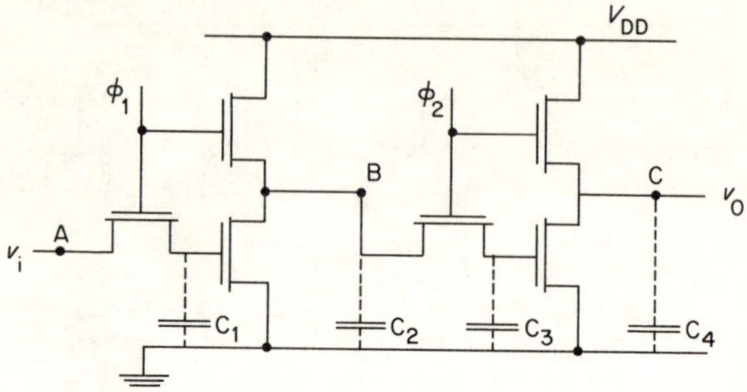

Fig. 11.14. Two-phase ratioless logic.

are shifted and inverted from B to C, i.e. C = B′ = A with 1 bit of delay from A to C. The scheme does *not* permit more than one inverter (with associated driver logic) in tandem on the same phase; this restricts the design to two levels of logic per bit time (one for each phase). Note also that when ϕ_2 goes on, C_2, if charged, has to share its charge with C_3. Voltage on C_3 after sharing must still exceed the threshold so that the driven device is ON. This consideration seriously limits *fan-out.* and reduces noise immunity.

An alternative approach to two-phase logic is very different from that just described. The only supplies are the clock lines ϕ_1 and ϕ_2 (non-overlapping) which also serve the functions of power and ground. The arrangement is shown in Fig. 11.15. During ϕ_1 time, input data reaches C_1 while C_2 is precharged via T_2 which behaves like a saturated load. When ϕ_1 goes off, the source of T_3 returns to ground and then C_2 either retains its charge (if C_1 is low) or discharges through T_3 (if C_1 is high). A further inversion occurs in the next half-state during and after ϕ_2 time so that, as usual, v_o is a replica of v_i delayed by 1 bit time. As there is no d.c. path, a clock line of voltage step height V at frequency f drives a purely capacitive load C; the power furnished by the clock is CV^2f. This expression for clock power is an important one. The topic is expanded and the CV^2f formula is verified in Detail 1 at the end of the chapter.

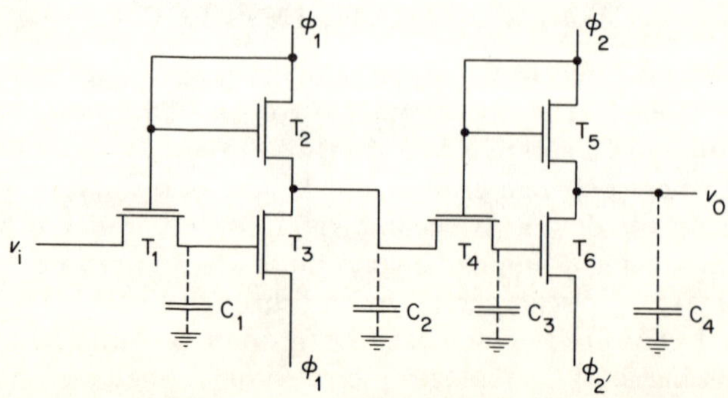

Fig. 11.15. Ratioless logic with no d.c. supply.

FOUR-PHASE LOGIC

While two-phase operation has wide acceptance it is a little restrictive for ratioless working. The limit of two inverters per bit time (T) can be circumvented by the use of a four-phase scheme. Several four-phase systems exist but one example will suffice to illustrate the most important features.

The four-phase scheme uses four non-overlapping clocks as in Fig. 11.16. There are four inverter types and, cascaded as shown, they incur 1 bit of delay from input to output. The general operation is such that gate type n is *precharged* during ϕ_{n-1} time. It *evaluates* (with inversion) the input data during ϕ_n time and for the next two clock phases it simply *holds* the output data valid. Any gates driven by gate type n must evaluate while gate type n is in hold; this restricts the driven gates to types $(n+1)$ and $(n+2)$. If $n=4$, $(n+1)$ becomes 1, and so on.

There is a half-bit delay from gate type n to type $(n+2)$ whether or not they are separated by a type $(n+1)$ gate. Thus, for example, gate type 2 may

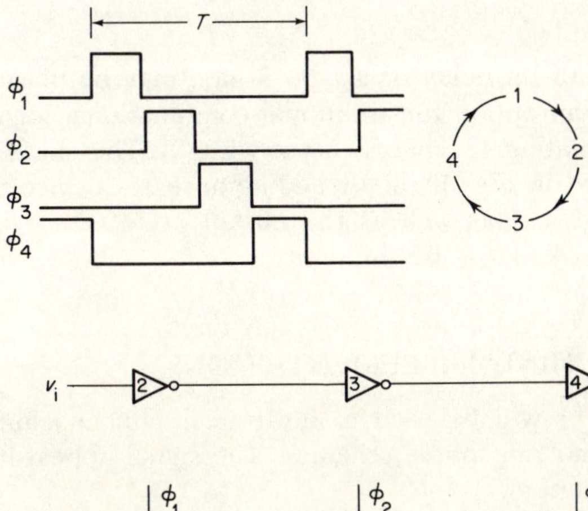

Phase → / ↓ Type ↓	1	2	3	4
1	E	H	H	P
2	P	E	H	H
3	H	P	E	H
4	H	H	P	E

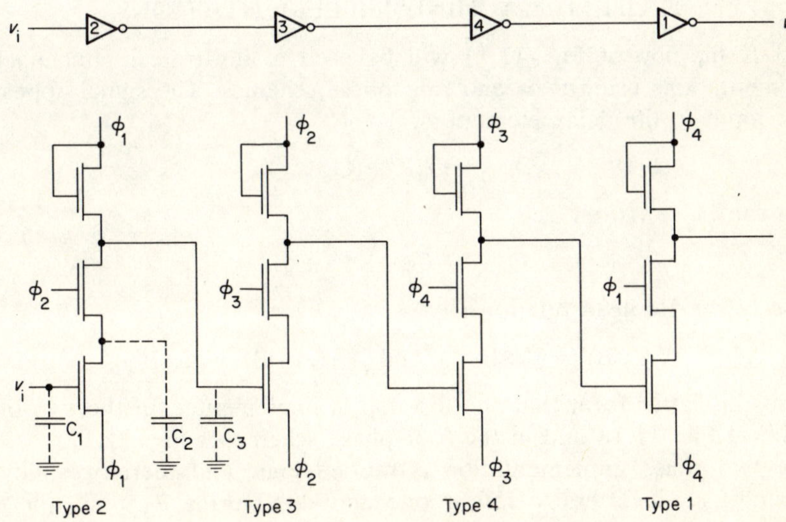

Fig. 11.16. Four-phase logic with non-overlapping clocks.

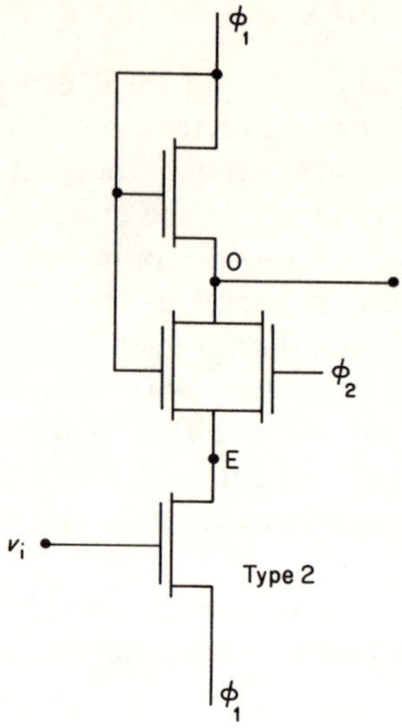

Fig. 11.17.

drive type 4 directly with a half-bit delay or a type 3 gate may be placed between them thus providing an option for additional combinational logic. Note that, when type 2 is evaluating, C_3 shares charge with C_2. This sharing can be eliminated as illustrated in Fig. 11.17 by adding an extra device to precharge the evaluation node (E) as well as the output node (O). This simplifies circuit design and increases noise immunity.

DIGITAL MACHINE IMPLEMENTATIONS

The J–K flip-flop of Fig. 11.11 will be used to illustrate digital machine implementations using two- and four-phase schemes. The signal appearing at the input to the delay element is:

$$Q_{n+1} = Q_n K' + (Q_n + J')'$$

which can be rewritten as:

$$Q_{n+1} = (Q_n' + K)' + Q_n' J$$

by use of the Boolean relations:

$$(A \cdot B)' = A' + B' \qquad \text{and} \qquad (A+B)' = A' \cdot B'$$

It is in this latter form that the flip-flop is implemented in the two-phase scheme of Fig. 11.18 and in the four-phase scheme of Fig. 11.19.

The two-phase implementation is ratioed logic and therefore allows a number of ϕ_2 inverters to follow one another. During ϕ_1 time, the data stored on C_1 are Q_n'. From this initial position, the movement of data around the loop are easily traced. C_1 preserves Q_n dynamically after ϕ_1 goes low.

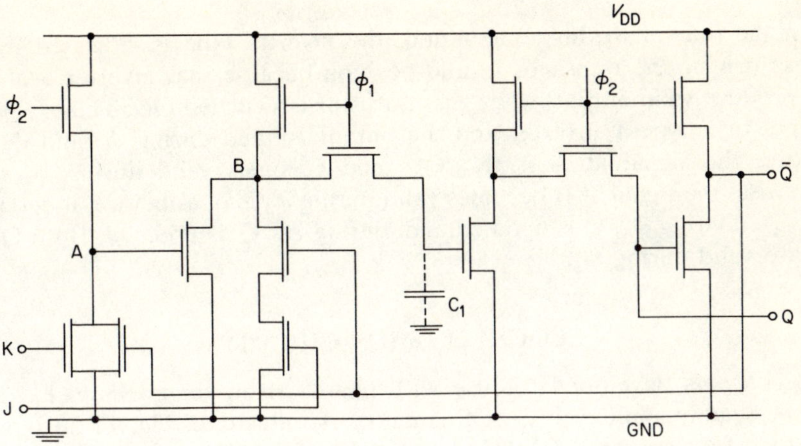

Fig. 11.18. J–K flip-flop (two-phase ratioed).

During ϕ_2 time, all inverters driven by ϕ_2 are activated. Q_n becomes the new output Q and its complement becomes available as Q′. The signal at A becomes $(K+Q'_n)'$ and the signal at B becomes the complement of:

$$(K+Q'_n)'+Q'_nJ \qquad \text{which is } Q_{n+1}$$

Therefore B acquires Q'_{n+1} during ϕ_2 time. Later, when ϕ_1 goes high, Q'_{n+1} is transferred to capacitance C_1. This completes the cycle. Notice that a ϕ_1 transmission gate and a ϕ_2 transmission gate are both necessary for data isolation, i.e. to ensure that the complete feedback loop is never closed.

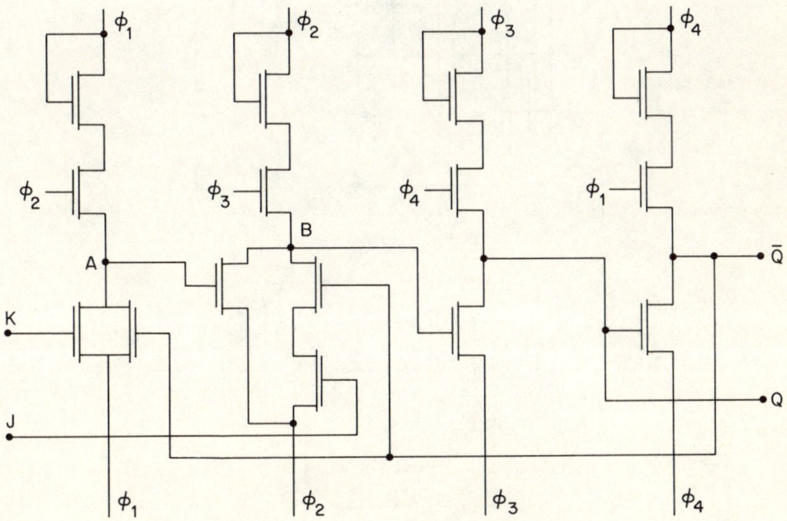

Fig. 11.19. J–K flip-flop (four-phase).

In the four-phase implementation, the inverter type is indicated by the phase driving the transistor in mid-position because that inverter evaluates (to produce valid output) when the transistor is switched ON. Thus, Q′ is the output of a type 1 inverter and the output is valid during ϕ_2 and ϕ_3. As before, the signal at A is $(K+Q'_n)'$ and becomes valid during ϕ_2 time. Likewise, the signal at B becomes valid during ϕ_3 and, as before, it describes $(Q_{n+1})'$. During ϕ_4, Q is updated and, during ϕ_1, Q′ is updated. Both Q and Q′ are valid during ϕ_2.

CLOCKED CMOS CIRCUITS

The concepts developed for use with phase-driven single-channel circuits can be readily applied to CMOS circuitry. To illustrate, Fig. 11.20 shows a portion of a semi-static CMOS shift register. Its action is better understood from the symbolic representation below and from the full register block diagram of Fig. 11.21. Each of the cascaded blocks that form the register contains circuits identical to that within the dotted square of the equivalent schematic. Alternate blocks are driven by ϕ_1 and by ϕ'_1. The blocks switch between the hold (H) and the evaluate (E) mode at the clock frequency. When ϕ_1 goes high, the ϕ_1 blocks go on HOLD and simultaneously permit

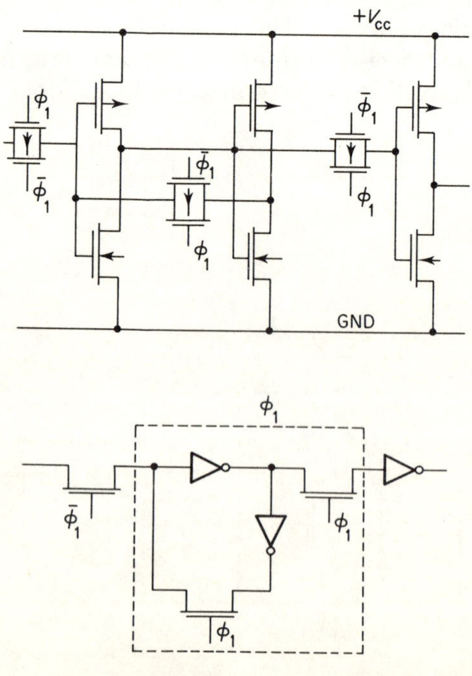

Fig. 11.20. Semi-static shift register and equivalent schematic.

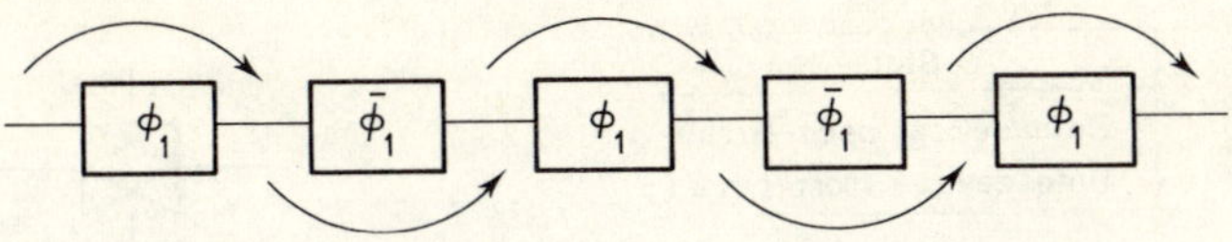

Fig. 11.21. Semi-static register (block diagram).

data through with inversion. Meanwhile, the intervening blocks open the feedback loops to evaluate and turn OFF their transmission gates. Because data are held in a static manner, ϕ_1 may stay high for any length of time so that low-frequency operation is possible. When ϕ_1 returns low, the data pass through the ϕ_1' blocks with a further inversion. Thus, each cycle of ϕ_1 passes data through a pair of blocks without inversion.

A dynamic CMOS shift register is depicted in Fig. 11.22. Note that two-phase working provides the interstage isolation. Data pass (with inversion) from A to B during ϕ_1 and from B to C during ϕ_2. Clock frequency must be sufficiently high to avoid data loss due to leakage from gate capacitance.

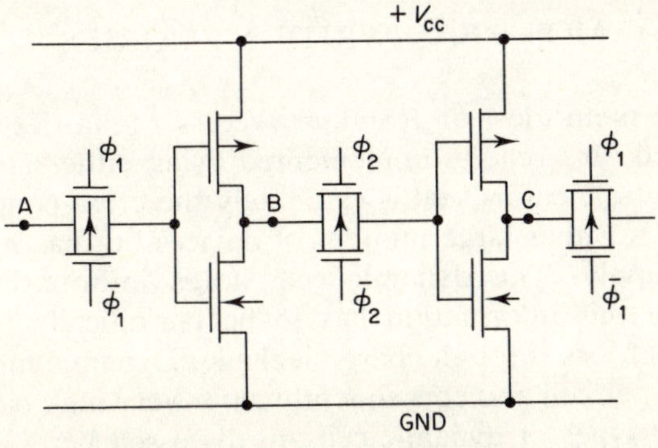

Fig. 11.22. Dynamic shift register.

As a final example, Fig. 11.23 shows a schematic of a flip-flop which utilizes the same principle as the semi-static shift register. It comprises two 'blocks' which alternately assume 'evaluate' and 'hold' modes and thereby

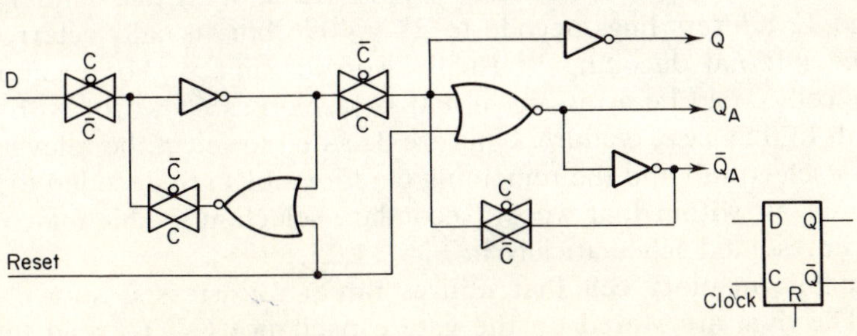

Fig. 11.23. A CMOS flip-flop (schematic).

C	State
0	Both devices open-circuit
1	Both devices short-circuit

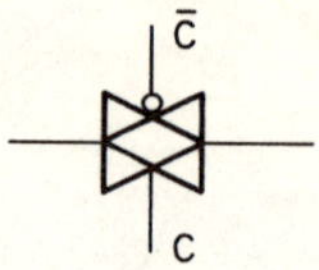

Fig. 11.24. CMOS dual-channel switch symbol and convention.

offers master–slave operation. A reset option is provided. A special symbol is used here to represent the CMOS transmission gates. The convention is given in Fig. 11.24.

It is not a difficult matter to convert Fig. 11.23 into a circuit schematic. The circuit is a straight mix of transmission gates, inverters, and a pair of two-input NOR gates. The most noteworthy feature is the very large number of devices which are required to build a semi-static element such as this.

MOS READ/WRITE MEMORIES

MOS read/write memories, or *R*andom *A*ccess *M*emories (RAM) as they are usually called, are readily implemented using either static or dynamic circuitry. The static memory cell is essentially the cross-coupled cell of Fig. 11.7. It requires a rather large number of devices but has the advantage of being easy to apply. The dynamic cell stores information on parasitic capacitances and this information has to be periodically rewritten or *refreshed* to prevent loss through charge leakage. Dynamic memories require very few devices per cell and consequently allow very high packing densities. A few important types of dynamic cell are discussed here.

The organization of large memories is usually such as to select one bit for output. If data are handled in 8-bit bytes, then every byte draws one bit from each of eight chips with identical addresses specified for each chip. The advantage of this scheme is that it minimizes the number of package pins required per chip. Each chip normally provides full address decoding. Thus, a 4k-bit memory organized in this way has 12 address lines and 1 output line. The 12 address lines decode to $2^{12} = 4096$ bits, usually referred to as '4k'. The internal decoding is usually word-organized. For example, the memory cells could be arranged in matrix fashion as 64 words with 64 bits per word. In that case, 6 address bits are decoded to select the relevant word ('primary selection') and the remaining 6 address bits are decoded to specify the desired bit within that word ('secondary selection'). This matrix structure is represented schematically in Fig. 11.25.

A dynamic memory cell that utilizes three transistors is shown in Fig. 11.26. The data are stored on the gate capacitance C_1. To read the data, primary selection identifies the desired WORD READ line and takes it

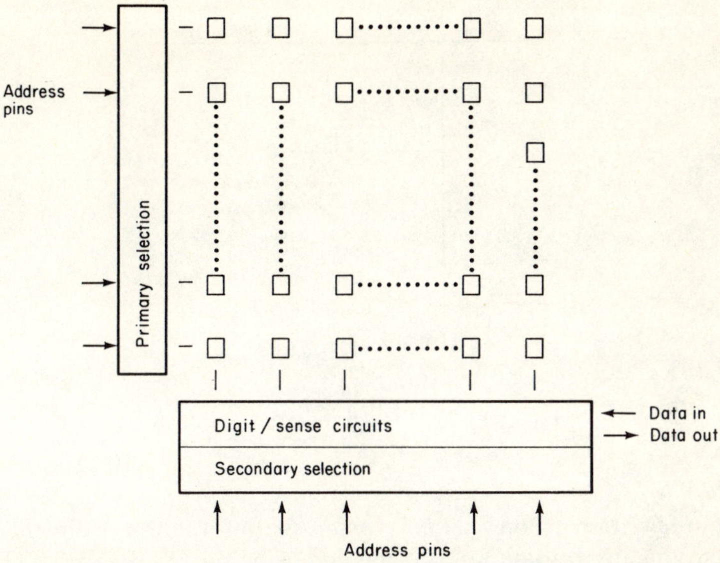

Fig. 11.25. A word-organized memory.

HIGH. This turns on T_2. If C_1 is uncharged, T_1 is OFF and the SENSE line remains LOW. If C_1 is charged, T_1 is ON and the sense line is charged to a HIGH level. The same activity is occurring simultaneously for all other bits in the selected word and for their associated sense lines. Secondary selection identifies the desired sense line and passes its contents to the DATA OUT line. The reading operation is clearly non-destructive.

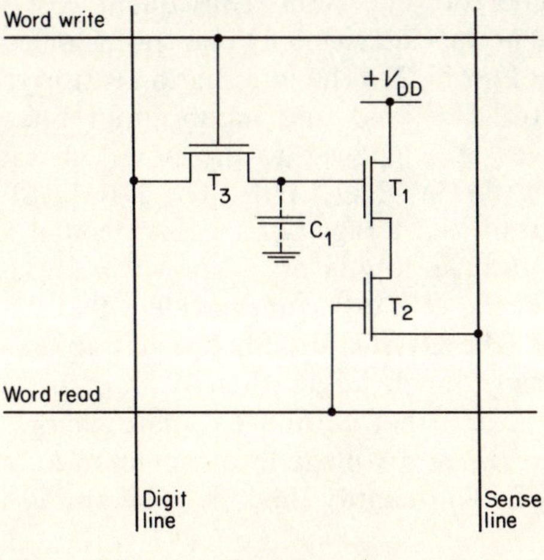

Fig. 11.26. A three-transistor memory cell.

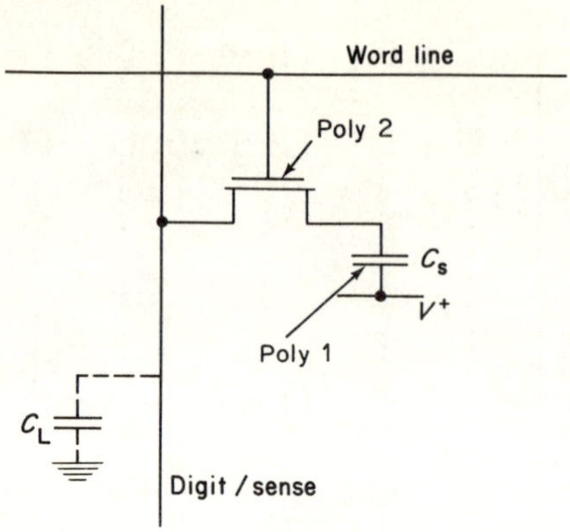

Fig. 11.27. One-transistor memory cell.

Data are written to the cell by passing the information from the DATA IN line to the appropriate DIGIT line and using the WORD WRITE line to turn on T_3. Notice that the corresponding transistors of all cells in the word are turned ON at the same time. Therefore, *all* DIGIT lines must contain correct data for inputting. For the selected bit, this data comes from the DATA IN line; for all other bits in the word, the data are replicas of existing contents which have just previously been read and temporarily held ready for re-entry. Therefore new data entry and refresh of old data are occurring simultaneously. Even if no new data are written, every cell must be refreshed at intervals of a few milliseconds. To achieve this, external circuitry must cycle through all of the 'primary selection' addresses within that time so that each word in turn gets refreshed.

Considerably higher densities (and consequent cost reduction) become possible with a scheme that utilizes only one transistor per cell! This kind of memory is shown in Fig. 11.27. The information is stored on the capacitance C_s. Only one control line and one input/output line are possible. The transistor itself serves as a gateway to the stored charge and this 'gate' is opened and shut by the WORD LINE. The DIGIT/SENSE line conveys data both in and out of the storage point C_s. A popular layout arrangement for a process using double poly layers is shown in Fig. (11.28).

On examination of the cell, it becomes evident that the reading operation is a destructive one. The DIGIT/SENSE line has an associated capacitance C_L which is inevitably much larger than C_s. Let us suppose that C_L is precharged to $V_{DD}/2$ and that C_s and C_L share charge. Because C_L is so much larger than C_s, the new voltage level, common to both, is only slightly different from $V_{DD}/2$. To quantify this, the total stored charge is simply:

$$Q = \frac{C_L V_{DD}}{2} + LC_s V_{DD}$$

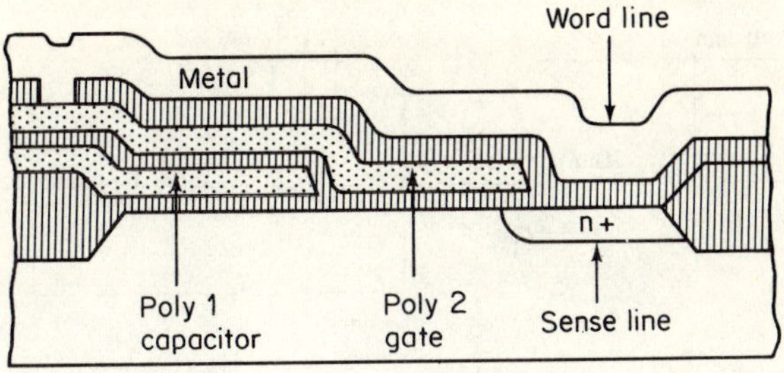

Fig. 11.28.

with $L=0$ or 1 depending on the logic level stored. After charge sharing:

$$V=\frac{Q}{C_L+C_s}=\frac{C_L}{C_L+C_s}\frac{V_{DD}}{2}+L\frac{C_s}{C_L+C_s}V_{DD}$$

For $L=0$: $$V=\frac{V_{DD}}{2}\left(1-\frac{C_s}{C_L+C_s}\right)$$

For $L=1$: $$V=\frac{V_{DD}}{2}\left(1+\frac{C_s}{C_L+C_s}\right)$$

Thus, the change in voltage is only $0.5V_{DD}C_s/(C_L+C_s)$ which may be merely tens of millivolts. It is necessary to detect the direction of the voltage change and to then rewrite the original data to the storage capacitor C_s. Both of these operations can be performed simultaneously using a special kind of sense amplifier which will now be described.

The sense amplifier is basically a flip-flop and, as such, has two stable states. But it also has an intermediate *metastable* state. An analogous mechanical situation is illustrated in Fig. 11.29 where a spherical ball can assume any of three positions in a shaped vessel. Positions 1 and 2 are clearly stable. Position 3 is such that a very small disturbance will displace the sphere to position 1 or 2, depending on the direction of the disturbance. This is the metastable state.

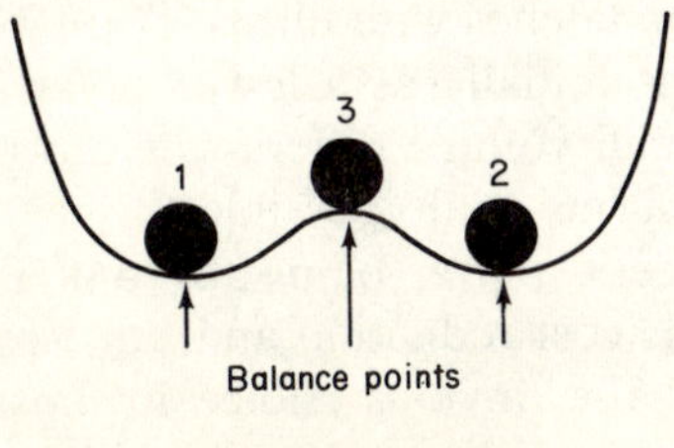

Fig. 11.29. Illustrating the metastable state.

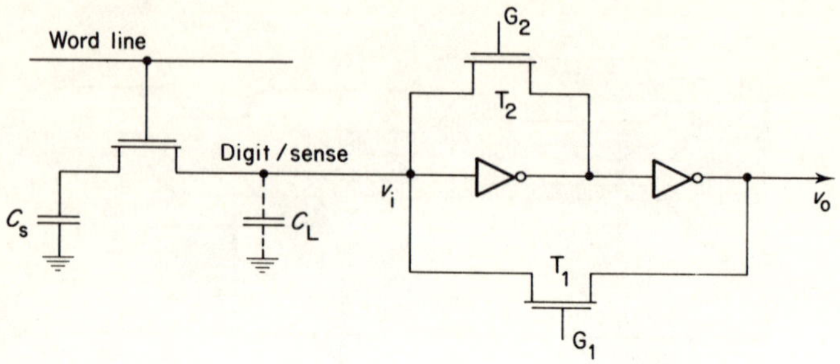

Fig. 11.30. Memory cell with sense amplifier.

Figure 11.30 shows the one-transistor cell complete with sense amplifier. The flip-flop is simply a pair of inverters (shown symbolically) for which the feedback loop may be closed via transistor T_1. Alternatively, T_1 can be left open and T_2 closed by raising G_2. This forces both inverters to the trigger level $V_{tr} = v_i = v_o$, which is also the metastable state. The procedure for a READ operation is as follows:

(a) Open T_1 and close T_2 to establish the metastable state.
(b) Open T_2.
(c) Raise WORD LINE. This causes a shift of $+\Delta V$ or $-\Delta V$ in the SENSE line voltage.
(d) Close T_1. This applies positive feedback and the movement initiated by the ΔV shift is accelerated towards a stable state.

The stable state reached is $v_i = v_o =$ HIGH for a positive ΔV and $v_i = v_o =$ LOW for negative ΔV where v_i and v_o are as shown in Fig. 1.30. On moving to a stable state, the output v_o recharges C_s through T_1 to its initial logic state. The WORD LINE can now be lowered. The data is available at v_o and has been restored on C_s. The READ operation has also served to perform a refresh of the stored data.

Writing to the one-transistor cell is accomplished very simply by connecting DATA IN (suitable buffered) to the DIGIT/SENSE line and raising the WORD LINE to admit the data.

Another form of the balanced sense amplifier is seen in Figure 11.31. T_1 serves simply as a switch which is closed when reading operations commence. Then the flip-flop latches with either T_2 on T_3 turned ON depending on whether point A was initially at a lower or at a higher potential than point B. In this way, small voltage differences can be detected.

The one-transistor cell has made possible the manufacture of very dense memories, the most recent arrival being the 64k RAM. The main consequence of high density is cost reduction and, for large memory systems, the dynamic MOS RAM is the obvious choice in most cases. For very small systems, static RAM requires less circuitry overall and it may be convenient to use RAM with several output lines rather than just one. In particular, a

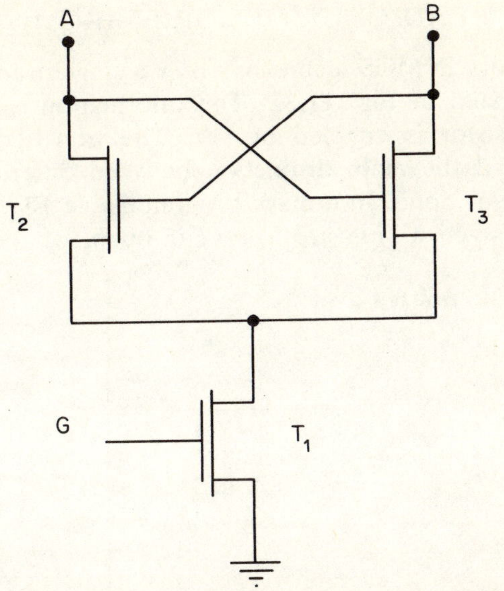

Fig. 11.31.

memory chip which outputs data in 8-bit bytes is very well suited to small microprocessor applications.

THE MNOS MEMORY

The MNOS cell is a transistor with a gate structure consising of *M*etal over *N*itride over *O*xide over the *S*ilicon substrate. The oxide layer is extremely thin. If a positive voltage pulse is applied to the gate, electrons tunnel through the oxide and are trapped near the oxide–nitride interface. This stored charge causes the threshold voltage to shift in the positive direction by as much as several volts. A negative gate pulse releases the stored charge and restores the original threshold voltage. If the threshold shift takes the device from the enhancement mode (normally OFF) to the depletion mode (normally ON), and vice versa, then the reading operation merely determines whether the transistor is conducting or not. This kind of cell is suitable as an *E*lectrically *A*lterable *R*ead *O*nly *M*emory (EAROM). It differs from RAM in that it is not volatile. The data are retained (in the form of a threshold voltage) after the power is switched off and are valid for months or even years. It also differs from RAM in that the WRITE time is relatively long. The MNOS device unfortunately suffers from a degradation mechanism which leaves some doubt regarding its long-term commercial acceptability.

READ ONLY MEMORIES (ROM)

A read only memory in MOS technology uses a single transistor in each bit position, as suggested in Fig. 11.32. The information stored depends on whether the transistor is enabled or not. The transistor is disabled by severing a link and the main distinction between different types is with regard to how this is done. In a mask-programmable ROM, the metallization mask specifies which links are to remain open.

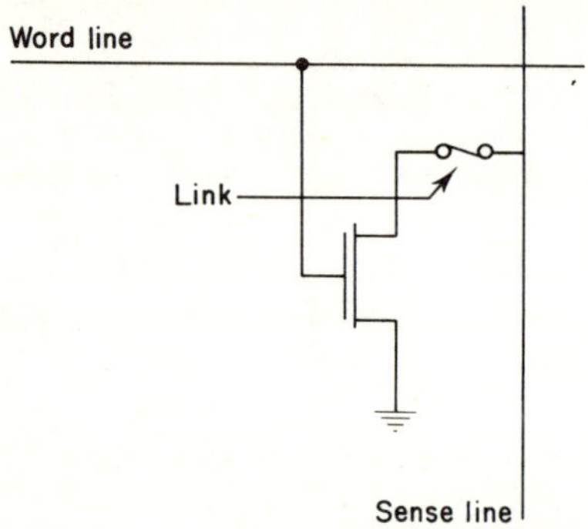

Fig. 11.32. MOS ROM element.

The method is suitable for large-volume production where the customer specifies the memory contents in advance. In a field-programmable ROM (a PROM), a fusible link is employed and the user programs the memory by injecting a high-current pulse to open-circuit the link. This is the most convenient form in most cases but has the disadvantage that the process is irreversible. Once blown, a link cannot be reestablished.

The FAMOS device permits fabrication of ROMs that do not have this restriction. FAMOS refers to a *F*loating gate *A*valanche injection *MOS*. The memory element is a transistor with a polysilicon gate that has no external connection—a floating gate. The gate is fully surrounded by the insulating silicon dioxide, as shown in Fig. 11.33.

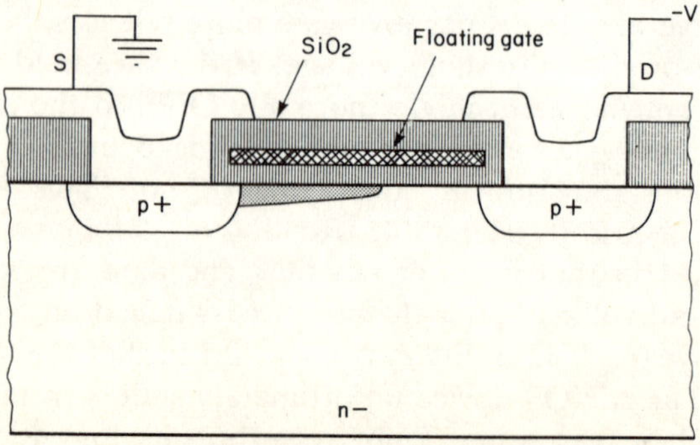

Fig. 11.33. The FAMOS transistor.

In the 'normal' state, the transistor is turned OFF. To turn it ON, the drain is reverse-biased to avalanche breakdown. This process generates high-speed electrons of which a small proportion penetrate the oxide and charge the floating gate. This negative charge on the gate causes surface inversion and the device goes ON. The gate charge is stable over very long periods of time. The charge can be released and the memory 'wiped clean' by exposing the chip to ultraviolet light. It is then ready to be programmed once again.

CHARGE-COUPLED DEVICES (CCDs)

In a charge-coupled device, the charge associated with an inversion layer is caused to move, in shift-register fashion, along the surface of a silicon wafer under the control of a three-phase (or similar) clocking system. These charge quantities can be viewed as a sequence of discrete analog levels which, in effect, pass through a clocked delay line from input to output. The special feature of the CCD is that no transistor structure is necessary. Its operation utilizes the inherent properties of the silicon surface.

The inversion layer charge can be recalled from Chapter 6 as:

$$q\int_0^{x_c} n\,dx = \text{gate charge} - V_{TO} \text{ inducing charge}$$

that is:

$$Q_I = C_{ox}V_G - C_{ox}V_{TO}$$

In the case of a transistor structure, this charge is readily supplied from the nearby source diffusion. In the illustration of Fig. 11.34, the channel is formed from source end electrons. These electrons are minority carriers from the standpoint of the p− bulk in which they exist. Inversion will still occur if no nearby diffusions exist. All that is necessary is a gate electrode over a thin oxide. In this case, however, there is no convenient source of electrons. They are supplied, in a relatively slow manner, by thermal generation. If the voltage on this isolated gate is suddenly stepped from zero

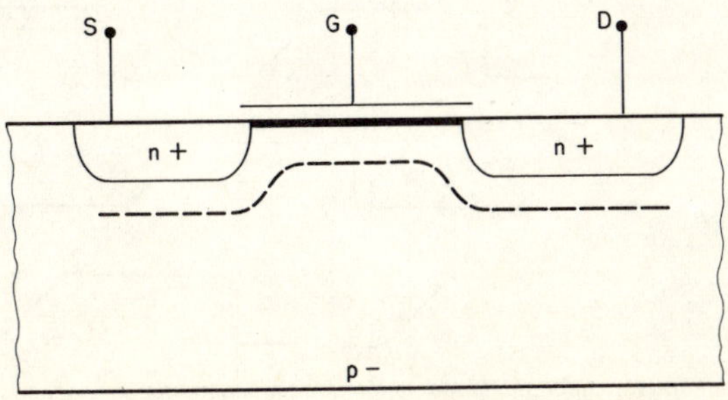

Fig. 11.34. Channel inversion.

to a value well above thresholding, the channel cannot form at once and, instead, the depletion layer is at first much wider than the expected value which corresponds to only $2\phi_F$ volts drop. Then, as the inversion charge grows, the depletion layer width falls back to the normal value. Similarly, if the gate voltage is then stepped back to zero, the inversion charge does not at once disappear. Rather, it gets 'covered' by the inflow of majority carriers from the bulk and recombination then gradually restores a normal equilibrium.

The inversion layer charge can be made to 'flow' along the surface if voltage is suitably applied to a succession of gates (Fig. 11.35). The charge

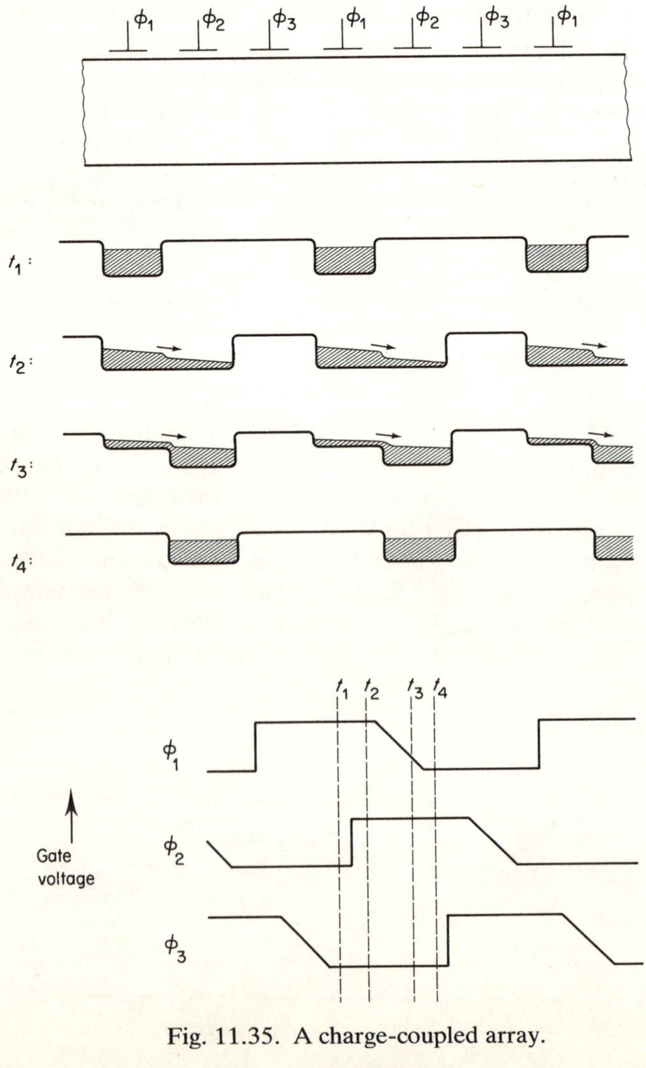

Fig. 11.35. A charge-coupled array.

Q_I can be viewed as the *capacity* of the inversion layer, which is varied by raising and lowering the gate potential V_G. The high capacity corresponding to a high potential V_G can be symbolized as a *deep potential well.* This kind of symbolism is used in Fig. 11.35 where an array of gates are driven from the three-phase system (ϕ_1, ϕ_2, ϕ_3) with clock voltage waveforms as shown. The potential wells, and the contained charge, are shown at successive points in time, t_1, t_2, t_3, and t_4:

t_1: The ϕ_1 wells are filled.

t_2: The ϕ_2 wells are 'sunk' by raising ϕ_2. This permits redistribution of the existing charge in the ϕ_1 wells.

t_3: Charge transfer to the ϕ_2 wells is encouraged by decreasing the capacity of the ϕ_1 wells.

t_4: The ϕ_1 well capacities have been reduced to zero and charge transfer to the ϕ_2 wells is complete.

During this sequence, charge packages have been shifted one step to the right. In the next time interval the same charge packets will be shifted from ϕ_2 wells to ϕ_3 wells. In purely electrical terms, the 'sinking' of a new well simply introduces a drop in electron potential energy level from left to right so that electrons fall downhill towards the lowest available level. As the process repeats, all charge packets are shifted to the right at a constant rate set by the clock frequency.

A charge input and output structure is illustrated in Fig. 11.36. The diffusions are, of course, reverse-biased as in a normal transistor structure. The input and output gates are used to regulate the flow of charge. It is frequently necessary also to measure the charge quantities at intermediate points along the line. All that is required is an electrode with thin dielectric separation from the point of measurement. If it is allowed to float, its voltage variations will reflect charge fluctuations nearby. Alternatively, if its potential is fixed, its own charge will vary and these variations can be detected.

Each step in the charge-transfer process can be modelled in the manner suggested by Fig. 11.37. The 'z^{-1}' block symbolizes unit delay in the notation of the Z transform. Of the total input charge Q_{IN}, a small portion δQ_{IN} gets left behind (and adds to the next sample) while a large proportion,

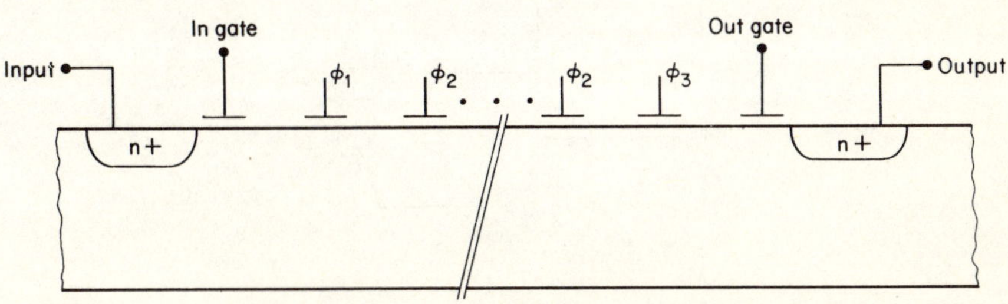

Fig. 11.36. CCD input/output structures.

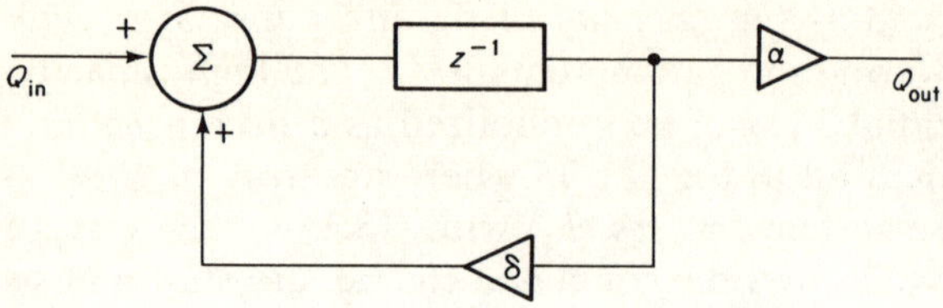

Fig. 11.37. A charge transfer model.

αQ_N, is transferred to the next stage. This representation is more than a mere flow chart. In fact:

$$\frac{Q_{OUT}}{Q_{IN}} = \frac{\alpha z^{-1}}{1 - \delta z^{-1}}$$

is a valid transfer function for this sampled-data system using Z-transform theory.

The properties of CCD arrays make them useful in a variety of digital filtering applications. They are also attractive as memory systems which store large blocks of data for *sequential* access. This is because of their high storage capability and low cost but limited data access capability.

DETAIL 1

This note calculates the power expenditure when a purely capacitive load is driven from a square-wave low-impedance voltage source. The result is

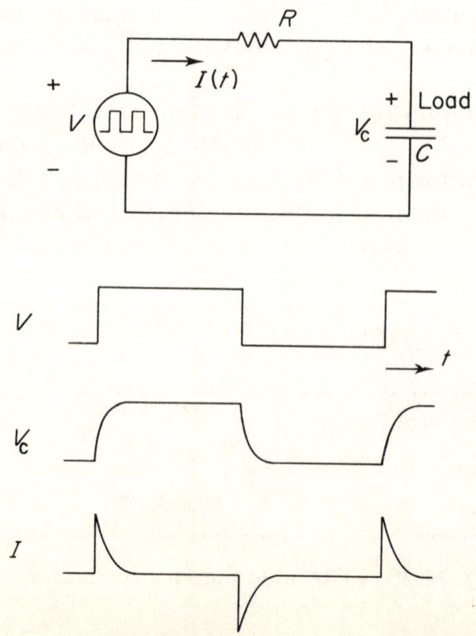

applicable to dynamic MOS circuits powered by two or more clock phases where no d.c. paths exist.

The square wave has clock frequency f and corresponding period $T = 1/f$. For low effective circuit resistance R, the time constant RC is small compared with the period T. Hence, the current transients are essentially complete in less that $T/2$ seconds.

$$I(t) = \frac{V}{R} e^{-t/RC}$$

for any half-cycle. The instantaneous power dissipation is I^2R and is integrated over time to give the energy loss E for one half-cycle:

$$E = \int I(t)^2 R \, dt = \frac{V^2}{R} \int e^{-2t/RC} \, dt = \frac{-CV^2}{2} [e^{-2t/RC}]_0^\infty$$

$$= CV^2/2 \text{ per half cycle}$$

$$\text{Power} = E/(T/2) = CV^2/T = CV^2 f \text{ Watts}$$

The source supplies energy at an average rate of CV^2f Watts and this energy is dissipated in the equivalent series resistance of the complete system.

CHAPTER 12

Digital System Design and Implementation

Up to recent times digital systems were built for the most part by the bringing together of a large number of integrated circuits onto one or more printed circuit boards. The engineer frequently completed his design at the electrical and logical level and passed the results to a group who would then design the layout of chips and interconnections on the printed circuit boards. The large inherent flexibility of the layout procedure permitted effective separation of electrical and logical design from topographical tasks.

That situation is changing. Nowadays, more and more logic fits on a single chip. It is becoming increasingly common for firms to design their own 'custom' chips and then send the designs to a fabrication facility for chip manufacture. Accordingly, designers are more frequently working at chip level and here the functional and topographical designs are not so conveniently separated. This chapter provides some broad guidelines on system design at chip level and these guidelines are as much concerned with layout as with functionality. Quite often, a design which looks optimum at circuit schematic level can be quite unattractive for implementation. Minimization of the number of active components cannot be equated to minimization of space requirements for a digital sub-system.

LAYOUT EXAMPLES AND DESIGN RULES

Actual fabrication of an integrated circuit requires the execution of a sequence of masking steps which collectively define the circuits which are to be built. The masks are first drawn on a much larger scale, typically 500× which means 500 times their final size. With this scale factor, a square circuit of 0.1 inch on the side would start at a set of drawings with a 50 inch side length. The first drawing to be made shows all geometric features simultaneously and it is from this drawing that the various masks are subsequently defined.

The generation of suitable artwork is best illustrated with reference to some examples. Figure 12.1 shows a layout for the realization of a two-input NOR gate in p-channel metal gate technology. The various features are distinguished as shown in Fig. 12.2. Note that all metal interconnects run across the full width of the drawing so that they can be reached from either side. This is good general practice as it facilitates the interconnection of different sub-systems on the chip. It is all the more important in metal gate

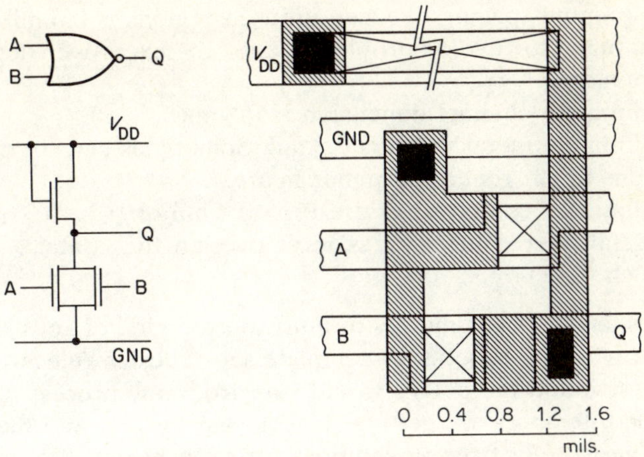

Fig. 12.1. Layout of a two-input NOR gate (aluminium gate processing).

technology which has only two interconnect levels, the diffused regions and the metal overhead.

There are many restrictions which should be observed in laying out an integrated circuit. These are necessary to assure consistent proper functioning of the system. They relate particularly to the tolerances that must be allowed for photographic work but, in some cases, to electrical constraints as well.

Mask-alignment tolerance must take account of several contributing error mechanisms. These include tolerance on photoresist edges due to variations in exposure, tolerance on oxide windows, tolerance on alignment by the operator, and many other factors. Collectively, they lead to a requirement for about 0.15 mil tolerance. This tolerance figure is the principal parameter underlying a set of *design rules* which will now be described.

Rule 1: Minimum width of a diffused region is 0.3 mil. This is necessary because of alignment tolerance.

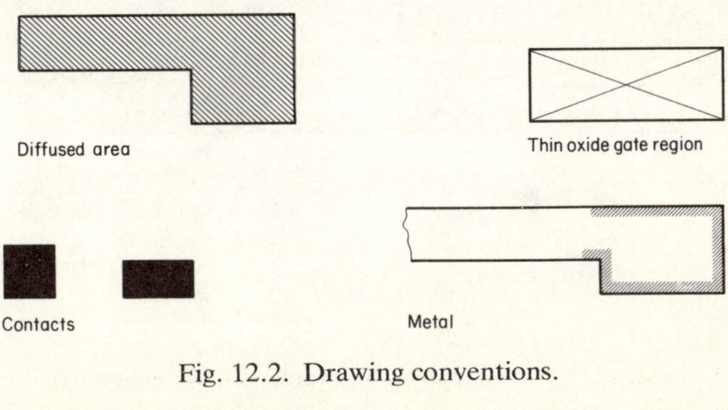

Fig. 12.2. Drawing conventions.

Rule 2: Minimum spacing between diffused regions is 0.4 mil. This is wide enough to avoid problems due to excessive depletion layer spreading.

Rule 3: Minimum channel dimension is 0.3 mil.

Rule 4: Minimum metal width is 0.3 mil. Sometimes current density limitations might require a higher figure.

Rule 5: Minimum contact sizes are 0.2×0.4 mil or 0.3×0.3 mil.

Rule 6: Metal over contact areas must overlap the contacts by 0.1 mil in each direction.

These rules are intended merely to illustrate the kind of constraints that are typical of any layout process. A complete set of design rules would be much more extensive and the details would vary from one process to another. As well as this, tolerances are getting smaller year by year and the design rules are changing to reflect the possibilities for even higher circuit densities.

In the layout of Fig. 12.1, the drivers are both 0.4 mil square. While the smaller size of 0.3 mil may be acceptable, better yield can be expected with the chosen dimensions. The load device might be 0.3×2.0 mil, mask dimensions. Such a large device is typical of conditions necessary to assure a sufficiently high beta ratio (and correspondingly low V_{OL}) after sideways diffusion has been taken into account.

An implementation in CMOS of the same logic function appears in Fig. 12.3. Notice the connection of the p− tub to ground together with the

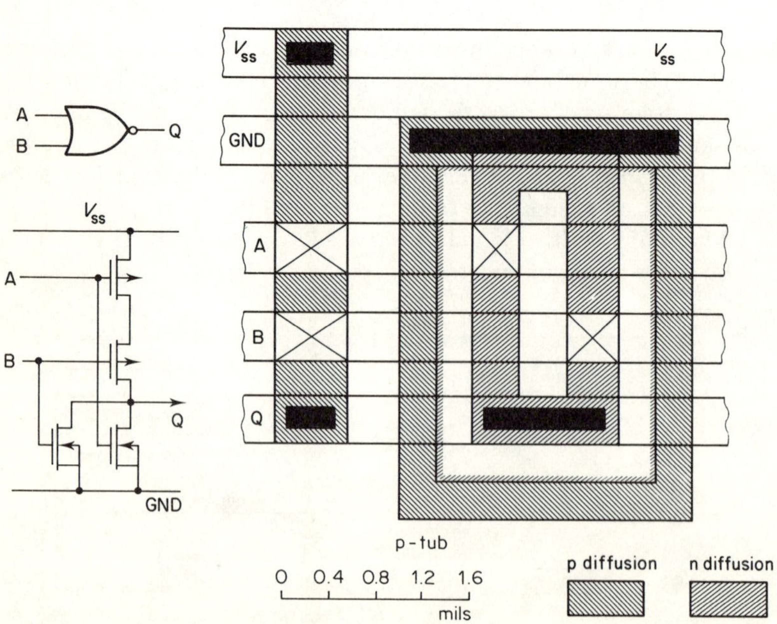

Fig. 12.3. Two-input NOR gate in CMOS technology.

common source diffusion for the two n-channel devices. It is noticeable, too, that the necessity for a p – tub makes large demands on chip area compared with the single-channel case. While this example and the one previous may constitute satisfactory layout solutions when viewed in isolation, the solution adapted may or may not be suited to the external interconnections that are required. It is desirable that, as far as possible, interconnection requirements be anticipated before detailed layout is undertaken. This process is assisted by the use of system blocks which are labelled according to the functions that they perform and the connections that must be made to them.

Silicon gate technology has layout procedures that are different in a few important respects. The polysilicon layer, which constitutes all gate regions, can also serve as a third interconnect medium. This is a large asset and greatly increases the flexibility of layout. It must, of course, be remembered that *crossovers* of polysilicon interconnects over diffused layers cannot exist. Wherever a polysilicon lay 'crosses' a diffused layer, a transistor is formed; the diffusion does not exist under the polysilicon because the polysilicon itself served as a *mask* while diffusions were being made. It is also important to remember that the poly interconnect medium is of fairly high resistance, frequently around 50 Ω/□ compared with 10 Ω/□ for diffusions and about 0.03 Ω/□ for metal interconnects.

Several additional design rules, appropriate to silicon-gate technology, are illustrated in Figure 12.4. This shows the layout of a transistor using different connection methods for source and drain. The left-hand contact is made by the poly medium, directly to the silicon surface. To allow this kind of contact, Mask 2 was used to make a window in the thin gate oxide before the polysilicon layer was deposited. The right-hand contact is a metal contact through a window in the oxide to the underlying silicon surface. The recommended minimum dimensions have been used and are multiples of 0.2 mils which describes a well-established and conservative set of design rules at the time of writing.

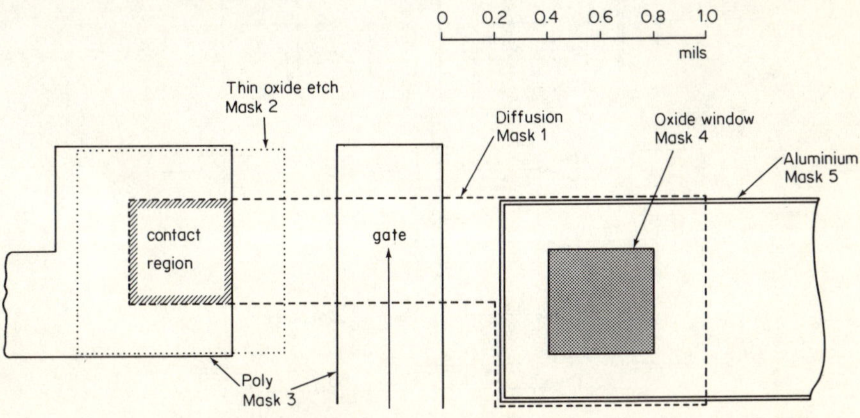

Fig. 12.4. Silicon gate design rules.

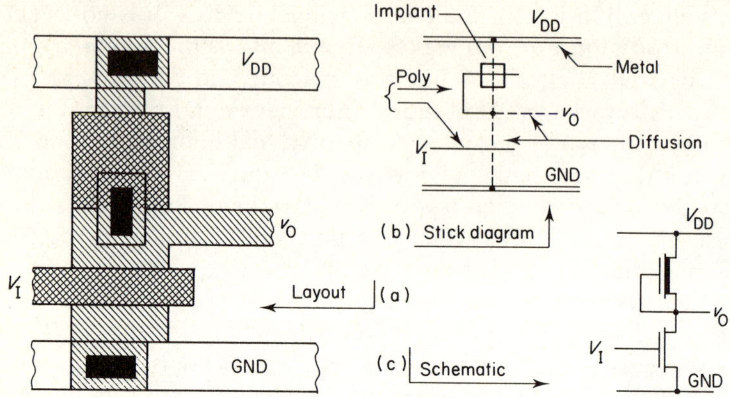

Fig. 12.5. Silicon gate inverter layout.

Figure 12.5 shows the layout for a simple inverter in silicon gate technology using a depletion load. It also shows, at (b), a much simplified representation known as a *stick diagram* as well as a circuit schematic at (c). The stick diagram is a convenient alternative to a full layout for most purposes prior to the final phase of the work. Most early designs can be sketched and

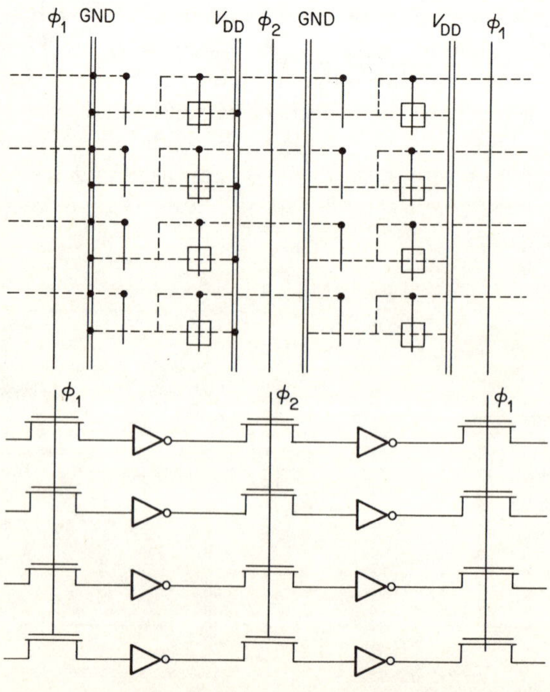

Fig. 12.6. Dynamic shift-register implementation.

examined in this form. Note that the output line v_o could be taken at any of the three levels since all are associated with the butting contact. The stick diagram representation will be used throughout the remainder of the chapter. As in Fig. 12.5(b), dotted lines will describe diffused areas while solid lines represent poly interconnects and double lines represent the metal overhead. Transistors are formed wherever poly crosses a diffused region.

A further example of circuit implementation in silicon gate technology is seen in Fig. 12.6. This shows the layout of a 4-bit wide dynamic shift register. Notice how easily the pass transistors are incorporated. In this particular arrangement, the control lines run vertically while the data are shifted horizontally. This is a common characteristic of well-ordered patterns. It is also true that a well-ordered pattern uses space much more effectively than some random grouping of functions. Thus, it would appear that the inherently repetitive structure of a shift register is far easier to implement than some arbitrary grouping of combinational logic circuits.

STRUCTURED DATA PATHS

It is, in fact, possible to adapt an organized approach to the implementation of combinational logic by the use of *P*rogrammable *L*ogic *A*rrays (PLAs). The manner in which these arrays are built will shortly be discussed. The use of the PLA to implement combinational logic in conjunction with two-phase working leads to a very general concept of digital machine structure. The broad structure is depicted in Fig. 12.7 where blocks of combinational logic are separated by dynamic register elements. Consecutive register banks are clocked by opposite phases. This provides for signal isolation and a delay of one half-cycle between successive registers. It is merely an enlargement of the concepts developed in Chapter 11 with reference to various two-phase and four-phase schemes. The registers are simply inverters which acquire data during one clock phase and hold the data during the next. The combinational logic may be purely static in itself. Notice that while one block of combinational logic is handling data acquired at some instant

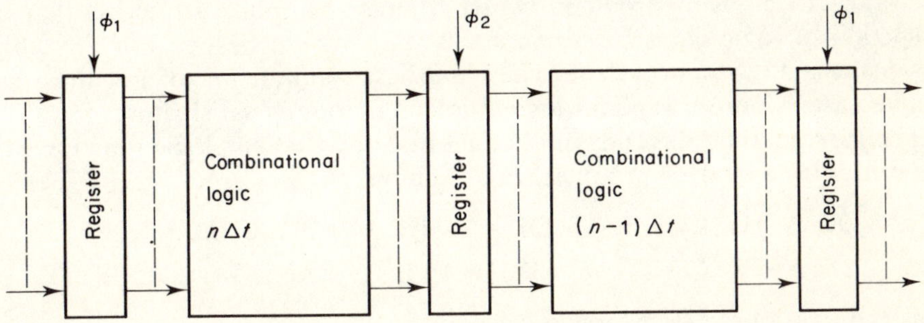

Fig. 12.7. Generalized data handling.

$(n-1)\Delta t$, the block to the left is processing the more recent data from $n\Delta t$. There are as many levels of simultaneous data handling as there are combinational blocks and all of these blocks are simultaneously in use. In this way, high efficiency of hardware utilization and high throughput are possible.

It was already seen that combinational logic can be fitted in place of inverter drivers. This approach is quite useful for very small systems and some examples of this kind were given in Chapter 11. It is a method whereby each new problem is different from the previous one and may require quite a different design strategy.

The combinational logic problem can be defined in general terms as that involving a system with n inputs and m outputs. There are a total of 2^n

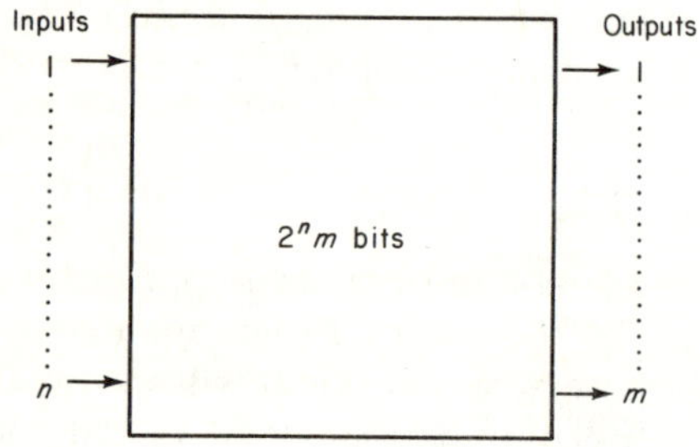

Fig. 12.8. Combinational logic system.

possible input combinations and, in principle, each combination can occur and can produce some specified output level. The output can be viewed as an m-bit word (Fig. 12.8). The system just described is readily implemented as a *R*ead *O*nly *M*emory (ROM) with n address lines and m output lines, i.e. an m-bit word size. The total number of bits in the memory is $2^n m$. It is a simple matter to program the ROM so that each combination of inputs (each address) gives the desired output. The ROM is frequently used in this way but the approach is inherently wasteful because it usually happens that only a very small proportion of all the possible input combinations are necessary to a given problem. A ROM may be more attractive than a collection of logic chips for some small-scale system design. For ON chip designs, the PLA approach provides the best solution for all but the very simple cases where a random-logic implementation is satisfactory.

The PLA method depends on the fact that any Boolean logic function can be written as a sum of products. For example:

$$Q = A \cdot [B \cdot C' + (B + C \cdot D)'] = A \cdot [B \cdot C' + B' \cdot (C \cdot D)']$$
$$= A \cdot [B \cdot C' + B' \cdot (C' + D')] = A \cdot B \cdot C' + A \cdot B' \cdot C' + A \cdot B' \cdot D'$$

This reduction used de Morgans' rules:

$$(X + Y)' = X' \cdot Y' \quad \text{and} \quad (X \cdot Y)' = X' + Y'$$

In this example, there were four inputs and one output. As many as sixteen product terms were possible but only three were required. The final form of the solution is known as a *sum of minterms.*

The PLA implementation differs from the ROM approach in that it implements the minterms of the problems rather than all possible input combinations. This is obviously much more efficient. The manner of implementation can be illustrated with respect to a system with inputs A, B, C, D and outputs Q_1, Q_2, Q_3 such that:

$$Q_1 = A \cdot B \cdot C' + A \cdot B' \cdot C' + A \cdot B' \cdot D'$$

$$Q_2 = A \cdot C + A \cdot B' \cdot D'$$

$$Q_3 = A \cdot B \cdot C'$$

These functions are implemented by the PLA of Fig. 12.9. Each horizontal line describes one of the minterms M_1, M_2, M_3, M_4. Each line uses a depletion load on the left and one or more 'drivers' or pull-down transistors in

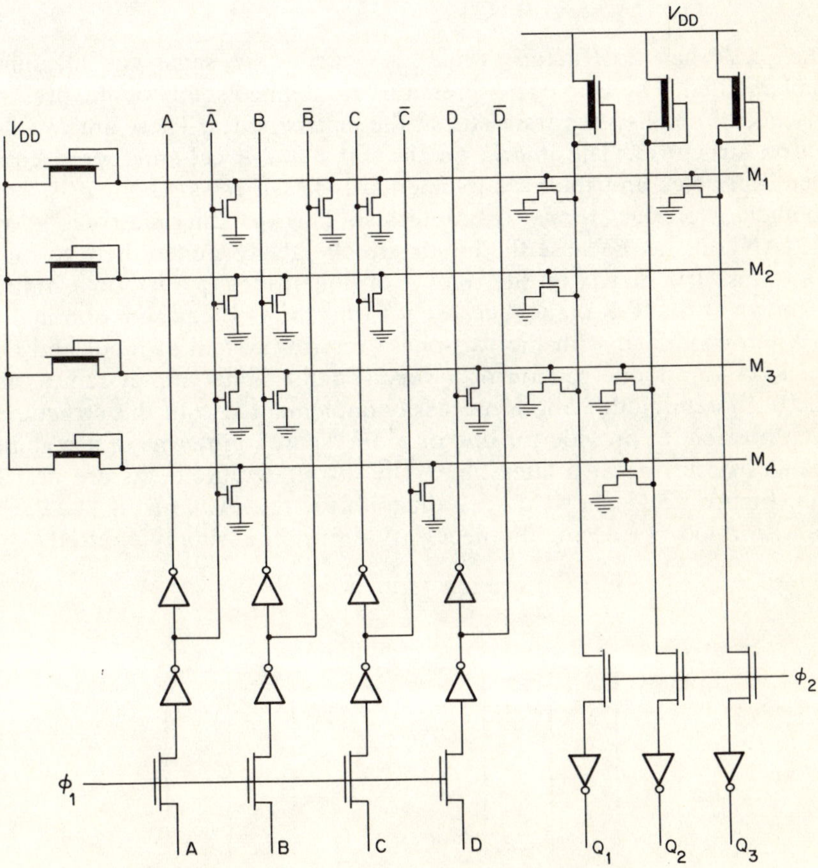

Fig. 12.9. A programmable logic array.

parallel. It is a multi-input NOR gate. Accordingly;

$$M_1 = (A' + B' + C)' = A \cdot B \cdot C'$$

In a similar way:

$$M_2 = (A' + B + C)' = A \cdot B' \cdot C'$$

$$M_3 = (A' + B + D)' = A \cdot B' \cdot D'$$

$$M_4 = (A' + C')' = A \cdot C$$

Note that there are, in this example, no more than four separate minterms required to describe all outputs. Each of the vertical lines on the right is a NOR gate with the various minterms available as inputs. Each NOR drives another inverter at output so that Q_1, Q_2, and Q_3 are in fact the logical OR of participating minterms. Thus:

$$Q_1 = M_1 + M_2 + M_3 = A \cdot B \cdot C' + A \cdot B' \cdot C' + A \cdot B' \cdot D'$$

$$Q_2 = M_4 + M_3 = A \cdot C + A \cdot B' \cdot D'$$

$$Q_3 = M_1 = A \cdot B \cdot C'$$

The PLA has a structure which is exactly the same for all functions implemented. The Boolean expressions are defined solely by the presence or absence of a pull-down transistor at the matrix nodes. There are two distinct matrix structures. The matrix on the left has $2m$ columns representing m input variables and their complements. It has p rows where p is the total number of product terms (or minterms). This $(p \times 2m)$ matrix is known as the 'AND plane' because the inputs are effectively 'anded' here to form the minterms. The matrix on the right is of dimensions $(p \times n)$ for n outputs. It is known as the 'OR plane' because it forms the Boolean sum of minterms.

When combined with the two-phase register system at input and output, the PLA forms one section of a very general digital sub-system as in Fig. 12.10. System modifications are easily implemented with this structure. It is even possible to provide for one or more 'spare' minterms to accommodate changes occurring at a later time. The layout requirements are also easily met. Figure 12.11 shows a stick diagram for the PLA of Fig. 12.9. Where transistors do not occur, the necessary diffusion is simply omitted.

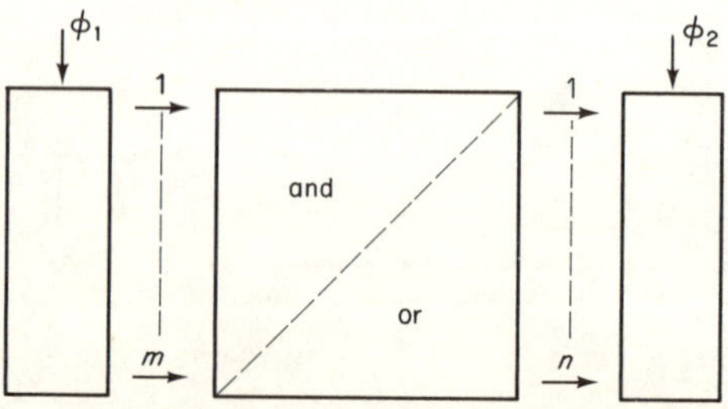

Fig. 12.10. A data path with PLA.

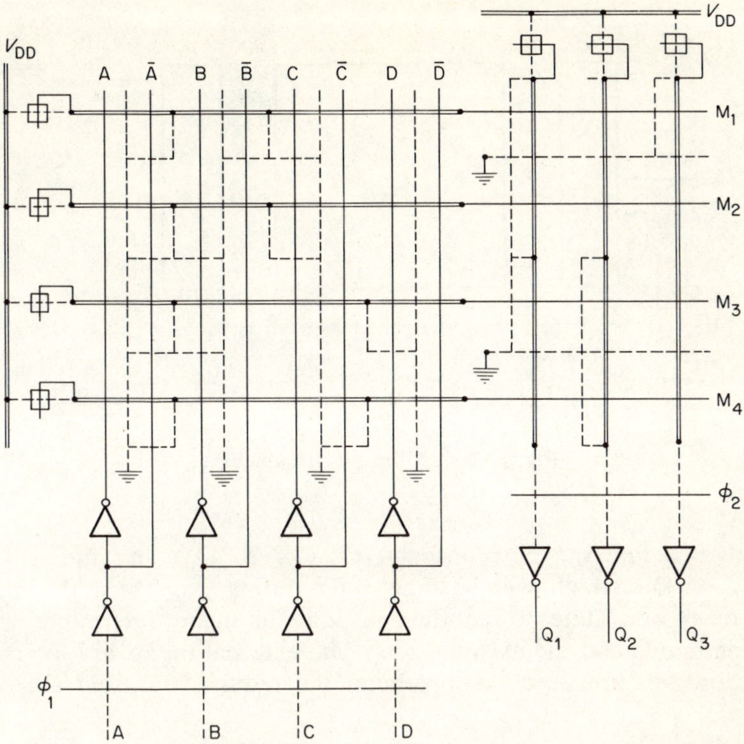

Fig. 12.11. Stick diagram for the PLA.

Since the inputs connect to gate terminals, they are defined in poly. Every matrix node has a grounded diffused region running nearby so that a pull-down transistor can easily be added wherever necessary. There is no limit on the permissible size and complexity of the PLA. It is a method of *hardware* implementation of any digital machine. This is in total contrast to the microprocessor where a general-purpose hardware is programmed by *software* to perform any desired function.

FINITE-STATE MACHINES

The data structure of Fig. 12.10 becomes a *finite-state machine* with the addition of feedback from output to input. The machine then has external inputs $I_1, I_2, \ldots, I_m$, outputs $O_1, O_2, \ldots, O_n$ and feedback signals $Q_1, Q_2, \ldots, Q_r$ (Fig. 12.12). Since the feedback lines serve as memory for the system, they can be thought of as defining the system *state* with 2^r possible states. During ϕ_1, the present inputs and the present state (i.e. the values $Q_1, Q_2, \ldots, Q_r$) are admitted to the combinational logic system.

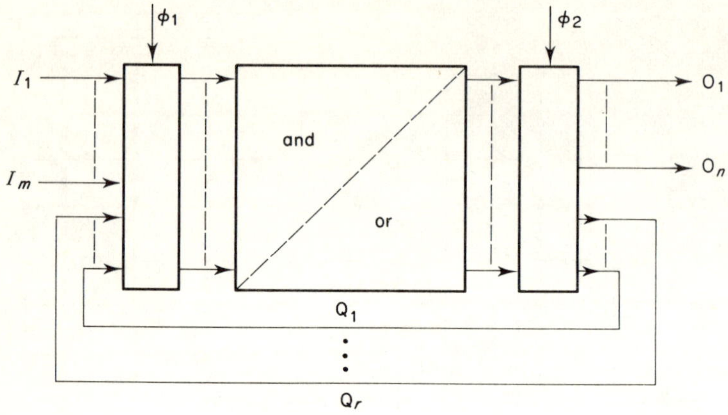

Fig. 12.12. A finite-state machine.

There they define the new outputs $(O_1, O_2, \ldots, O_n)$ and the new state $(Q_1, Q_2, \ldots, Q_r)$ which will become valid during ϕ_2. Thus, the machine moves from one state to another in some manner determined by the prevailing inputs and the existing state. The external inputs are the controls and the outputs are used to implement the required tasks in the outside world.

The introduction to feedback greatly complicates the behaviour of this data-processing device. An extensive body of theory exists for predicting the behaviour of finite-state machines. A general mathematical approach is facilitated by the use of the Z-transformation. This is particularly useful for the design of digital filters. A discussion of digital filtering is beyond the scope of this book but it will be possible to illustrate the action of a finite-state machine in a much simpler role where it merely cycles repeatedly through certain state sequences.

The example chosen relates to the different display modes of a digital watch. It is a simplified example where a single push-button P serves as input. $P=1$ when the button is depressed and $P=0$ otherwise. The watch has three display modes: TIME (hours, minutes), DATE (month, day), and SECS (shows seconds only). Depressing the button once in the TIME mode causes transfer to the DATE mode. It returns automatically to the TIME mode after 4 sec have elapsed. If the button is pressed again before 4 sec have elapsed, it moves to the SECS mode. This mode is maintained until the button is again pressed, upon which it returns to the TIME mode.

A state diagram which adequately describes the problem is shown in Fig. 12.13. The system is shown as having six states rather than three. This provision is necessary to prevent more than one state transition from occurring when the button is pressed. For example, the button is depressed to move from the regular time mode TIME/2 to DATE/1 and the system

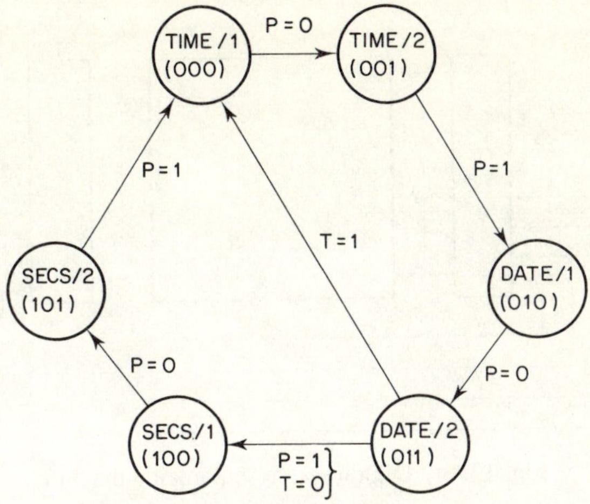

Fig. 12.13. State transition diagram.

moves to DATE/2 when the button is released. Note that DATE/1 and DATE/2 give identical displays and similarly for TIME and for SECS. The display options are selected by two output bits (O_2, O_1) encoded as follows:

MODE	O_2	O_1
TIME	0	0
DATE	0	1
SECS	1	0

Three bits are necessary to encode the six possible states and the coding convention is given in brackets for all states in Fig. 12.13. A timer is necessary to count the 4 sec interval. It has an input ST (start timer) which causes counting to start when it goes from 0 to 1 (positive edge triggering). While counting, the output T is held low. After 4 sec have elapsed, T returns high and stays there until the next positive edge of ST. The timer is started on entry to the DATE/2 state. If the button is pressed again within 4 sec, the SECS/1 state is accessed, otherwise the TIME/1 state is established when T returns high.

Inputs, outputs, and feedback signals are defined in Fig. 12.14. It is best to view T as an external input and ST as an output. The feedback signals are those that define the state of the system. The outputs (O_2, O_1) are used to select one of the three display options.

The state sequences defined by Fig. 12.13 can be presented in tabular form as shown in Fig. 12.15. All parameters to the left of the vertical dividing line are inputs to the PLA and become valid during ϕ_1. Parameters to the right are PLA outputs which become valid during ϕ_2 and are determined

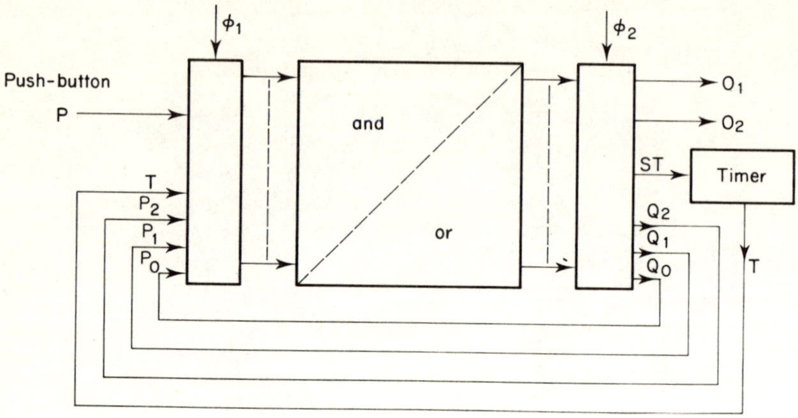

Fig. 12.14. Digital watch sequencing machine.

solely by the inputs which were established during ϕ_1. Each row of the table defines one of the several state transitions that can occur. Several of these entries are for circumstances where the new state and the old state are identical.

The PLA design follows directly from the transition table. Each row defines a minterm involving PLA inputs (and their complements) for that row. The PLA outputs are defined by selected minterms. To illustrate this,

Entered during ϕ_1					Sent out during ϕ_2							
Inputs		Present state			Next state			Outputs				
P	T	P_2	P_1	P_0	Q_2	Q_1	Q_0	ST	O_2	O_1		
1	X	0	0	0	0	0	0	0	0	0	M_1	
0	X	0	0	0	0	0	1	0	0	0	M_2	TIME 1 → TIME 2
0	X	0	0	1	0	0	1	0	0	0	M_3	
1	X	0	0	1	0	1	0	0	0	1	M_4	TIME 2 → DATE 1
1	X	0	1	0	0	1	0	0	0	1	M_5	
0	X	0	1	0	0	1	1	1	0	1	M_6	DATE 1 → DATE 2
0	0	0	1	1	0	1	1	0	0	1	M_7	
0	1	0	1	1	0	0	0	0	0	0	M_8	DATE 2 → TIME 1
1	X	0	1	1	1	0	0	0	1	0	M_9	DATE 2 → SECS 1
1	X	1	0	0	1	0	0	0	1	0	M_{10}	
0	X	1	0	0	1	0	1	0	1	0	M_{11}	SECS 1 → SECS 2
0	X	1	0	1	1	0	1	0	1	0	M_{12}	
1	X	1	0	1	0	0	0	0	0	0	M_{13}	SECS 2 → TIME 1

Fig. 12.15. State transition table.

the minterms are defined $M_1, M_2, \ldots, M_{13}$ in the table. Thus, for example:

$$M_4 = P \cdot P_2' \cdot P_1' \cdot P_0$$
$$M_8 = P' \cdot T \cdot P_2' \cdot P_1 \cdot P_0$$

and so on for the other terms. To illustrate the output formations:

$$Q_1 = M_4 + M_5 + M_6 + M_7$$
$$ST = M_6$$
$$O_2 = M_9 + M_{10} + M_{11} + M_{12}$$

The resulting PLA has an AND matrix of (13 × 10). The ten columns are for

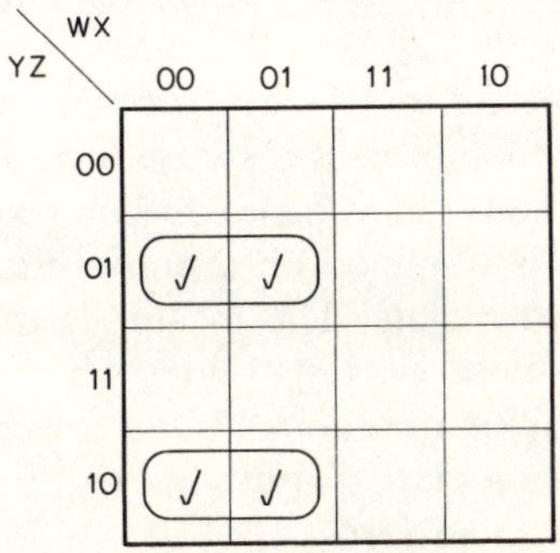

Fig. 12.16. Karnaugh map.

(P, T, P_2, P_1, P_0) and their complements. The OR matrix has (13 × 6) entries to accommodate the six PLA outputs (Q_2, Q_1, Q_0, ST, O_2, O_1).

The approach just given is a valid solution mechanism but frequently a wasteful one. Each output is a sum of minterms and it happens quite often that the number of terms can be reduced. To take an example:

$$Q = W' \cdot X' \cdot Y' \cdot Z + W' \cdot X' \cdot Y \cdot Z' + W' \cdot X \cdot Y' \cdot Z + W' \cdot X \cdot Y \cdot Z'$$

can be written much more simply as:

$$Q = W' \cdot Y' \cdot Z + W' \cdot Y \cdot Z'$$

The simplification can be assisted in a number of ways. The Karnaugh map of Fig. 12.16, which was used to achieve this reduction, is a popular method. Whatever the method, the main point is that some simplification is generally possible by reduction of the number of minterms required. When this is done, the system is implemented in exactly the same way as before but on a somewhat smaller chip area.

COMPUTER-ASSISTED GRAPHICS

A number of layout examples were given earlier in this chapter to illustrate the first step in the generation of masking plates for integrated circuit fabrication. The next operation is to produce drawings for each of the masks required by the process. These drawings follow directly from the original layout. Each drawing simply defines 'black and white' areas, i.e. areas which must be masked and areas which must not. It is now necessary to reduce the drawings photographically to final size. The first step is to define each drawing in a form suitable for photographic work. This is very often done on a sheet of Rubylith. Rubylith is a clear mylar material with a red overlay. The overlay is cut away in certain areas leaving a clear pattern on a red background. In subsequent photographic work, the red areas look opaque to the light used for exposure. The Rubylith cutting is generally done by a computer-controlled drafting machine. This means that the drawings have to be specified in the form of numerical coordinates which tell the computer where the red overlay is to be removed.

The initial reduction brings the drawing to 10× or 10 times the final size on a photographic plate. The final reduction uses a step-and-repeat camera which reproduces the same mask as many times as it can be accommodated on a wafer surface. The final pattern is obtained on a photosensitized glass plate which is referred to as the *master plate.* The master is then used to generate several *working plates.* It is these working plates which are placed in contact with the wafers during masking operations. Due to scratches and other imperfections, the life of a working plate is quite limited, in the region of thirty operations without excessive yield loss.

As already mentioned, the 'drawings' which define the masking plates are specified for the computer in some convenient numerical form. Rubylith cutting is not the only option at this point. An alternative method is to use a pattern generator. This is a device which exposes a photographic plate in a sequence of small rectangular elements. The length, width, and orientation of each rectangle are variable and are specified by the computer which directs the actions of the generator. The position of each rectangle is defined by moving the photographic plate in the xy plane beneath the generator under computer control. The mask is thereby built up as a large number of small rectangles which are defined in numerical form for the computer. The resulting plate is usually 10× size. The final reduction is performed as usual with step-and-repeat operations to replicate the mask throughout the entire surface.

Various software packages exist to simplify the mask definition for the user. These provide some convenient language medium with statements that define features of a drawing in the simplest way possible. For example:

RECT M2 (120, 44) 12 8 0

could define a rectangle on mask number 2 with its centre on the coordinate position (120, 44), with a width of 12 units, height of 8 units at orientation of 0^0. Another example might be:

PATH 6 (30, 20) (30, 80) (100, 80)

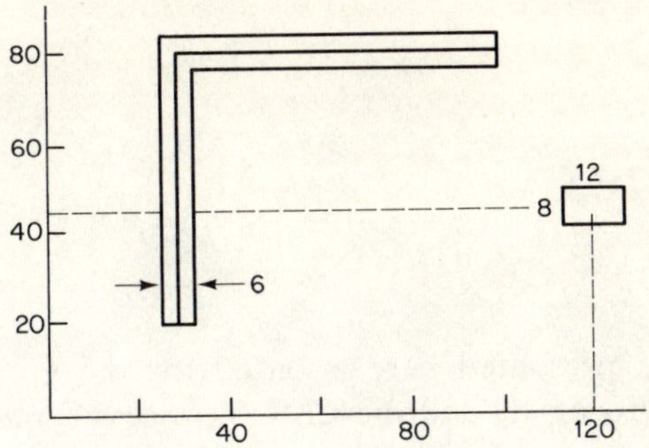

Fig. 12.17. Illustrating graphics software.

This would define a path 6 units wide linking the points specified in coordinate form. Both of the above examples are illustrated in Fig. 12.17.

Undoubtedly, computer assistance is best utilized in an interactive mode where portions of the work are displayed on a screen in graphical form. The 'drawing' which is now stored magnetically can be altered at will and very powerful editing options exist. To take one simple example, it is a straightforward software operation to rotate an entire drawing through any desired angle, to translate it to a new location, or to do both. Scale-change features are also readily available to assist the operator. Colour displays can also be used to great advantage by making it easier for the operator to interpret the work and thereby to identify errors or to avoid them in the first place. The manual development and checking of integrated circuit design files is a very tedious and time-consuming business. There is little doubt therefore that computer-aided graphics will be heavily utilized in a future which holds great promise for circuit integration at ever-increasing densities.

Case Studies

The set of problems presented here is based for the greater part on the n-channel process (Process 1) and the CMOS process (Process 2) for which essential data are given on page 203. Answers are given on page 204.

Q1. Consider a metal track of rectangular cross-section (0.2×0.1 mil) by 150 mil in length. Find the resistance between end faces if the metal is:
(a) Intrinsic silicon at 25 °C
(b) Intrinsic silicon at 60 °C
(c) Phosphorus-doped silicon, $N_B = 10^{15}$ atoms/cm^3 at 25 °C
(d) As in (c) at 60 °C
Assume $\mu_n = 1400$ and $\mu_p = 500$ cm^2/V-sec in intrinsic silicon.

Q2. A p–n junction approximates to the ideal with $N_D = 10^{19}$ and $N_A = 10^{15}$. Find the contact potential ψ_0 at 35 °C. Hence determine depletion layer widths X_p and X_n on the p side and n side respectively and the junction capacitance C in femtofarad per square mil for (a) 0 V applied and (b) 15 V of reverse bias. Find the avalanche breakdown voltage for the junction.

Q3. An n-type wafer has uniform resistivity of $0.5\ \Omega \cdot$ cm. A boron predeposition has 10^{16} atoms/cm^2. Find the required duration of the subsequent (Gaussian) drive-in diffusion to achieve a junction depth of 1.5 μm at a diffusion temperature of 1200 °C. Find also the final N_s atoms per cubic centimetre at the surface and $N(x)$ at a depth of 0.5 μm if these symbols refer to net concentrations of impurity atoms.

Q4. Referring to the n-channel Process 1, use appropriate data from the Appendices to determine the bulk doping level N_B, the net surface concentration N_s for a Gaussian profile, and ϕ_F at 300 K. Find also $C_{ox(G)}$, C_{pn-5V}, and C_{pn-10V} in femtofarads per square mil. (The voltages refer to external reverse bias.) Enter your results on the table of parameters.

Q5. Use the basic parameters for the CMOS process together with the data in the Appendices to determine the various concentrations (N atoms per cubic centimetre), the ϕ_F values, the Q_B values, and the listed capacitances. Enter your results in the Process 2 parameter list.

Q6. For the n-channel Process 1, determine the body-effect factors M_G and M_F and the gate and field thresholds with and without substrate bias. What is the long-term influence on the threshold of the ionic contaminant $Q_{Na^+}/q = 1.5 \times 10^{11}$? Assume aluminium gate processing.

Q7. For the CMOS Process 2 determine all listed body-effect factors and threshold voltages. Assume aluminium gate processing.

Q8. For the CMOS Process 2 determine L' for p-channel and for n-channel devices if each device is separately biased into saturated operation ($V_{GS} = V_{DS}$) using a 12 V supply. Find the current I_{DS} in each device assuming that the β' data provided in Appendix K is appropriate to this process. Assume $W_{eff} = 10$ mil and $L_{eff} = 0.24$ mil under non-saturated conditions.

Q9. Use the data given below to develop a curve of I_{DS} versus V_{DS} at $V_{GS} = +5$ V and at $V_{GS} = +15$ V for an n-channel MOS transistor. Use the L' effect on the effective W/L to define a finite slope in the saturated region and give a 'linear resistance' approximation to the slope in each case:

$N_B = 1.6 \times 10^{15}$ $\quad$ $\beta'_{(ave)} = 6.0 \times 10^{-6}$

$t_{ox} = 1500$ Å $\quad$ $W_{eff} = 1.2$ mil

$V_{TO} = -0.5$ V $\quad$ $L_{eff} = 0.3$ mil (full length)

$V_{BS} = -5$ V $\quad$ $2\phi_F = 0.6$ V

Q10.

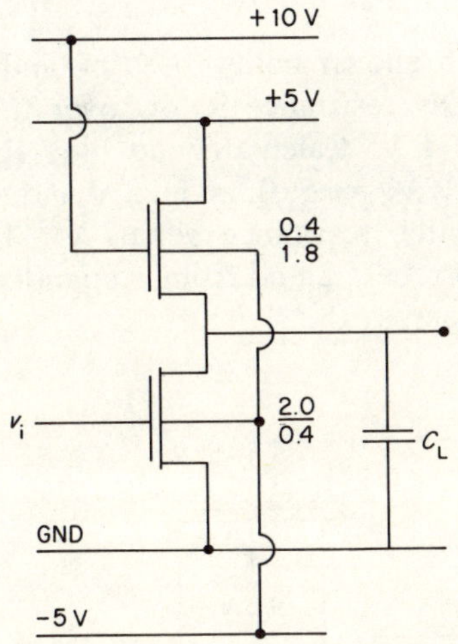

Parameters for the circuit shown can be drawn from the n-channel Process 1 data. Assume the given aspect ratios W/L with $\Delta L =$ 0.1 mil and use β' information from Appendix K. Use the static transfer characteristics and iterative methods to find the trigger level $V_{tr} = v_i = v_o$.

Q11. The circuit in Q10 is driven by a square-wave input with levels of 0 and +5 V. Find the output rise and fall times between +1.0 and +4.0 V if the output-node capacitance is 1.5 pF and the β' data of Appendix K (n-channel) are appropriate.

Q12.

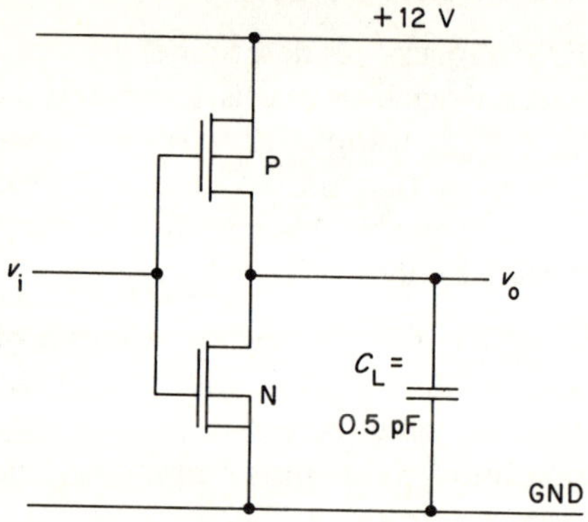

The simple CMOS inverter shown has devices with $(W/L)_{eff} = 0.4/0.2$ (in mils) in non-saturated operation. For step input changes between 0 and +12 V, find the time required to charge and discharge the output-node capacitance between the limits of +2 and +10 V. Use data from the CMOS Process 2 and the β' information from Appendix K.

Q13. The analog switch shown below uses p- and n-channel devices in parallel so that ON resistance is low over the entire analog signal range of +5 to −5 V. Calculate and plot R_{on} for each device at analog signal levels $V_S = +5$, 0, and −5 V and then calculate and plot the composite switch resistance versus V_S. Take appropriate data from the CMOS Process 2 and from Appendix K. Assume a negligible voltage drop across the switch.

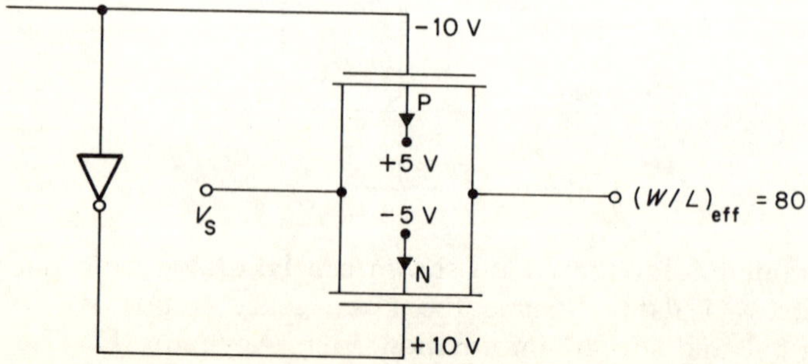

Process 1	N-channel
Parameters	Value
Bulk resistivity $\bar{\rho}_B(\Omega \cdot \text{cm})$	10
x_j (μm)	1.7
Sheet resistance $R_s(\Omega/\square]$	50
ϕ_F	
$t_{ox(G)}$(Å)	1500
$t_{ox(F)}$(Å)	14,000
Q_{ss}/q (charges/cm^2)	1.1×10^{11}
$C_{ox(G)}$ (fF/mil^2)	
C_{pn-5V} (fF/mil^2)	
C_{pn-10V} (fF/mil^2)	
V_{TOG}	
V_{TOF}	
M_G	
M_F	
V_{TG-5V}	
V_{TF-5V}	
N_s	
N_B	

Process 2		CMOS	
Parameters	P-channel	Parameters	N-channel
Wafer resistivity	3 Ω · cm	Tub x_j	10 μm
N_B		Tub R_s	3000 Ω/□
		Tub N_s	
		Tub N(2.5 μm)	
p-diffusion x_j	2.0 μm	n-diffusional x_j	2.5 μm
p-diffusion R_s	50 Ω/□	n-diffusion R_s	5 Ω/□
p-diffusion N_s		n-diffusion N_s	
ϕ_{fn}		ϕ_{fp}	
Q_{ss}/q	1.25×10^{11}	Q_{ss}/q	1.25×10^{11}
Q_B		Q_B	
t_{oxG}	1200 Å	t_{oxG}	1200 Å
V_{TOG}		V_{TOG}	
M		M	
V_{TG-12V}		V_{TG-12V}	
L'		L'	
$C_{MOS(G)}$		$C_{MOS(G)}$	
C_{pn-OV}		C_{pn-OV}	
C_{pn-12V}		C_{pn-12V}	
		C_{tub-OV}	
		$C_{tub-12V}$	

SOLUTIONS TO PROBLEMS

Q1. (a) $8.15\times10^{11}\ \Omega$, (b) $5.81\times10^{10}\ \Omega$, (c) 14.17 MΩ, (d) 14.17 MΩ.

Q2. $\psi_0=0.803$ V. By the single-sided step-junction approximation, $X_p=1.02\ \mu$m, $X_n=1.02\times10^{-4}\ \mu$m, $C_{-0V}=65.5$ fF/mil^2, $C_{-15V}=14.7$ fF/mil^2.

Q3. $t_d=406$ sec, $N_s=2.36\times10^{20}$ atoms/cm^3, $N(0.5\ \mu\text{m})=7.85\times10^{19}$ atoms/cm^3. (Bulk doping has a negligible influence on these figures.)

Q4. Answers tabulated on page 205.

Q5. Answers tabulated on page 205.

Q6. Answers tabulated on page 205. Positive gate field causes Q_{Na^+} to drift down to the silicon dioxide–silicon interface thereby maximizing its effect. V_{TG} can drift in the negative direction by up to 1.04 V. Depletion mode operation is marginally possible.

Q7. Answers tabulated on page 205.

Q8. L' values on page 205. $I_{DS}=14.0$ mA (p-channel) and 37.7 mA (n-channel). The 'gain' β' is calculated as the average of estimates at both ends of the channel.

Q9. The following data points were determined:
$V_{GS}=+5$ V: 0.222 mA at 10 V on drain
0.297 mA at 20 V on drain giving 117 kΩ approx.
$V_{GS}=+15$ V: 2.0 mA at 12.08 V on drain
3.0 mA at 20 V on drain giving 7.9 kΩ approx.

Q10. $V_{tr}=1.90$ V.

Q11. Rise time = 120 nsec. Fall time = 4.5 nsec.

Q12. Discharge time = 4.0 nsec. Charge time = 12.0 nsec.

Q13.

Volts	p-ohms	n-ohms	Combined ohms
+5	222	1298	189
0	373	117	89
−5	2445	59	57

Process 1	N-channel
Parameters	Value
Bulk resistivity $\bar{\rho}_B(\Omega \cdot cm)$	10
x_j (μm)	1.7
Sheet resistance $R_s(\Omega/\square)$	50
ϕ_F	0.295 V
$t_{ox(G)}$ (Å)	1500
$t_{ox(F)}$ (Å)	14,000
Q_{ss}/q (charges/cm^2)	1.1×10^{11}
$C_{ox(G)}$(fF/mil^2)	148
C_{pn-5V} (fF/mil^2)	27.1
C_{pn-10V} (fF/mil^2)	20.6
V_{TOG}	−0.38 V
V_{TOF}	−0.99 V
M_G	0.90
M_F	8.42
V_{TG-5V}	+1.05
V_{TF-5V}	+12.45
N_s	2.5×10^{19} atoms/cm^3
N_B	1.3×10^{15} atoms/cm^3

Process 2		CMOS	
Parameters	P-Channel	Parameters	N-Channel
Wafer resistivity	$3\ \Omega \cdot cm$	Tub x_j	10 μm
N_B	1.7×10^{15}	Tub R_s	3000 $\Omega/\square$
		Tub N_s (nett)	7.8×10^{15}
		Tub N(2.5 μn) (nett)	6.8×10^{15}
p-diffusion x_j	2.0 μm	n-diffusion x_j	2.5 μm
p-diffusion R_s	50 $\Omega/\square$	n-diffusion R_s	5 $\Omega/\square$
p-diffusion N_s	3.8×10^{19}	n-diffusion N_s	3.5×10^{20}
ϕ_{fn}	0.30	ϕ_{fp}	0.337
Q_{ss}/q	1.25×10^{11}	Q_{ss}/q	1.25×10^{11}
Q_B (C/cm^2)	1.84×10^{-8}	Q_B (C/cm^2)	3.9×10^{-8}
t_{oxG}	1200 Å	t_{oxG}	1200 Å
V_{TOG}	−2.24 V	V_{TOG}	0.41 V
M	−0.825	M	1.65
V_{TG-12V}	−4.53 V	V_{TG-12V}	4.92 V
L' (μm)	1.74	L' (μm)	0.86
$C_{MOS(G)}$(fF/mil^2)	185	$C_{MOS(G)}$(fF/mil^2)	185
C_{pn-OV}(fF/mil^2)	83.8	C_{pn-OV}(fF/mil^2)	161.0
C_{pn-12V}(fF/mil^2)	21.3	C_{pn-12V}(fF/mil^2)	39.0
		C_{tub-OV}(fF/mil^2)	27.4
		$C_{tub-12V}$(fF/mil^2)	13.5

APPENDIX A

Average Conductivity of Diffused Layers (after J. C. Irvin)

The set of graphical data presented here gives average conductivity values for layers extending from a depth x to the semi-conductor junction at x_j (see

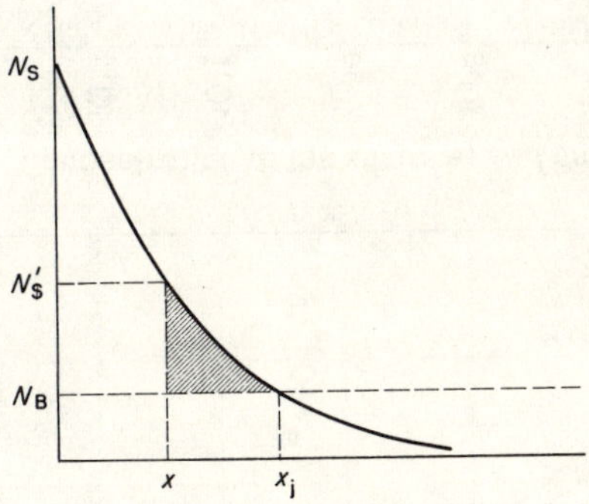

Fig. A.1.

Fig. A.1). Then:

$$R_s = \frac{1}{\bar{\sigma}(x_j - x)}$$

is the sheet resistance of the layer with the top slice removed. A convenient utilization guide follows.

Select diffusion profile:	Erfc—Figs A.2 and A.3 Gaussian—Figs A.4 and A.5
Select dopant type:	N-type diffusion—Figs A.2 and A.4 P-type diffusion—Figs A.3 and A.5
Select N_B:	Graphs are given for $N_B = 10^{14}$, 10^{15}, 10^{16}, 10^{17}. Use the graph offering the closest N_B value on a log scale.
Parameters are:	N_s, $\bar{\sigma}$, x/x_j. Use any two to find the third.

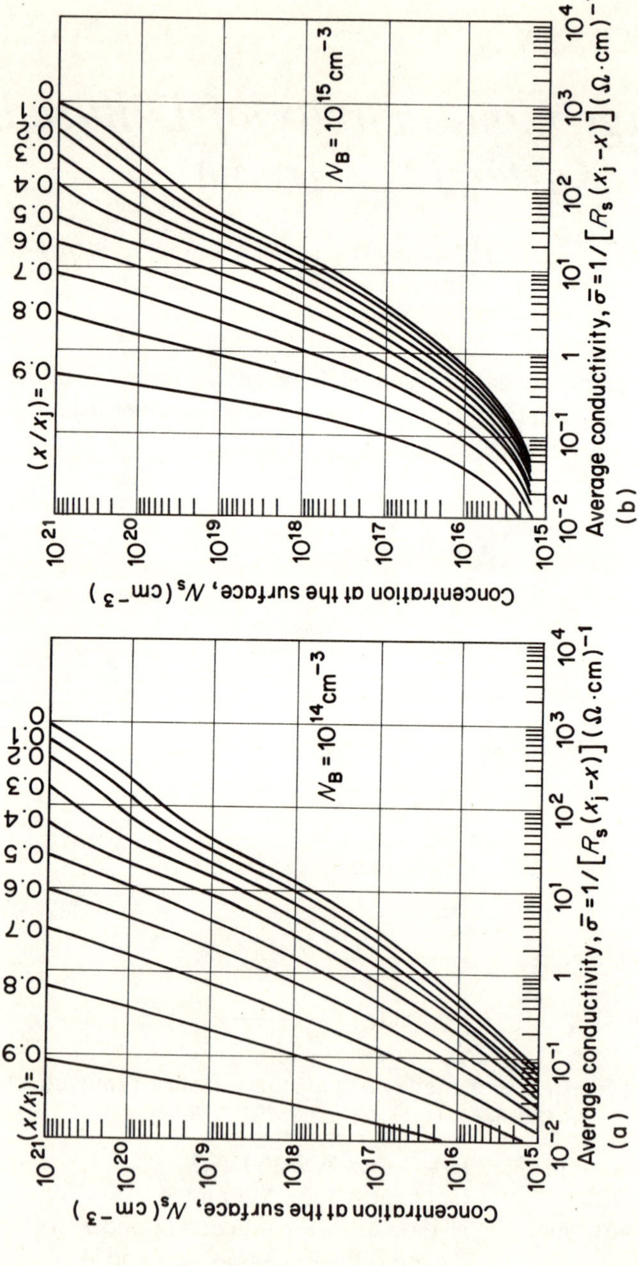

Concentration at the surface, N_s (cm^{-3})
Average conductivity, $\bar{\sigma} = 1/[R_s(x_j - x)]$ ($\Omega \cdot$cm)$^{-1}$
$(x/x_j) = 0.9$ 0.8 0.7 0.6 0.5 0.4 0.3 0.2 0.1 0
$N_B = 10^{14}$cm^{-3}
$N_B = 10^{15}$cm^{-3}
10^{15} 10^{16} 10^{17} 10^{18} 10^{19} 10^{20} 10^{21}
10^{-2} 10^{-1} 1 10^{1} 10^{2} 10^{3} 10^{4}
(a)
(b)

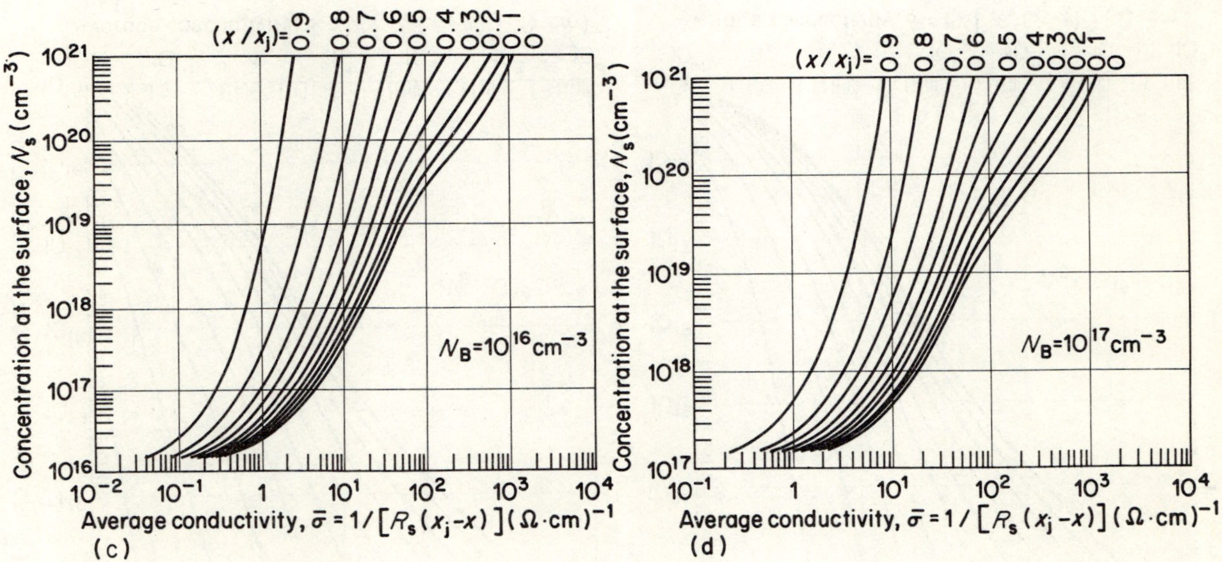

Fig. A.2. Average conductivity of n-type complementary error function layers in silicon.

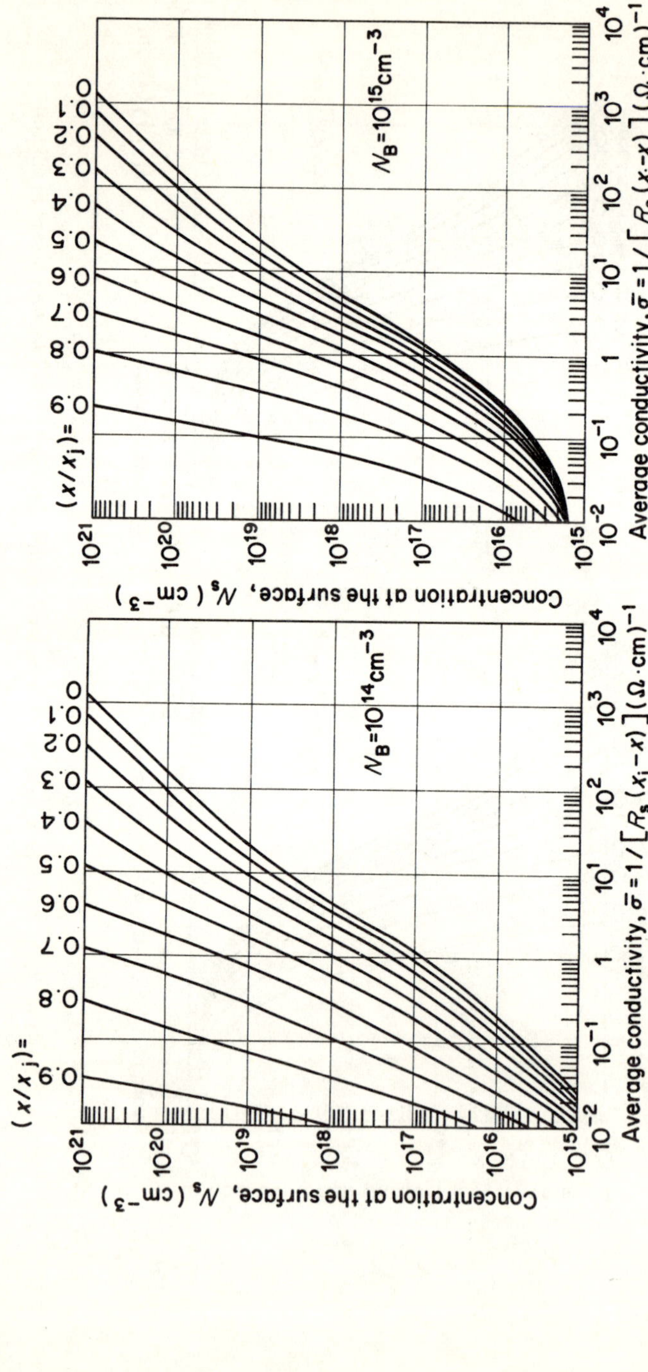
(a)
$N_B = 10^{14}\,cm^{-3}$
(x/x_j) =
0.9
0.8
0.7
0.6
0.5
0.4
0.3
0.2
0.1
0
Concentration at the surface, N_s (cm^{-3})
Average conductivity, $\bar{\sigma} = 1/[R_s(x_j - x)]$ $(\Omega \cdot cm)^{-1}$
(b)
$N_B = 10^{15}\,cm^{-3}$
Concentration at the surface, N_s (cm^{-3})
Average conductivity, $\bar{\sigma} = 1/[R_s(x_j - x)]$ $(\Omega \cdot cm)^{-1}$

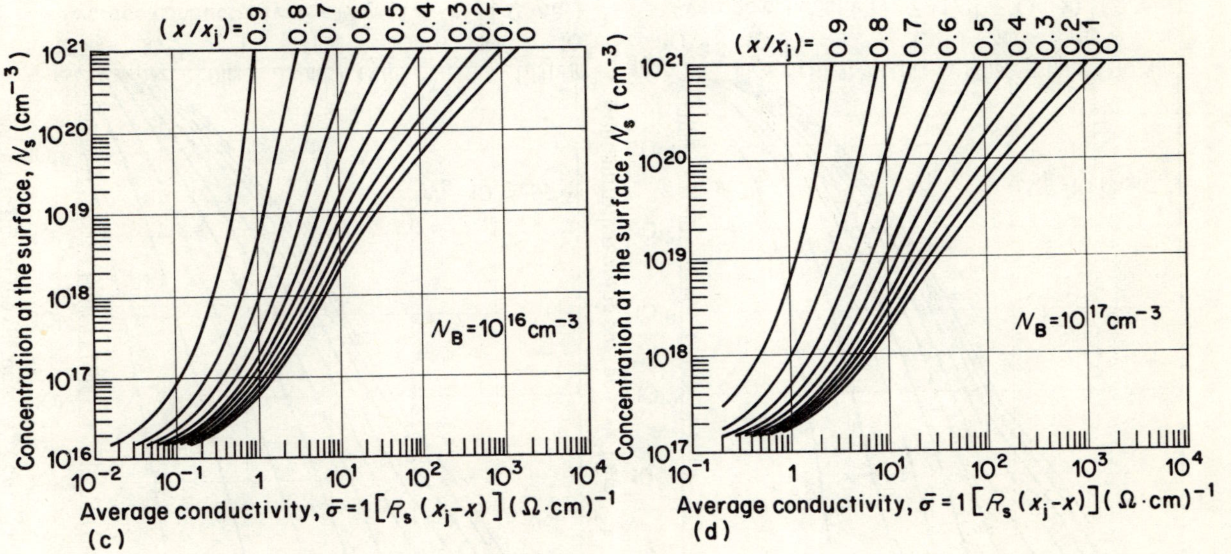

Fig. A.3. Average conductivity of p-type complementary error function layers in silicon.

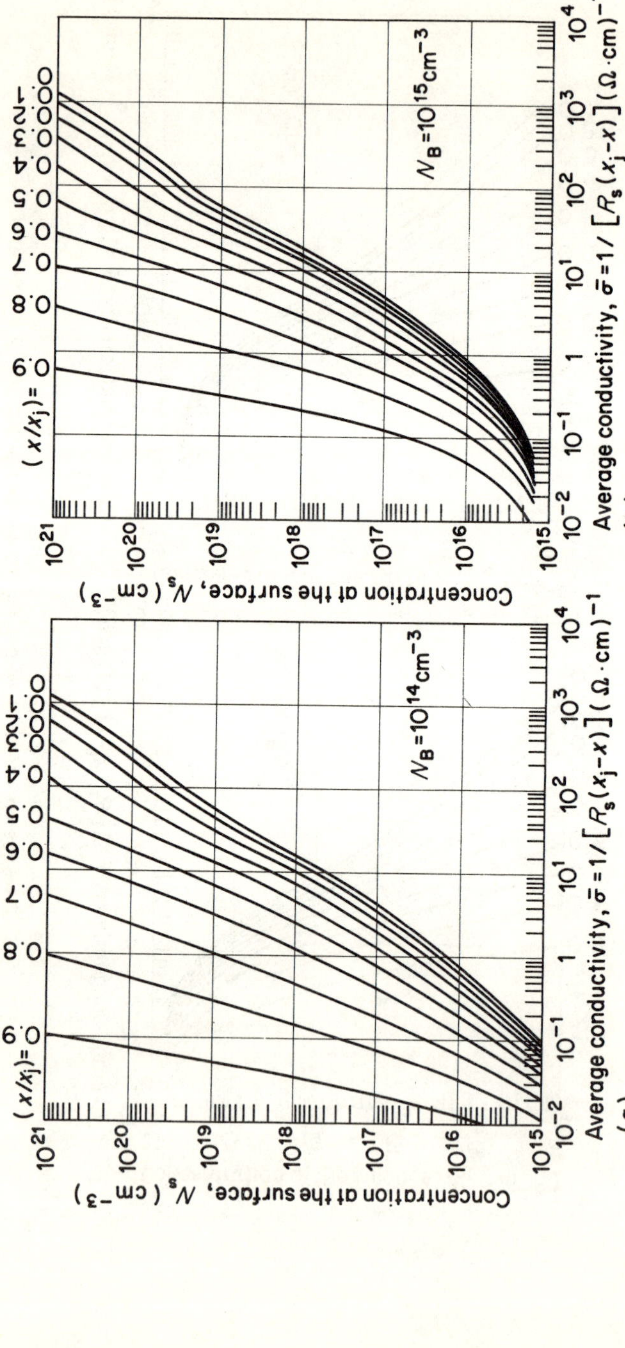

(x/xj)=0.9
0.8
0.7
0.6
0.5
0.4
0.3
0.2
0.1
0.0
NB=10^14 cm^-3
Concentration at the surface, Ns (cm^-3)
10^15
10^16
10^17
10^18
10^19
10^20
10^21
10^-2
10^-1
1
10^1
10^2
10^3
10^4
Average conductivity, σ̄=1/[Rs(xj-x)] (Ω·cm)^-1
(a)
(x/xj)=0.9
0.8
0.7
0.6
0.5
0.4
0.3
0.2
0.1
0.0
NB=10^15 cm^-3
Concentration at the surface, Ns (cm^-3)
10^15
10^16
10^17
10^18
10^19
10^20
10^21
10^-2
10^-1
1
10^1
10^2
10^3
10^4
Average conductivity, σ̄=1/[Rs(xj-x)] (Ω·cm)^-1
(b)

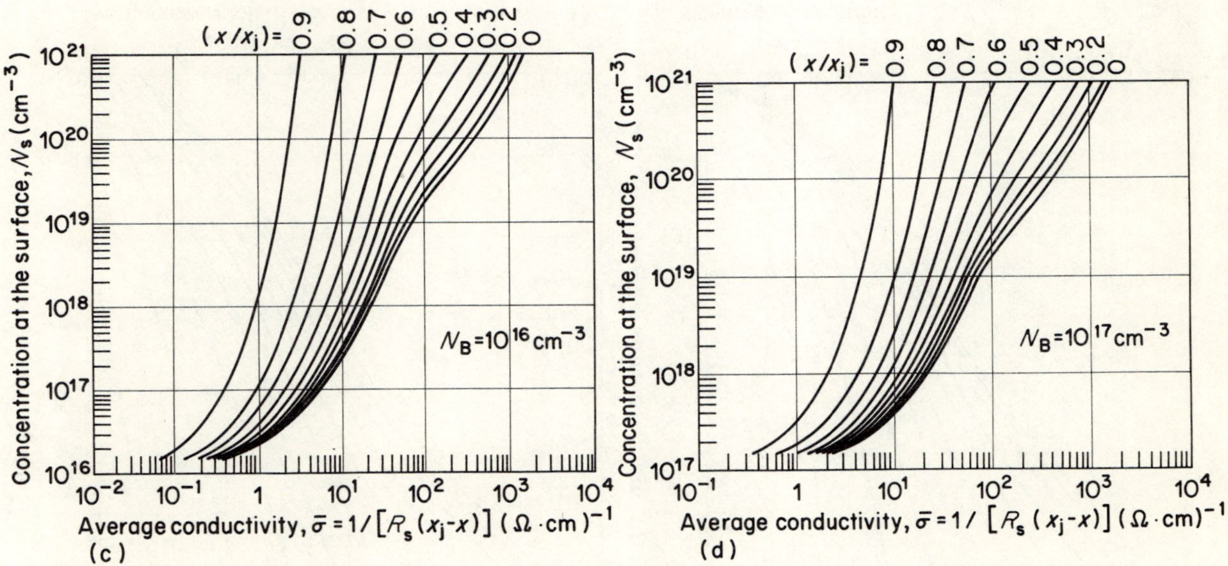

Fig. A.4. Average conductivity of n-type Gaussian layers in silicon.

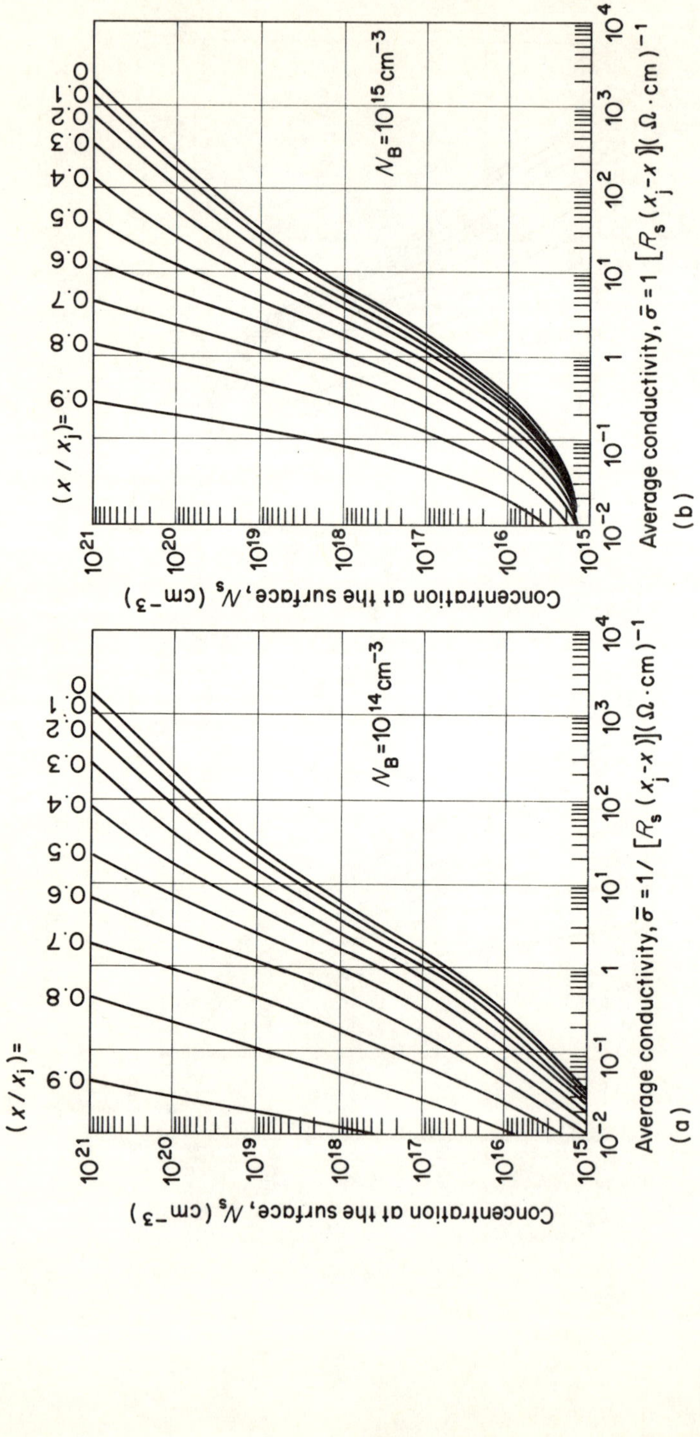
Concentration at the surface, N_s (cm^{-3})
(x/x_j)= 0.9 0.8 0.7 0.6 0.5 0.4 0.3 0.2 0.1 0
$N_B=10^{14}$cm^{-3}
Average conductivity, $\bar{\sigma}=1/[R_s(x_j-x)](\Omega\cdot$cm$)^{-1}$
(a)
Concentration at the surface, N_s (cm^{-3})
(x/x_j)= 0.9 0.8 0.7 0.6 0.5 0.4 0.3 0.2 0.1 0
$N_B=10^{15}$cm^{-3}
Average conductivity, $\bar{\sigma}=1[R_s(x_j-x)](\Omega\cdot$cm$)^{-1}$
(b)

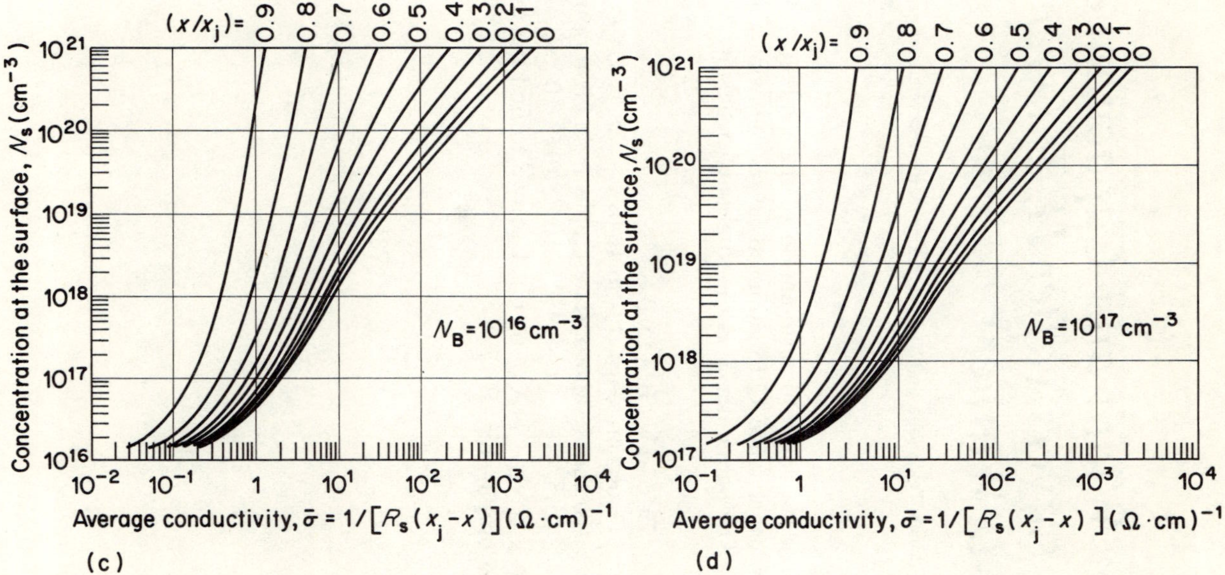

Fig. A.5. Average conductivity of p-type Gaussian layers in silicon.

APPENDIX B

Depletion Width and Capacitance of Diffused Layers (after H. Lawrence and R. M. Warner)

These curves provide accurate determination of the depletion layer width X_{mL} and junction capacitance C for diffused layers following the erfc and Gaussian profiles. The parameter V is the total barrier height. A utilization guide follows.

Select profile: Erfc—Figs B.1 to B.8
Gaussian—Figs B.9 to B.16
Select N_B/N_S: To choose appropriate graphs from eight options: 10^{-1}, 10^{-2}, 10^{-3}, 10^{-4}, 10^{-5}, 10^{-6}, 10^{-7}, 10^{-8}
Graph parameters: V/N_B, x_j, (X_{mL} or C). Use two to find the third.
Peak electric field graph—Fig. B.17

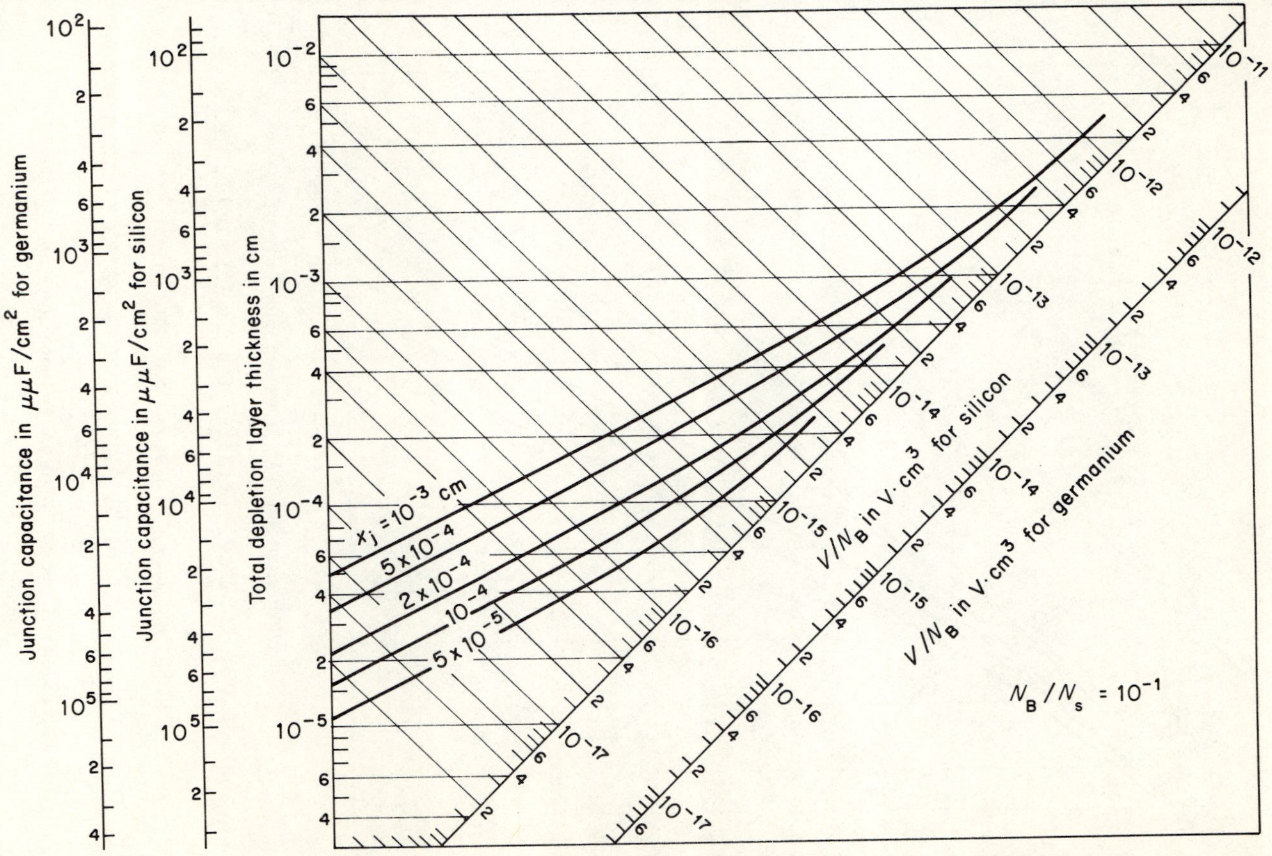

Fig. B.1. Chart for use in range 3×10^{-2} to 3×10^{-1}, erfc distribution.

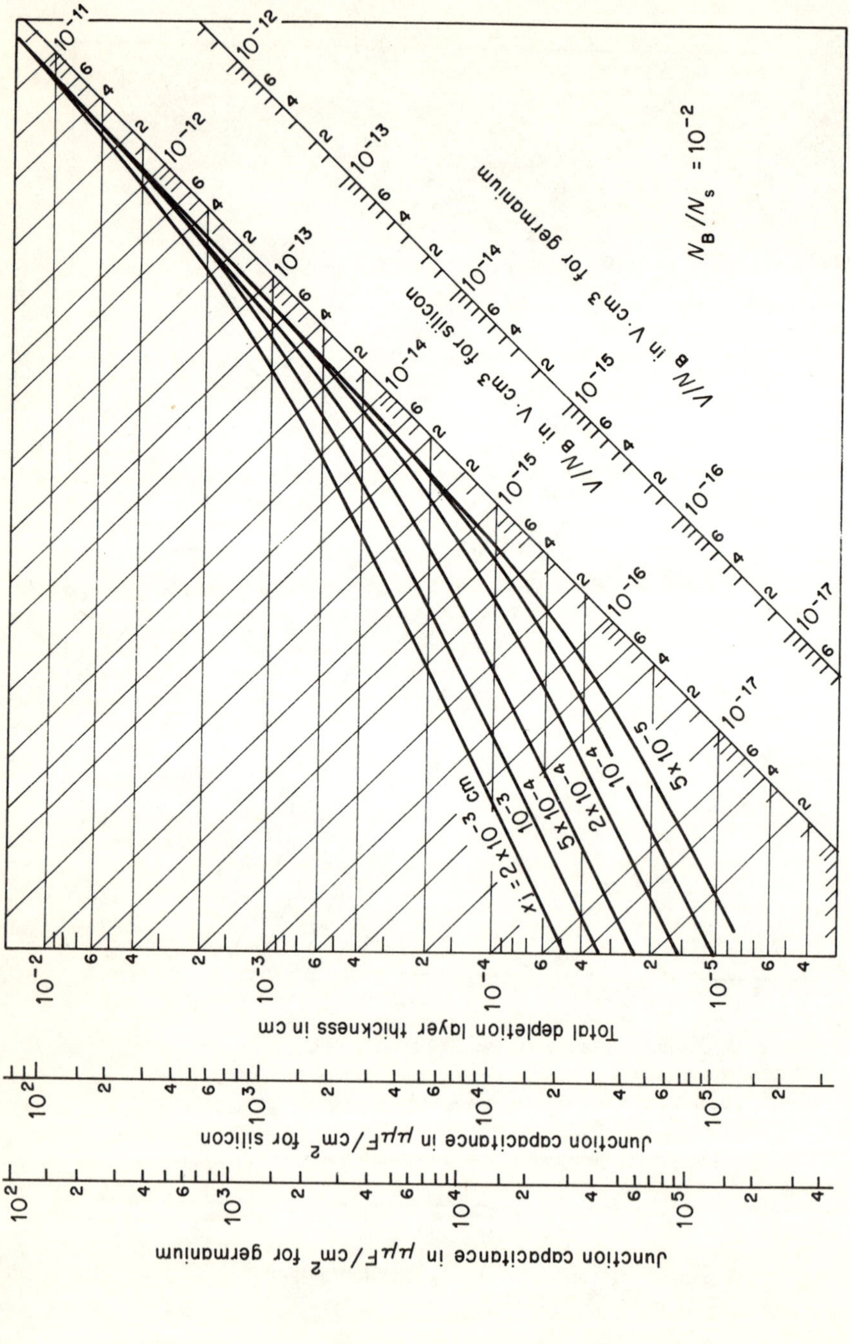

Fig. B.2. Chart for use in range 3×10^{-3} to 3×10^{-2}, erfc distribution.

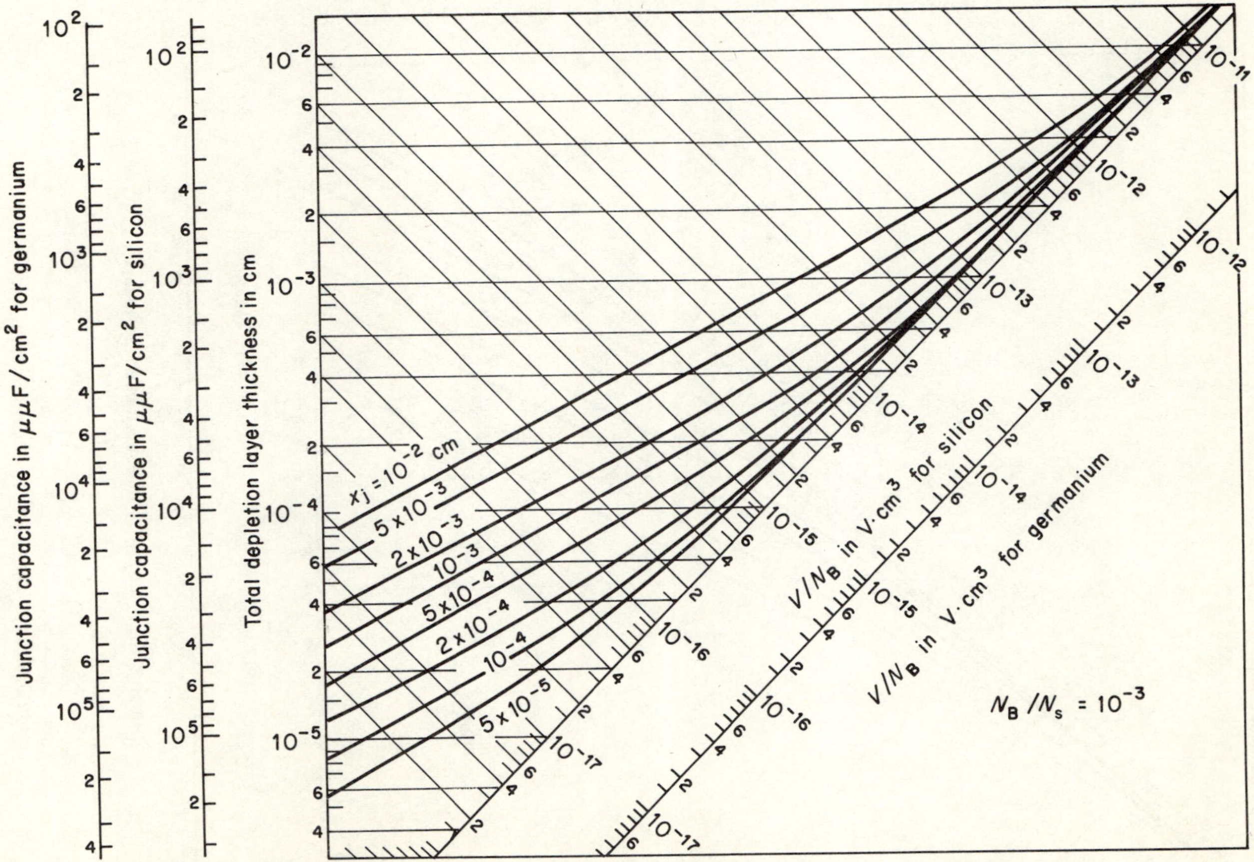

Fig. B.3. Chart for use in range 3×10^{-4} to 3×10^{-3}, erfc distribution.

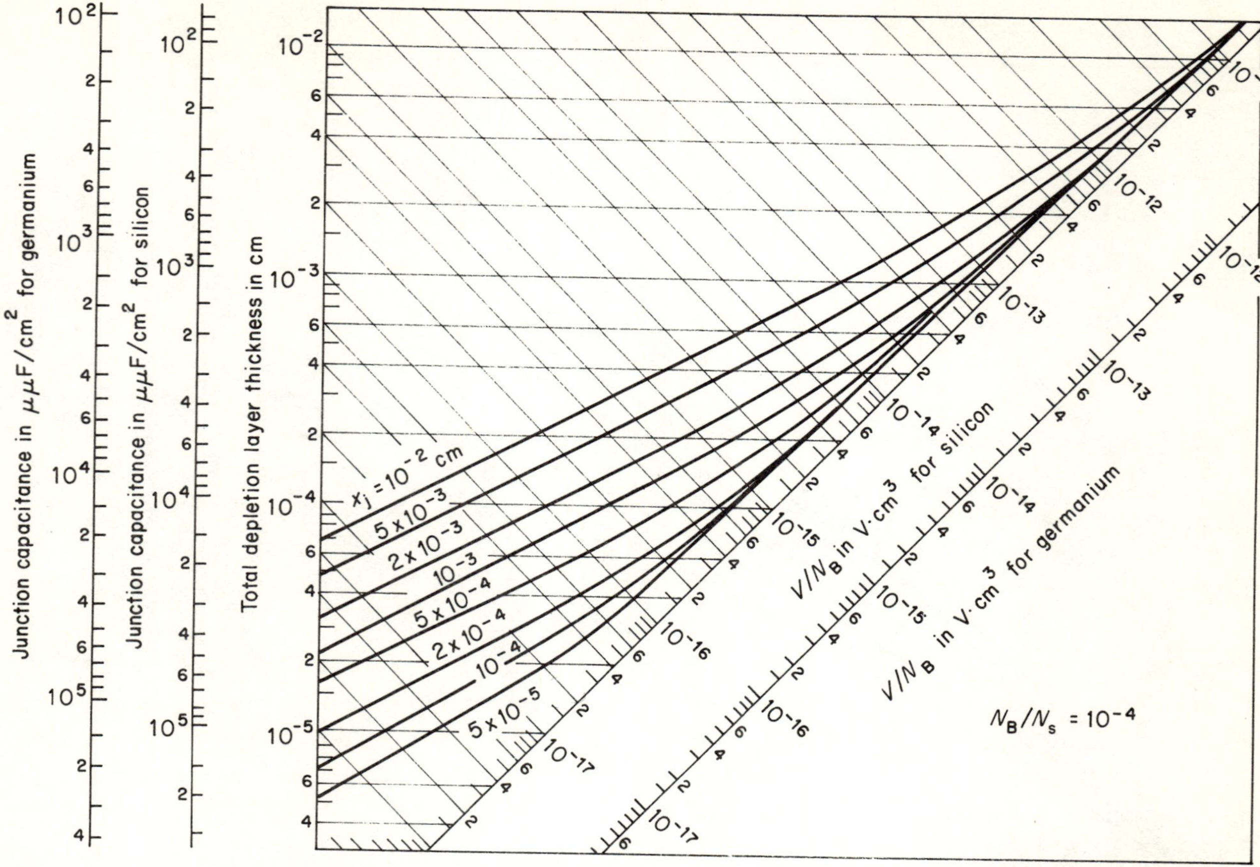

Fig. B.4. Chart for use in range 3×10^{-5} to 3×10^{-4}, erfc distribution.

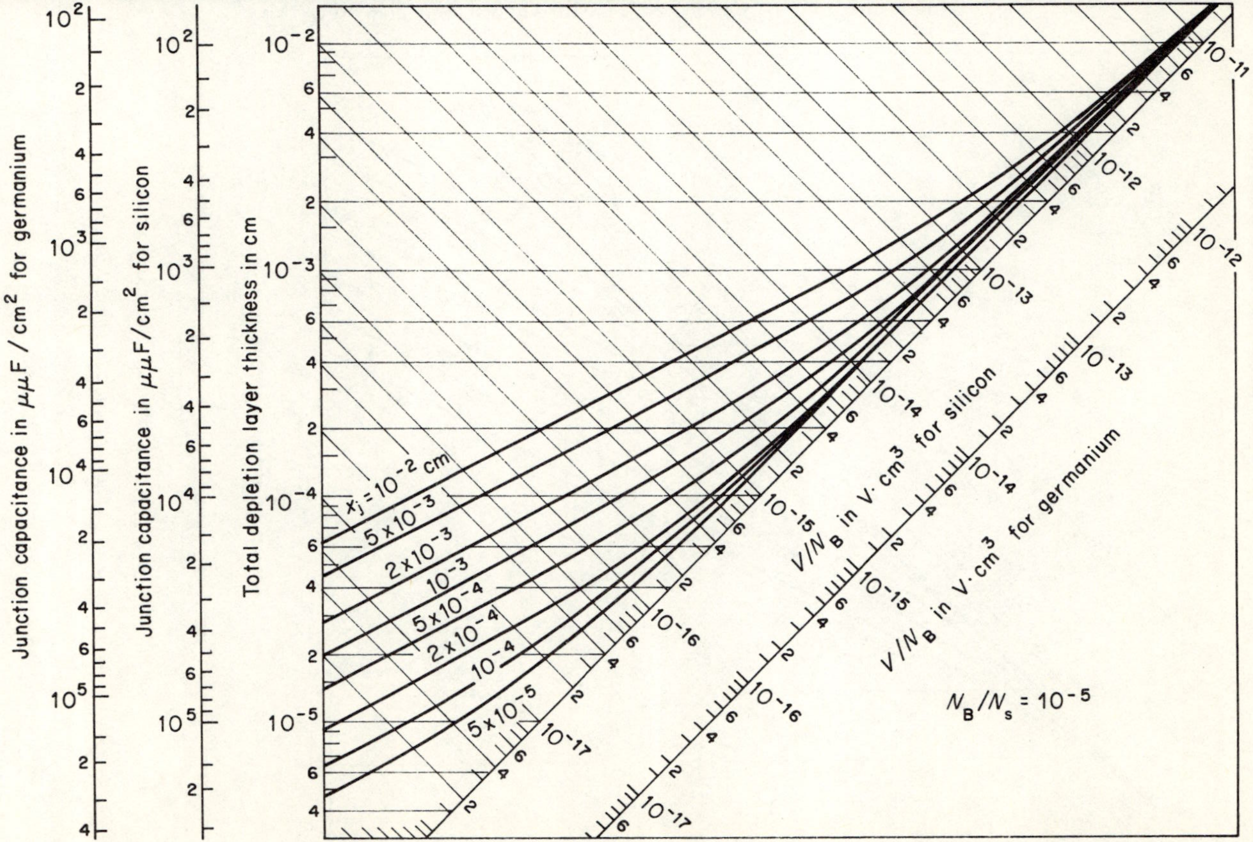

Fig. B.5. Chart for use in range 3×10^{-6} to 3×10^{-5}, erfc distribution.

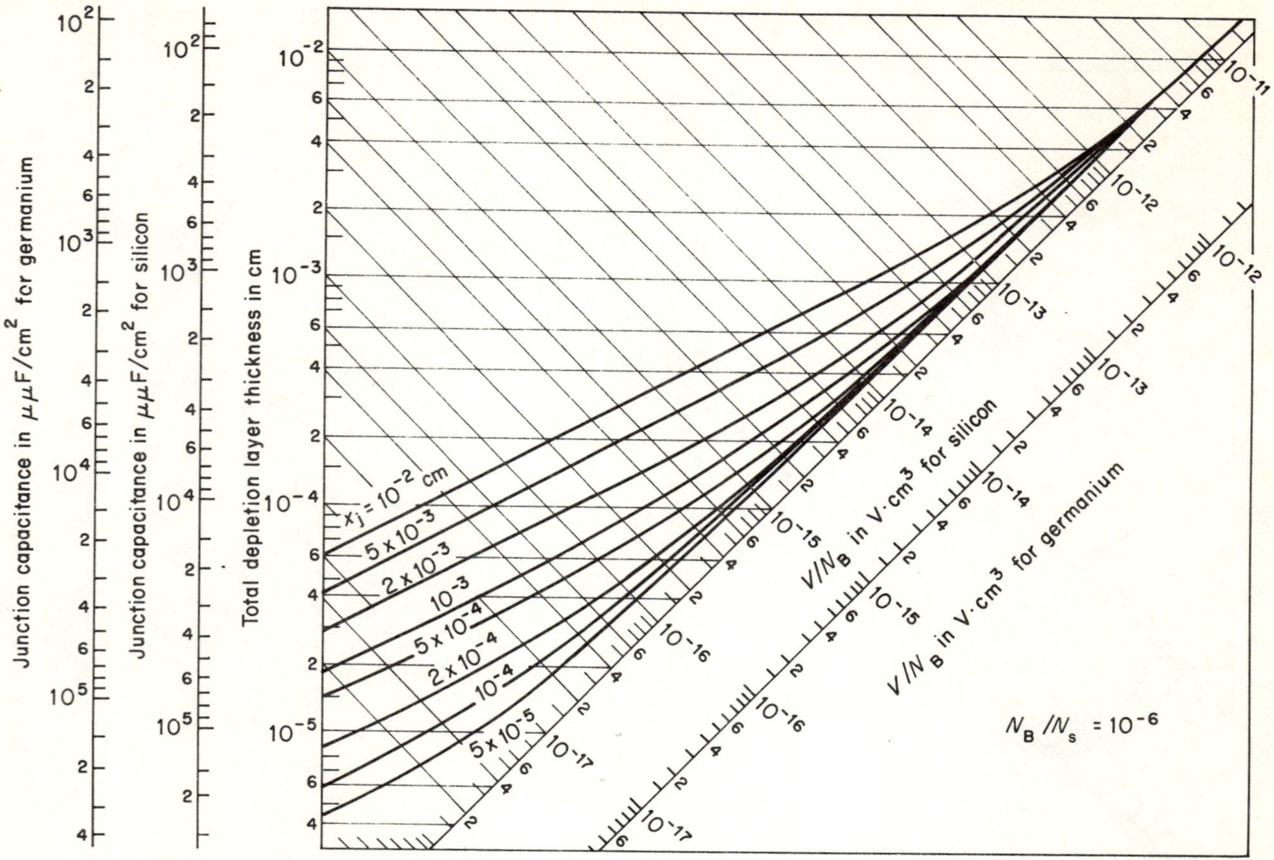

Fig. B.6. Chart for use in range 3×10^{-7} to 3×10^{-6}, erfc distribution.

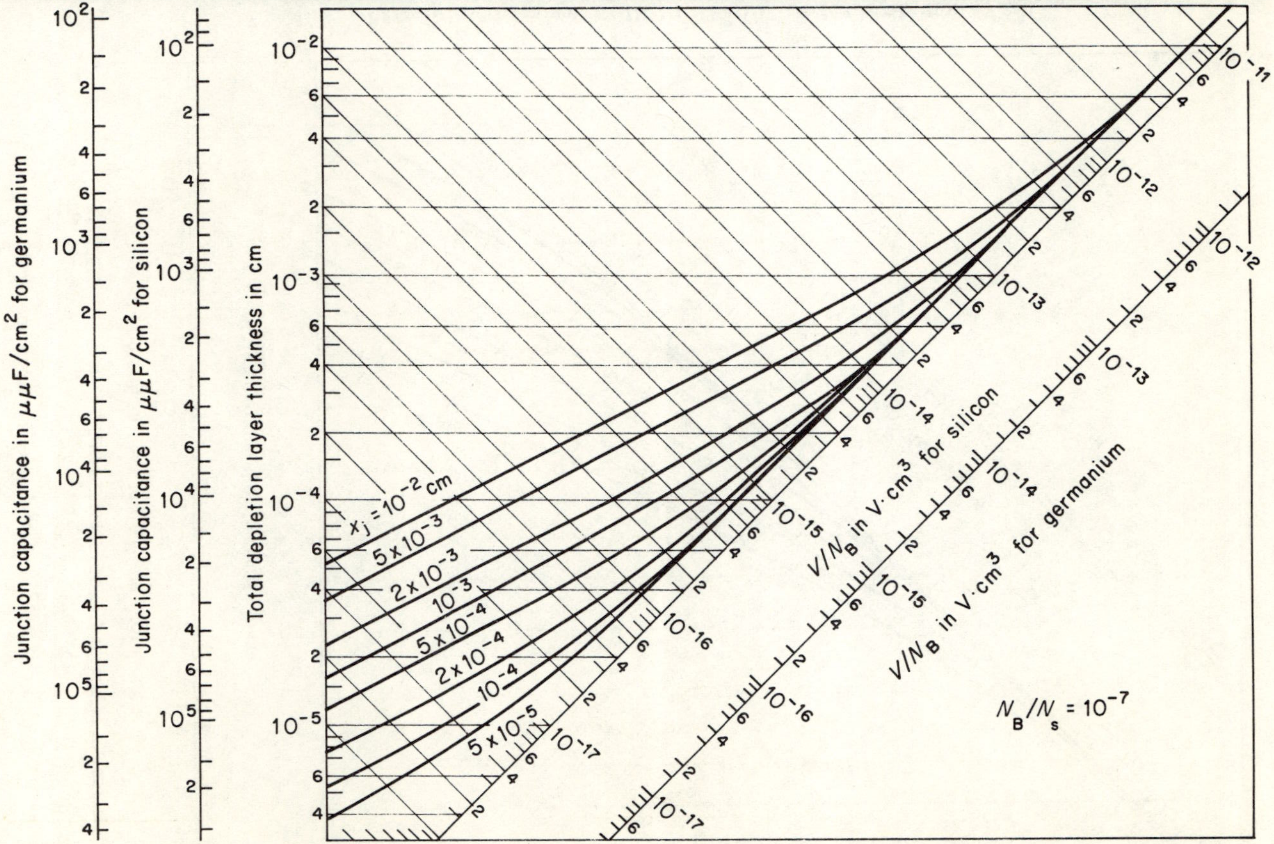

Fig. B.7. Chart for use in range 3×10^{-8} to 3×10^{-7}, erfc distribution.

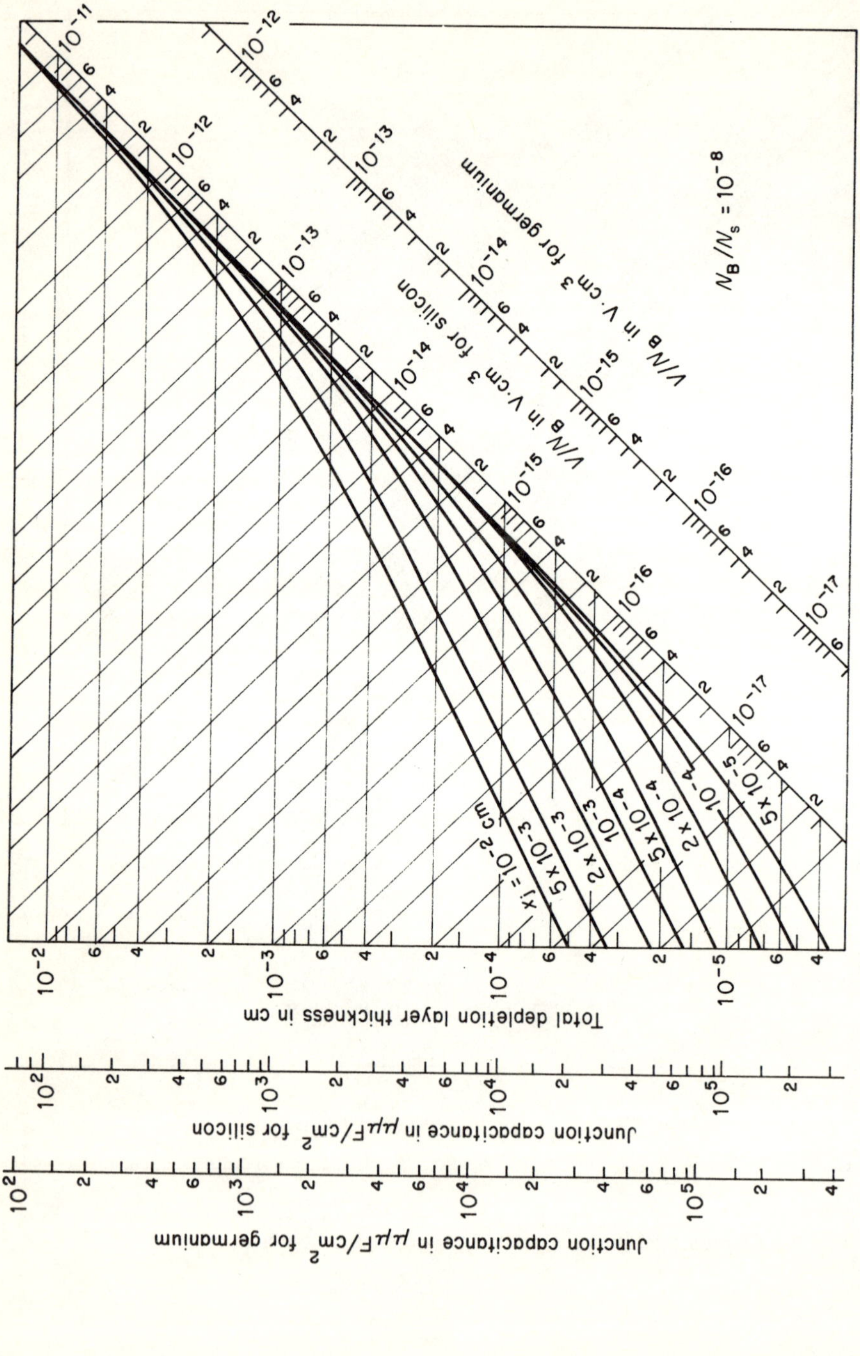

Fig. B.8. Chart for use in range 3×10^{-9} to 3×10^{-8}, erfc distribution.

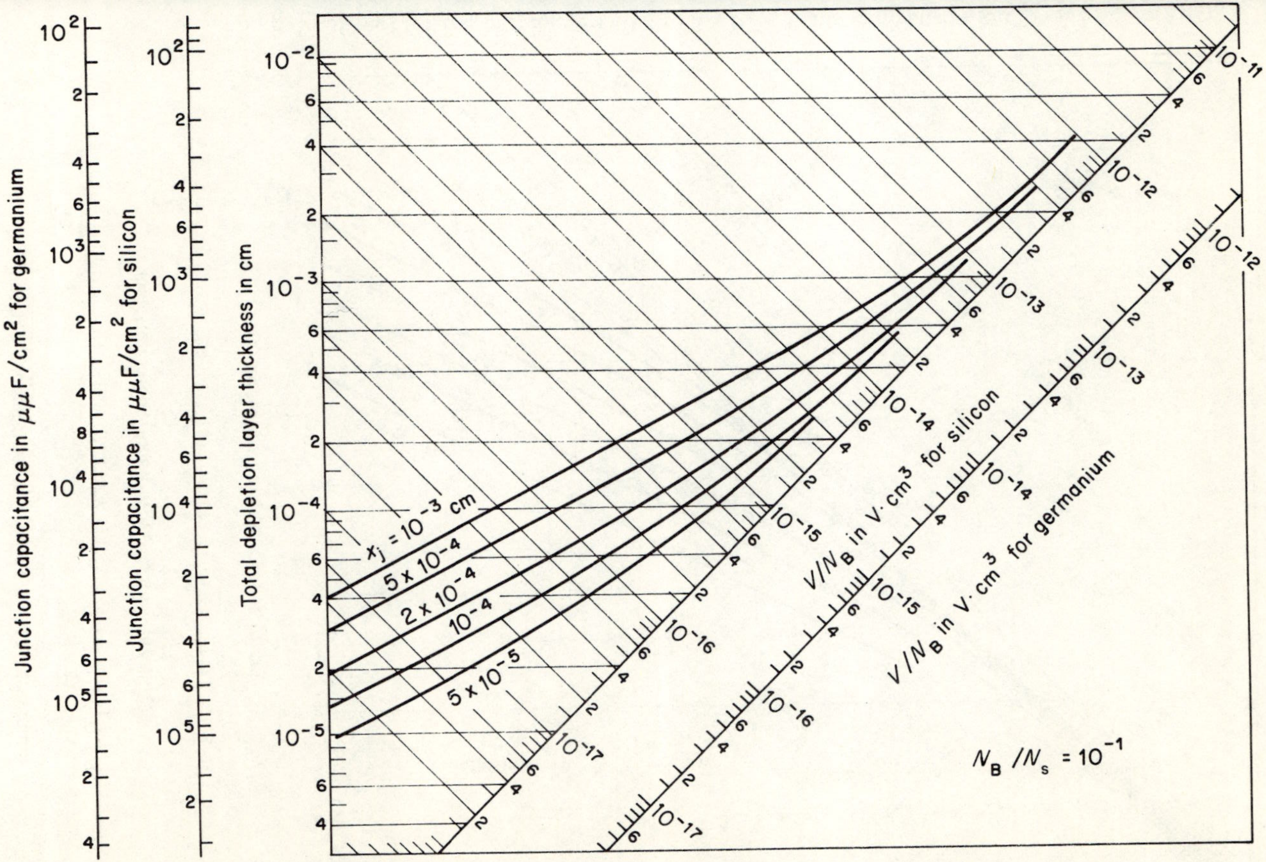

Fig. B.9. Chart for use in range 3×10^{-2} to 3×10^{-1}, Gaussian distribution.

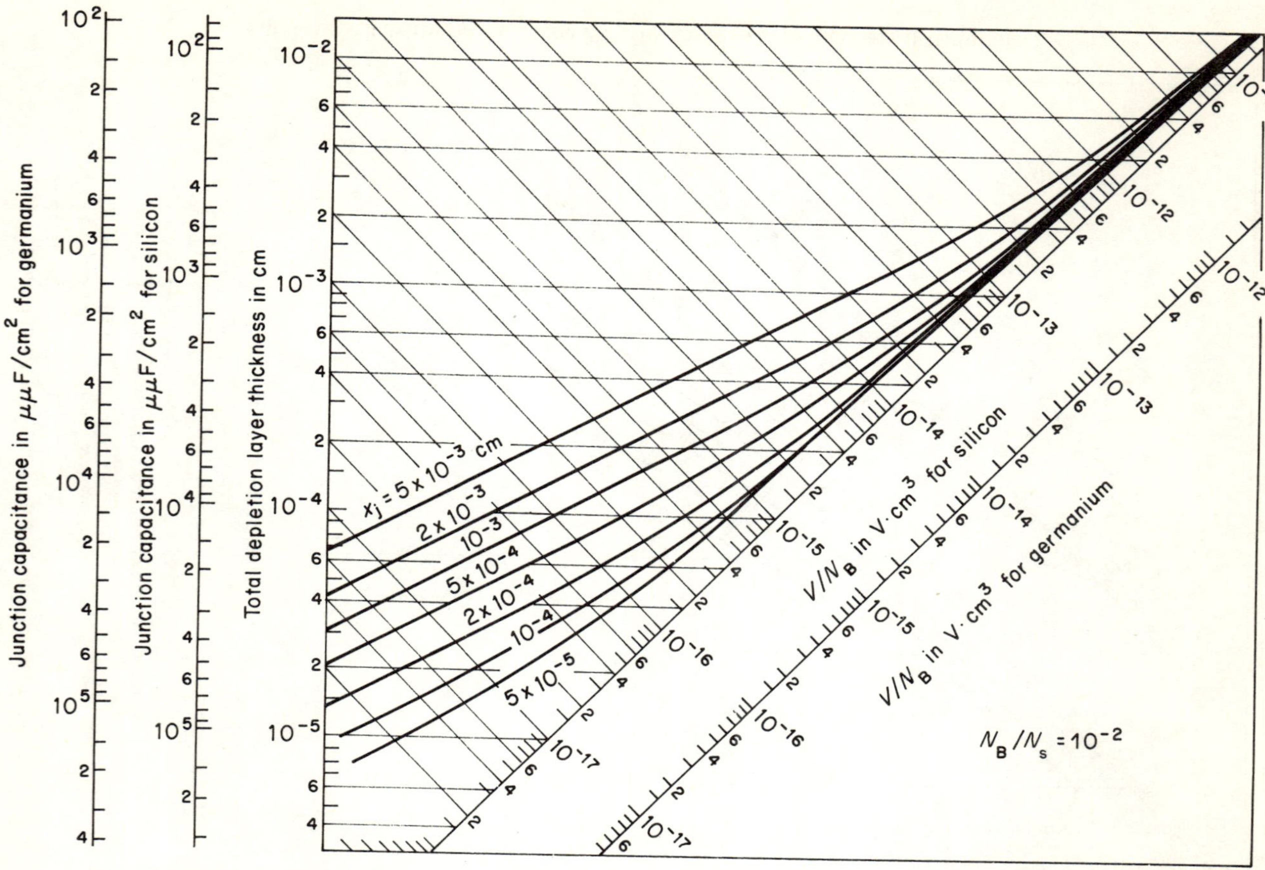

Fig. B.10. Chart for use in range 3×10^{-3} to 3×10^{-2}, Gaussian distribution.

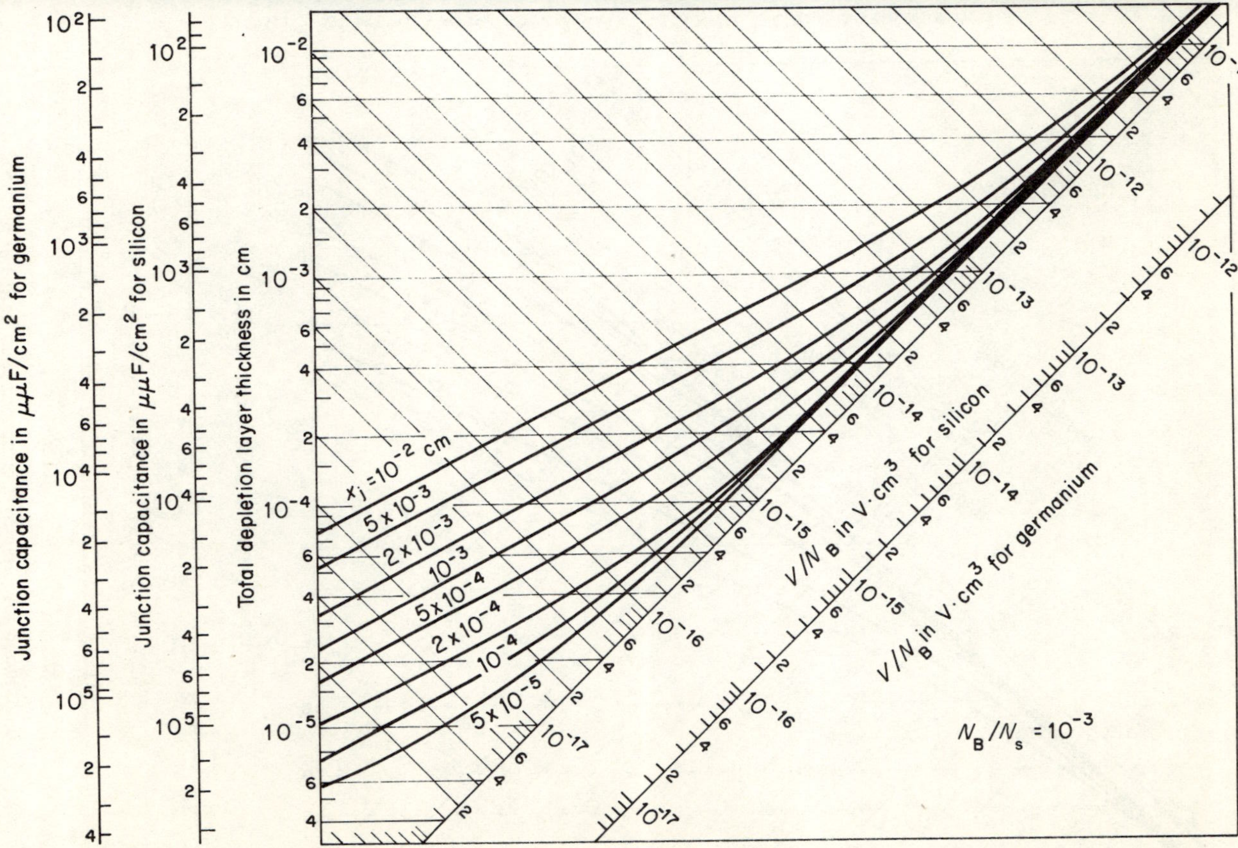

Fig. B.11. Chart for use in range 3×10^{-4} to 3×10^{-3}, Gaussian distribution.

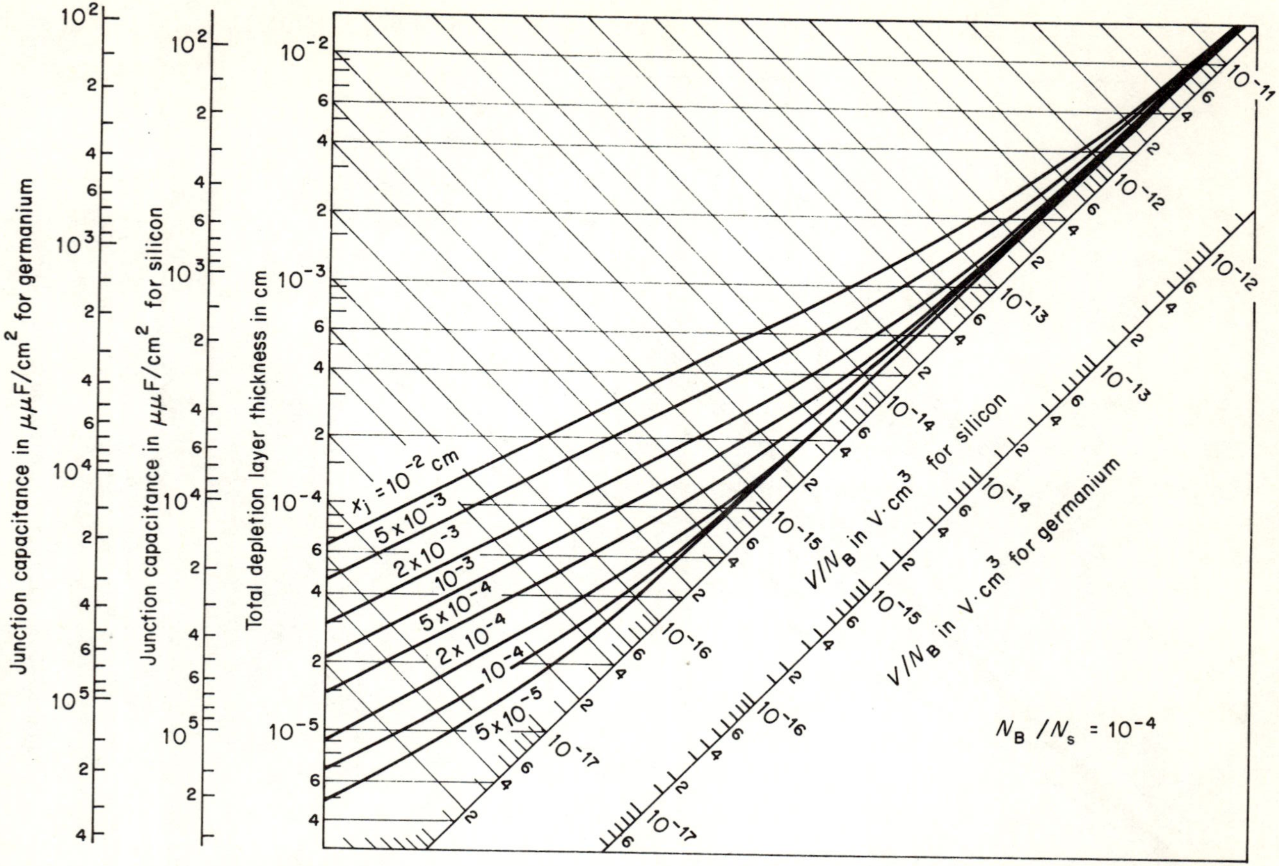

Fig. B.12. Chart for use in range 3×10^{-5} to 3×10^{-4}, Gaussian distribution.

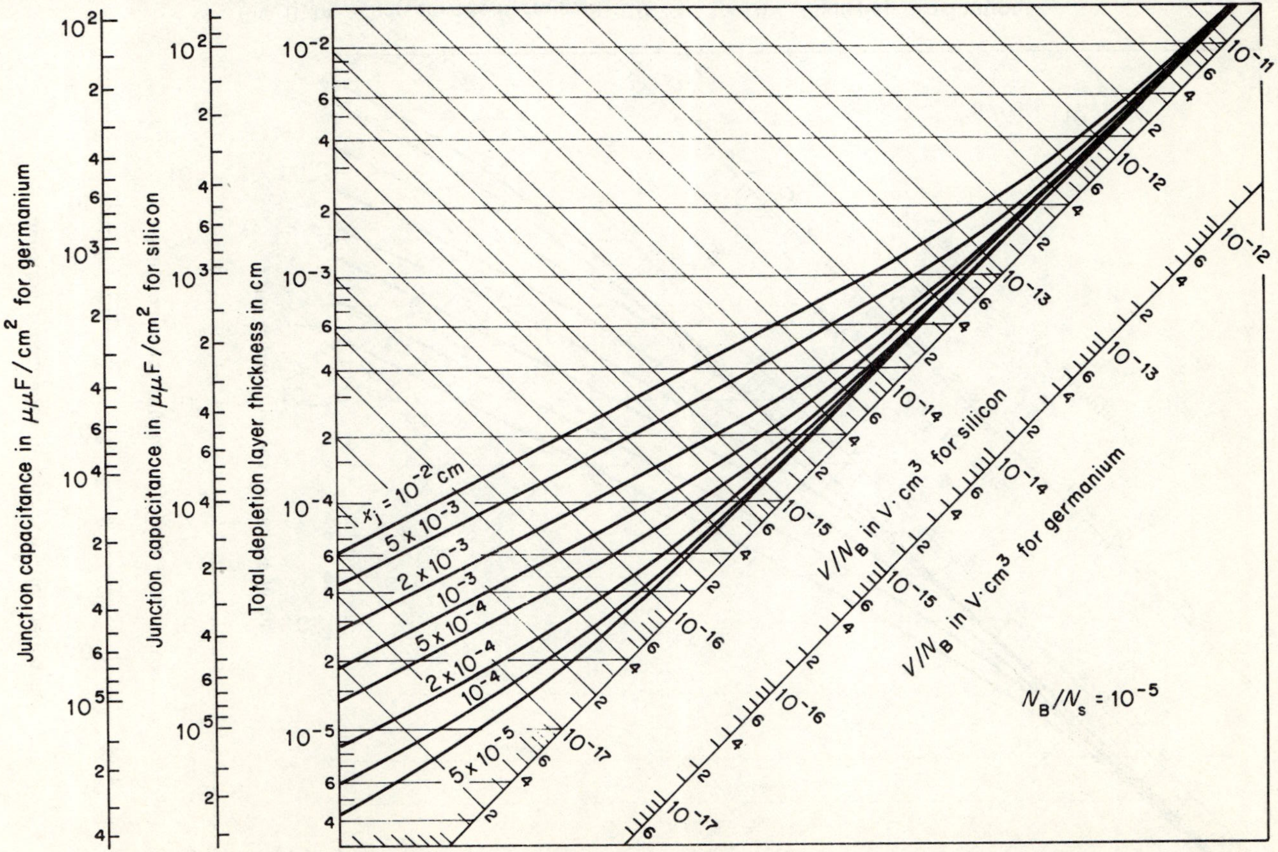

Fig. B.13. Chart for use in range 3×10^{-6} to 3×10^{-5}, Gaussian distribution.

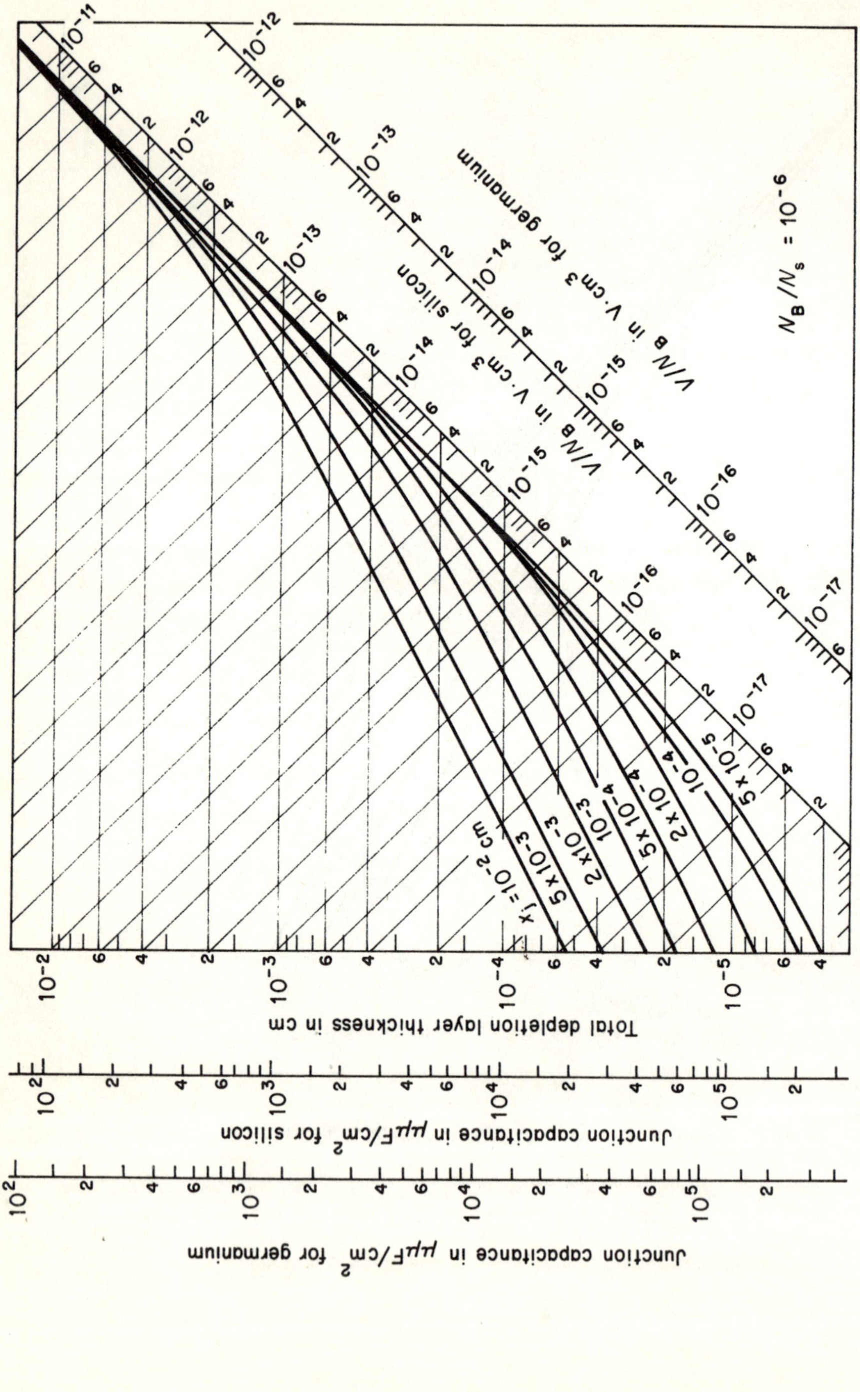

Fig. B.14. Chart for use in range 3×10^{-7} to 3×10^{-6}, Gaussian distribution.

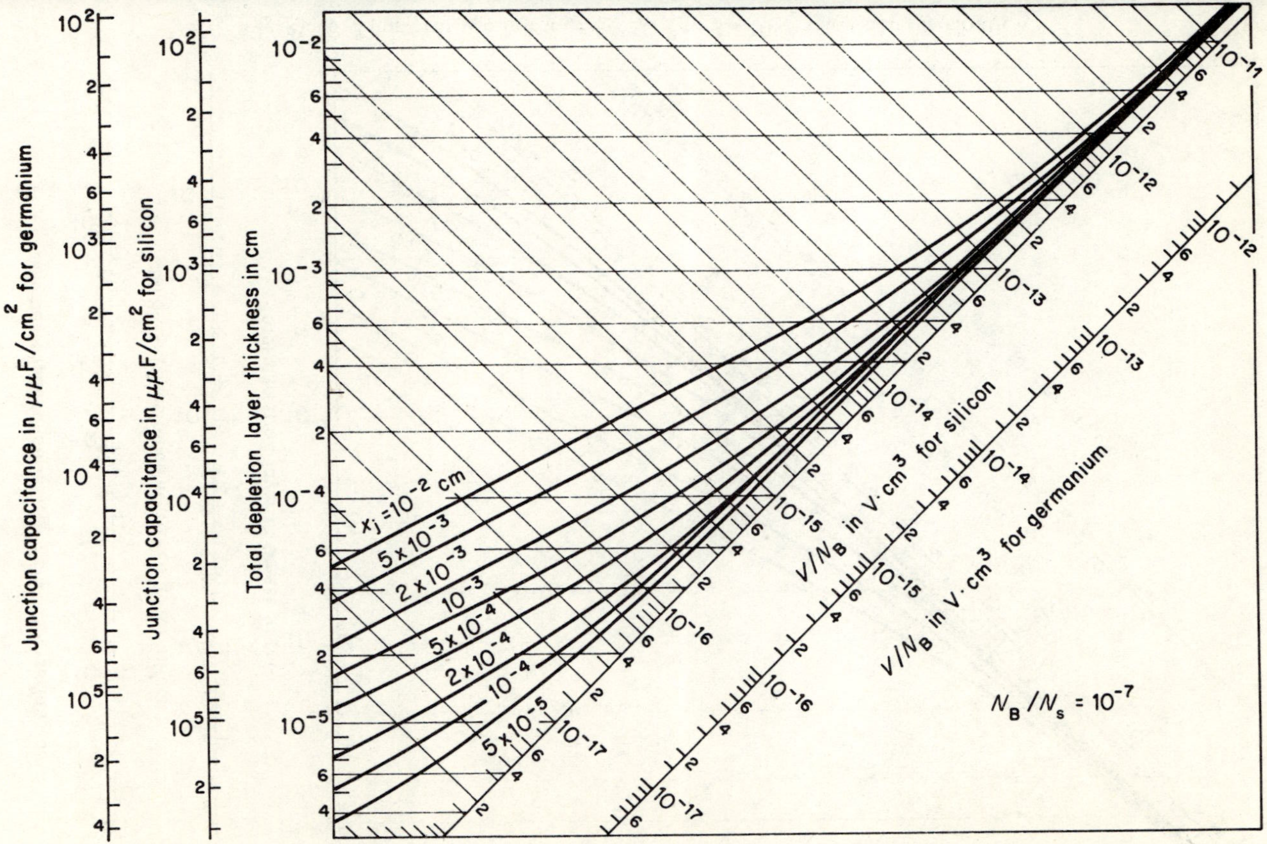

Fig. B.15. Chart for use in range 3×10^{-8} to 3×10^{-7}, Gaussian distribution.

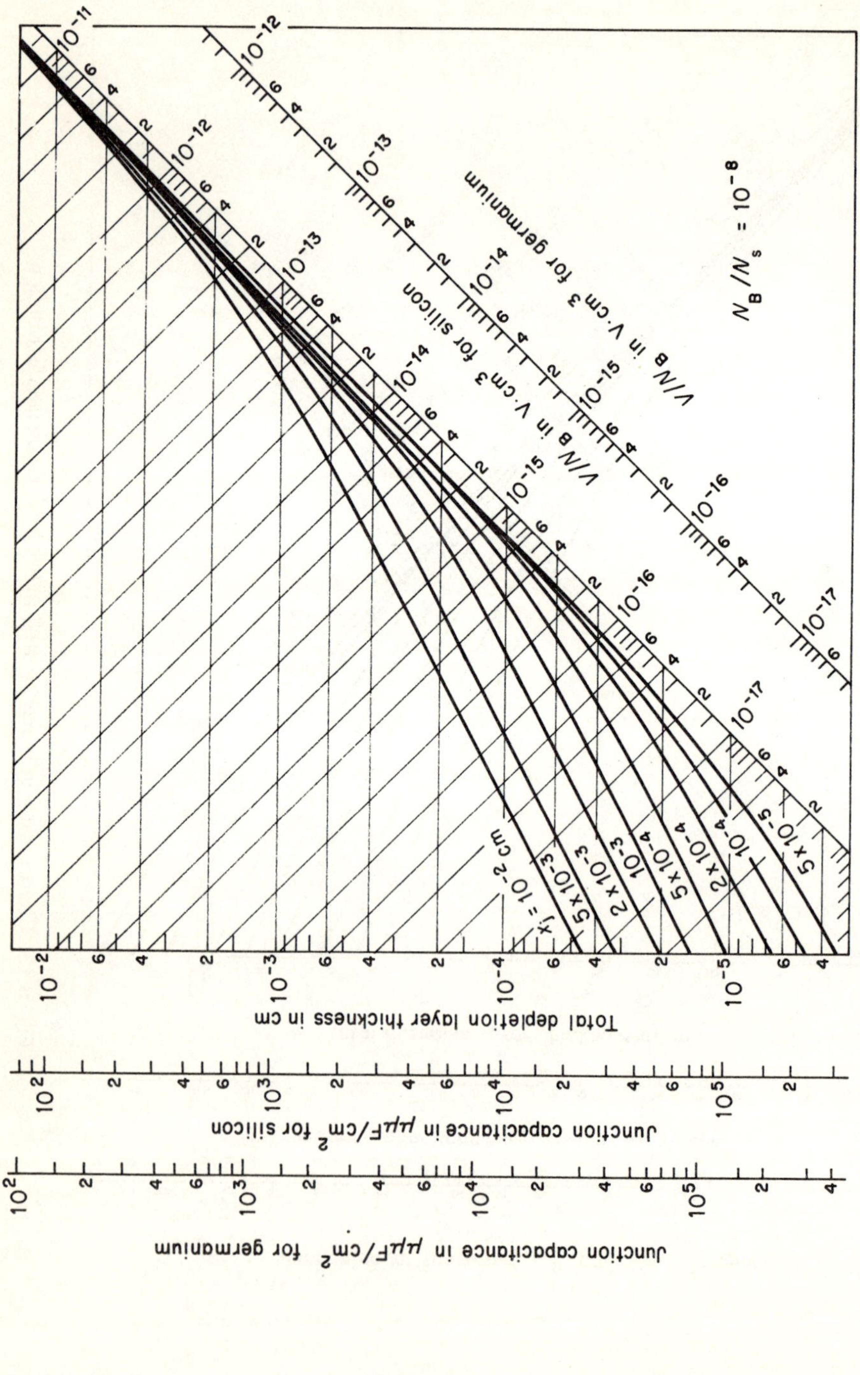

Fig. B.16. Chart for use in range 3×10^{-9} to 3×10^{-8}, Gaussian distribution.

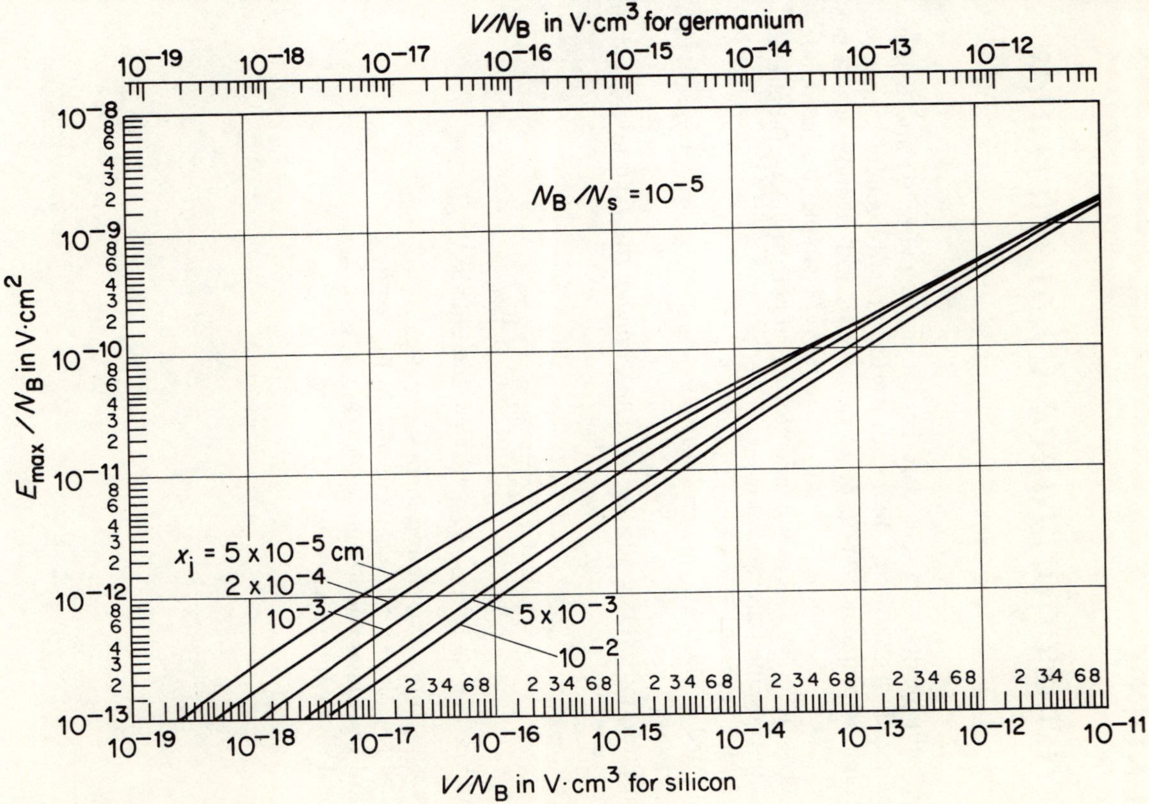

Fig. B.17. Peak electric field divided by N_B versus V/N_B, for $N_B/N_s = 10^{-5}$. For erfc distribution, use in range 10^{-4} to 10^{-8}. For Gaussian distribution, use in range 10^{-4} to 10^{-6}, $N_B = 10^{14}$.

APPENDIX C

Inverter Performance Characteristics

This appendix includes various data pertaining to the static and dynamic performance characterisics of different inverter types. In general, these data are a graphical implementation of certain theoretical relationships established in the text and are based on relatively simple device models. Nevertheless, it serves as a useful guide for initial performance evaluations.

The data presented are as follows:

Static transfer curves: enhancement loads—Fig. C.1

Saturated MOS loads use $m = 1.0$ but $V_{OH} = (V_{GG} - V_{TL})$ varies with V_{GG}. Also, $(V_{GG} - V_{TL}) < V_{DD}$.

Non-saturated MOS loads use m values ranging from 0.0 to 1.0 and m approaches 0.0 for large V_{GG}. Thus:

$$m = \frac{V_{DD}}{2(V_{GG} - V_{TL}) - V_{DD}}$$

and $(V_{GG} - V_{TL}) > V_{DD}$. V_{OH} is fixed and equals V_{DD}.

Static transfer curves: depletion loads—Fig. C.2

Theoretical rise times: enhancement loads—Fig. C.3

Theoretical rise times: depletion loads—Fig. C.4

Theoretical discharge times—Fig. C.5

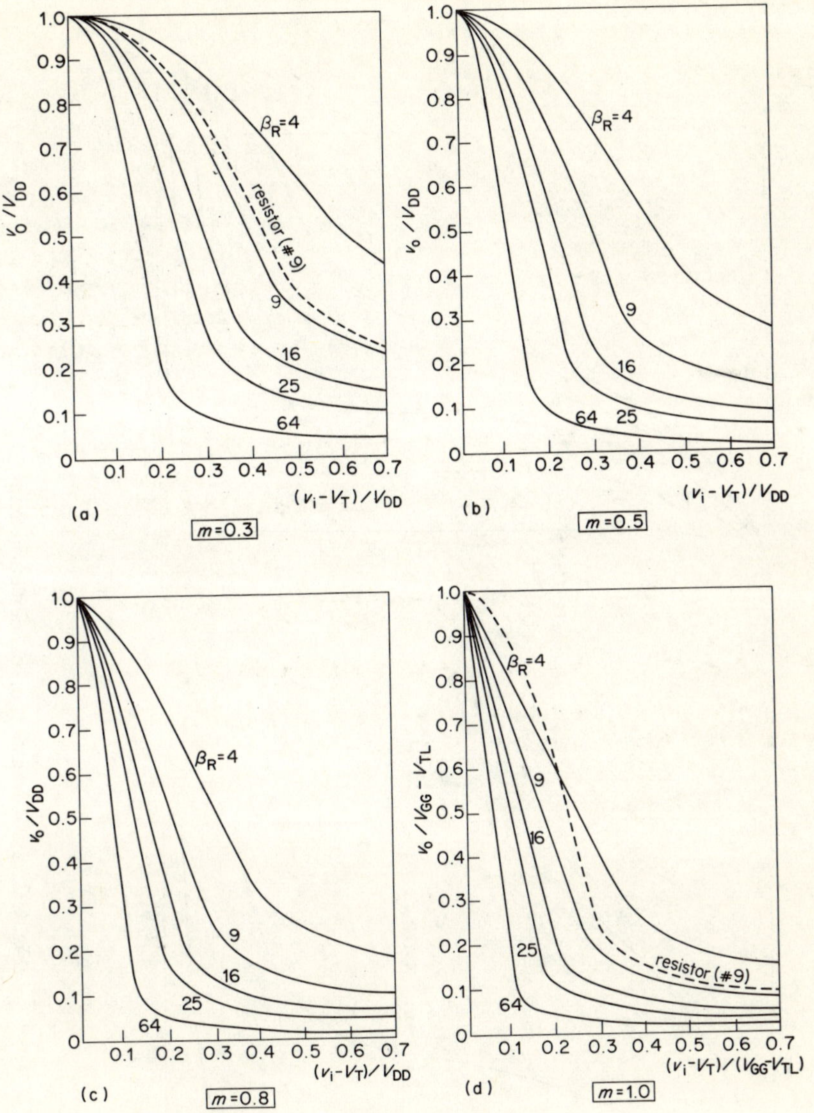

Fig. C.1. Static transfer curves: Enhancement loads.

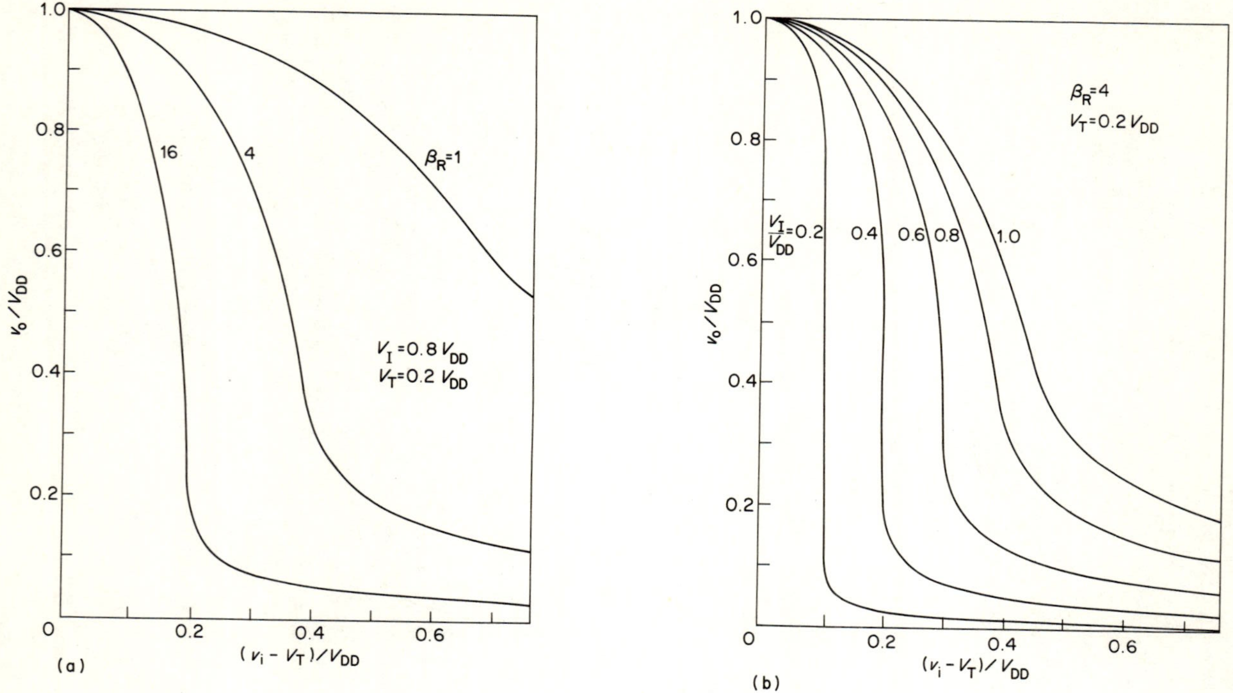

Fig. C.2. Static transfer curves: depletion loads.

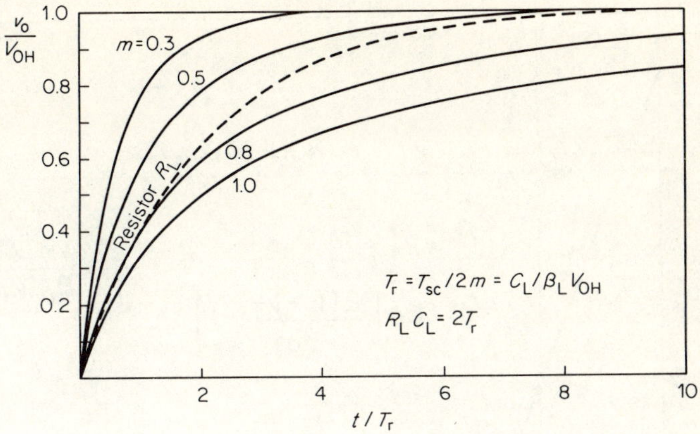

Fig. C.3. Theoretical rise times: enhancement loads.

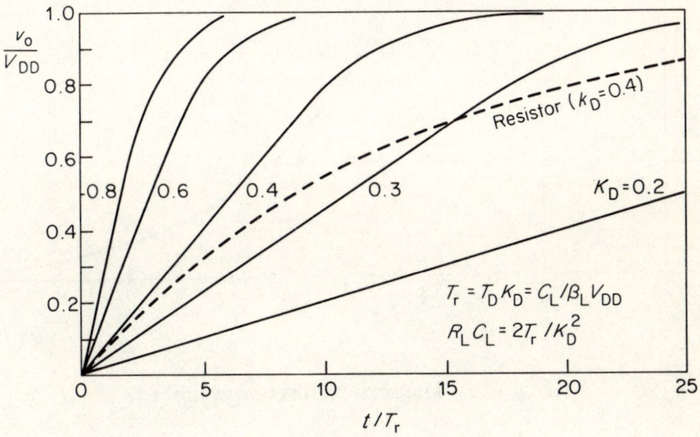

Fig. C.4. Theoretical rise times: depletion loads.

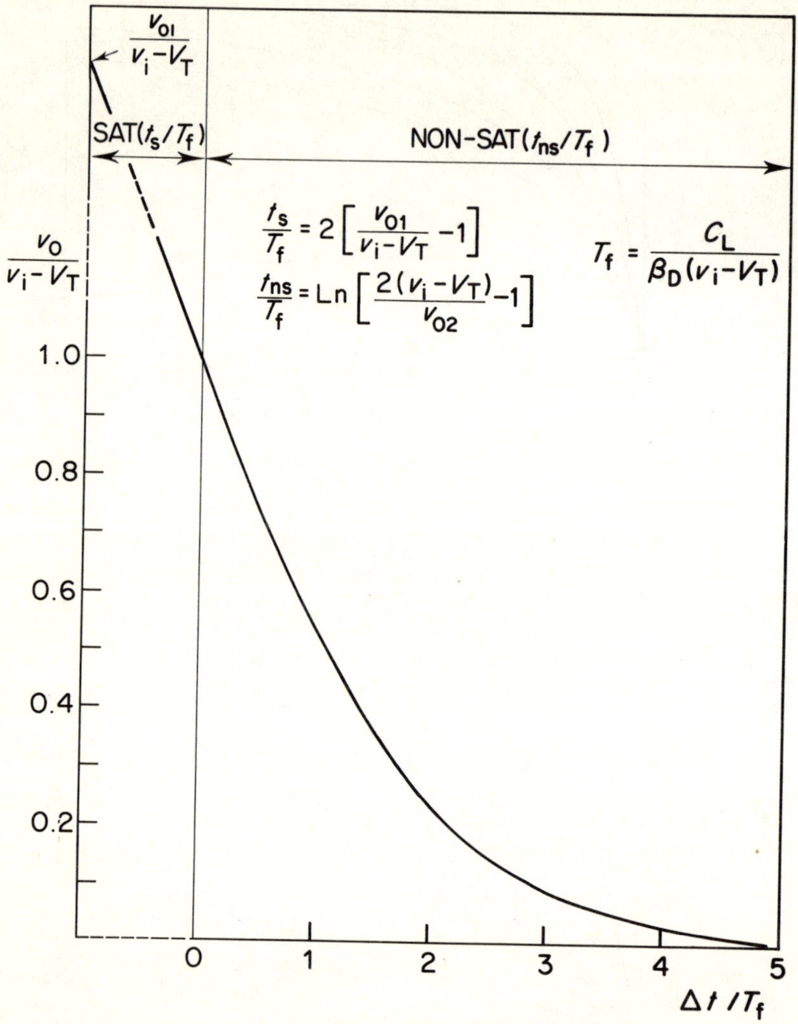

Fig. C.5. Theoretical discharge times.

APPENDIX D
Resistivity of Uniform Layers

Graph showing resistivity versus concentration for *uniformly doped* p-type and n-type materials, typically for substrate doping (after J. C. Irvin).

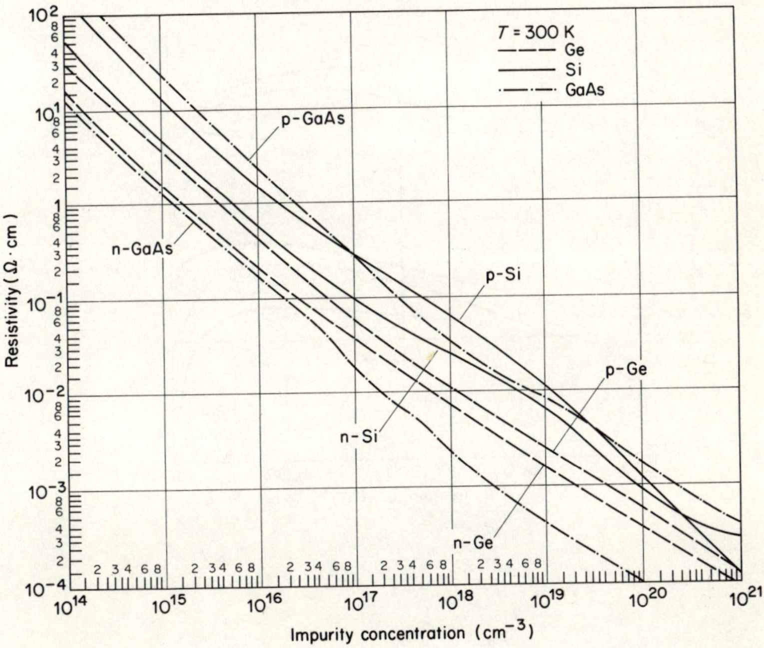

APPENDIX E

Solid Solubility of Various Dopants in Silicon (after F. A. Trumbore)

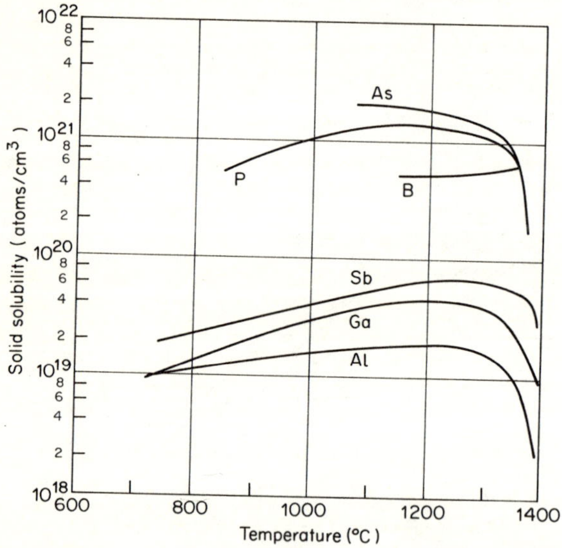

APPENDIX F

Gaussian and erfc Functions

Graphical presentation of the *Gaussian function* $\exp(-z^2)$ and the *complementary error function* $\text{erfc}(z)$.

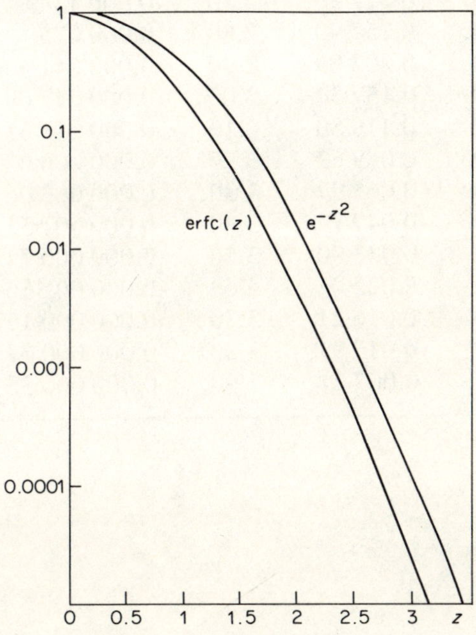

Tabular presentation of erfc(z).

z	erfc(z)	z	erfc(z)
0	1.000 00	2.00	0.004 68
0.10	0.887 54	2.10	0.002 98
0.20	0.777 30	2.20	0.001 86
0.30	0.671 37	2.30	0.001 14
0.40	0.571 61	2.40	0.000 689
0.50	0.479 50	2.50	0.000 407
0.60	0.396 14	2.60	0.000 236
0.70	0.322 20	2.70	0.000 134
0.80	0.257 90	2.80	0.000 075
0.90	0.203 09	2.90	0.000 041
1.00	0.157 30	3.00	0.000 022 09
1.10	0.119 80	3.10	0.000 011 65
1.20	0.089 69	3.20	0.000 006 03
1.30	0.065 99	3.30	0.000 003 06
1.40	0.047 72	3.40	0.000 001 52
1.50	0.033 90	3.50	0.000 000 743
1.60	0.023 65	3.60	0.000 00 356
1.70	0.016 21	3.70	0.000 000 167
1.80	0.010 91	3.80	0.000 000 77
1.90	0.007 21	3.90	0.000 000 35

APPENDIX G

Diffusion Coefficients in Silicon

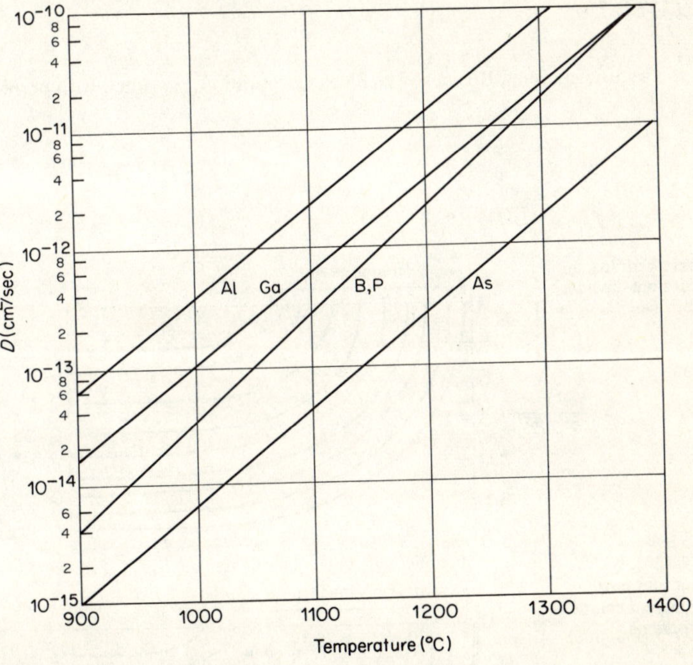

APPENDIX H

Lateral Diffusion Profiles (after D. P. Kennedy and R. R. O'Brien)

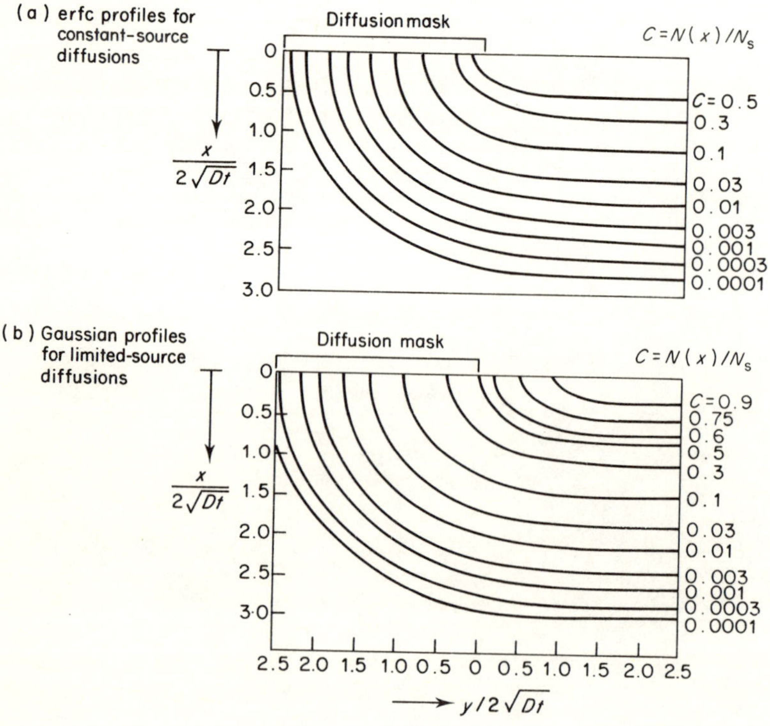

APPENDIX I

Oxide Growth Parameters

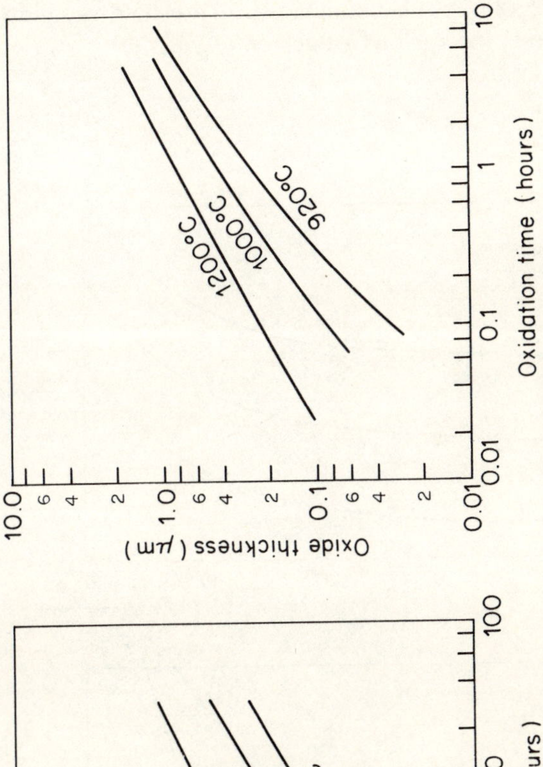

(b) Wet oxidation

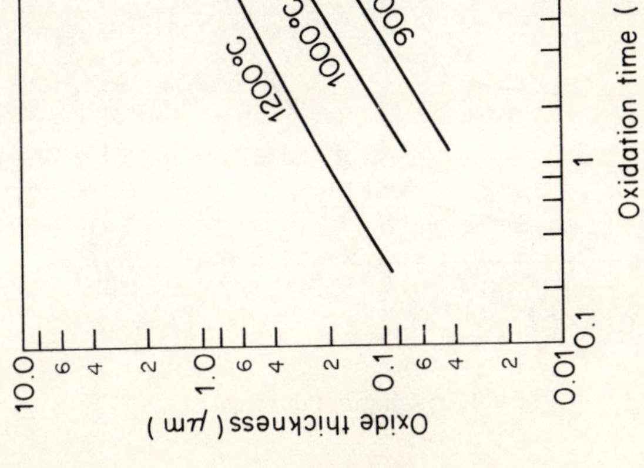

(a) Dry oxidation

APPENDIX J

Theoretical Threshold Voltage Graphs

Theoretical threshold voltages for an aluminium gate process using 1000 Å gate oxide and 15,000 Å field thickness. Threshold is plotted against substrate doping level for various values of Q_{ss}/q.

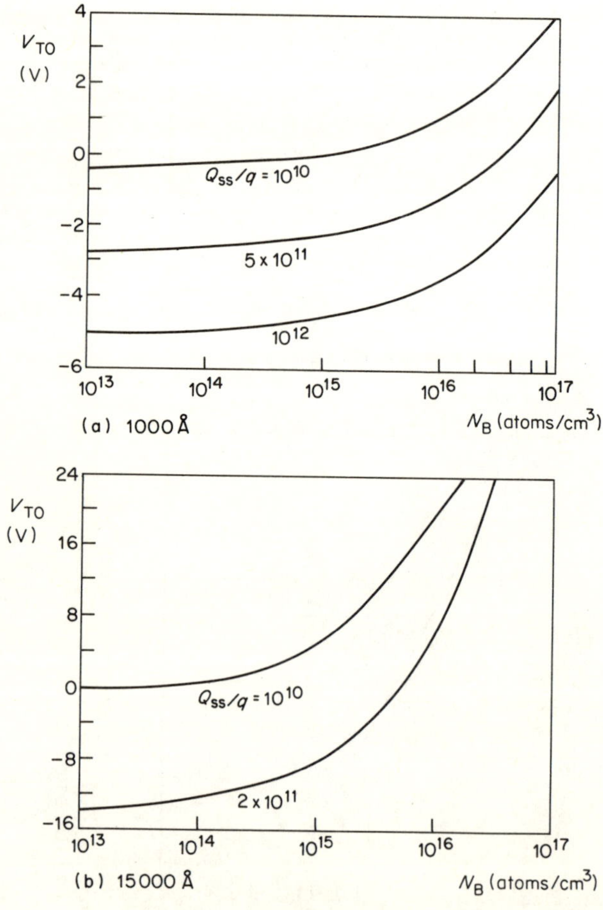

Fig. J.1. n-Channel devices.

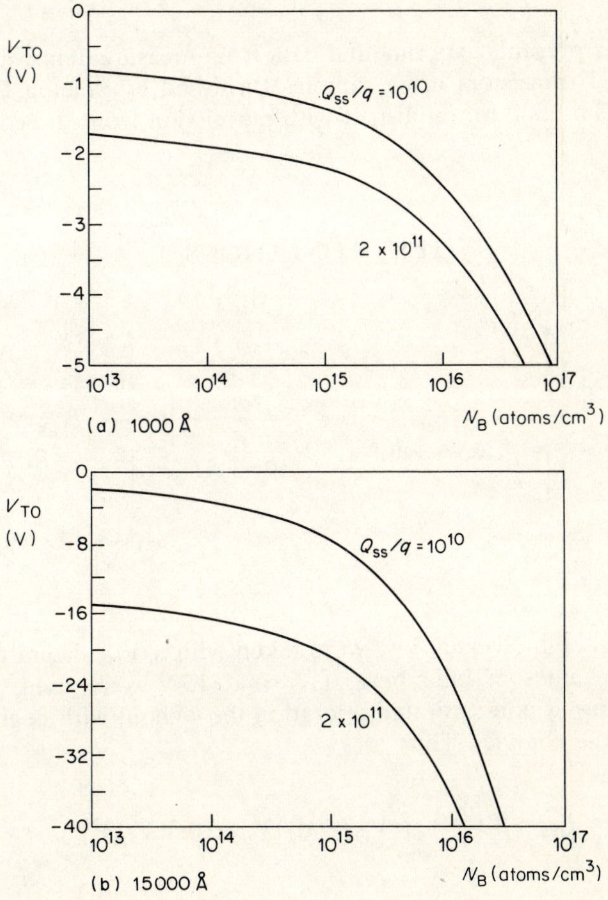

Fig. J.2. p-Channel devices.

APPENDIX K

Evaluation of Device Parameters

This appendix presents experimental data from measurements on p-channel and n-channel transistors using samples furnished by Analog Devices BV (Limerick). The data are published with permission from the company.

TEST CONDITIONS

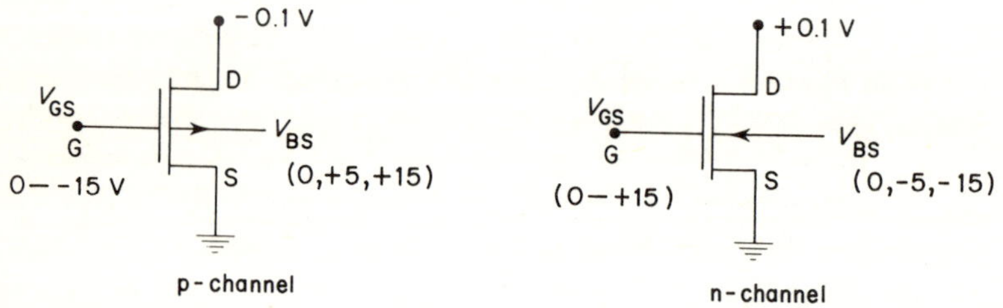

Measurements of I_{DS} versus V_{GS} were taken with a V_{DS} magnitude of only 0.1 V. Three values of bulk bias, 0, 5, and 15 V were used. These test conditions allowed non-saturated operation throughout with negligible variations along the channel. Thus:

$$I_{DS} = \beta'\left(\frac{W}{L}\right)_{\text{eff}} [(V_{GS} - V_T)V_{DS} - 0.5(V_{DS})^2]$$

Mask dimensions were $W_M = 40$ mil, $L_M = 0.4$ mil, and an estimate for ΔL was 0.08 mil. This gave $(W/L)_{\text{eff}} = [40/(0.4 - 0.16)] = 166.6$. Then, with $V_{DS} = 0.1$ V:

$$I_{DS} = \beta'(16.66)[V_{GS} - V_T - 0.05]$$

Treatment of p-channel and n-channel devices was identical except for polarity changes. The test circuit is shown below. It offers a stable V_{DS} independent of current level by placement of the microammeter inside the feedback loop of an operational amplifier. Gate bias was provided from a separate variable power supply unit.

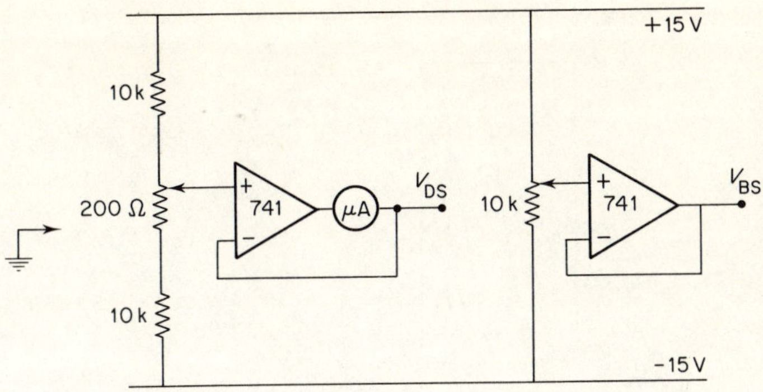

V_T varied with V_{BS} and was estimated graphically by linear extrapolation of the early portion of each I_{DS} V_{GS} curve. This estimate was then used in the above current expression to determine β' which was then plotted as a function of $(V_{GS} - V_T)$, i.e. of the overhead field E_G, for different values of V_{BS} where V_{BS} represents the field E_B under the channel. m values were estimated from the observed magnitude of body effect for both device types.

The use of very low V_{DS} ensured negligible variations along the channel and minimal heating effects. Note also that the current-flow equation, although approximate, gives very accurate results under these conditions.

The test results are presented in the pages that follow.

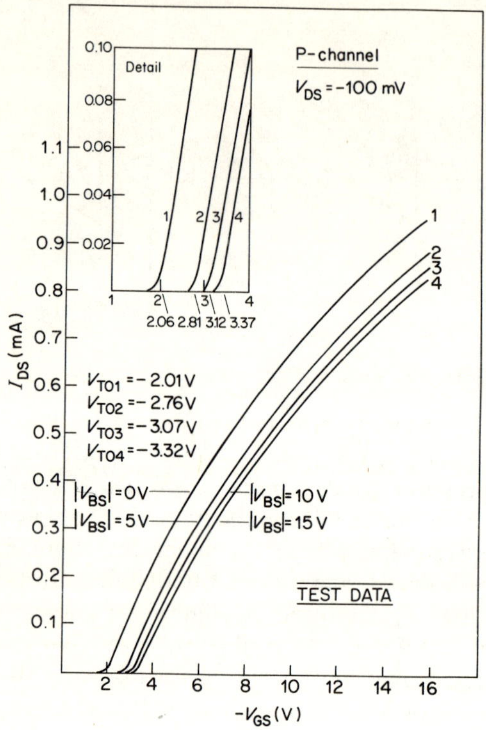
P-channel
V_{DS} = −100 mV
Detail
0.10
0.08
0.06
0.04
0.02
1 2 3 4
2.06 2.81 3.12 3.37
1.1
1.0
0.9
0.8
0.7
0.6
0.5
0.4
0.3
0.2
0.1
I_{DS} (mA)
V_{T01} = −2.01 V
V_{T02} = −2.76 V
V_{T03} = −3.07 V
V_{T04} = −3.32 V
$|V_{BS}|$ = 0 V
$|V_{BS}|$ = 5 V
$|V_{BS}|$ = 10 V
$|V_{BS}|$ = 15 V
TEST DATA
2 4 6 8 10 12 14 16
$-V_{GS}$ (V)

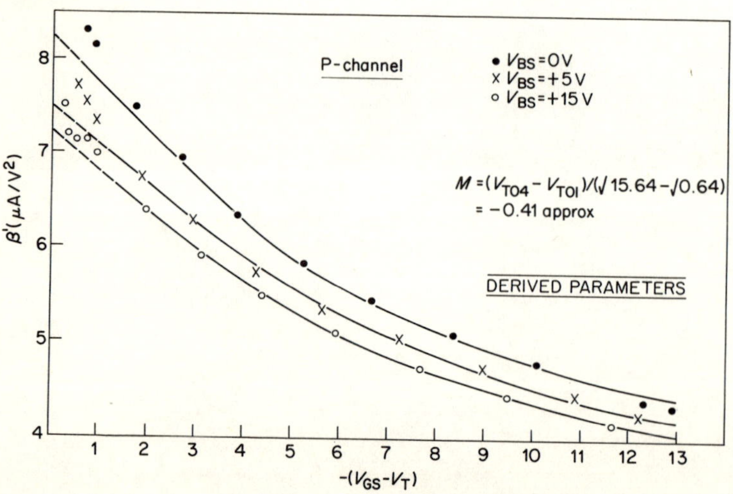
P-channel
● V_{BS} = 0 V
× V_{BS} = +5 V
○ V_{BS} = +15 V
$M = (V_{T04} - V_{T01})/(\sqrt{15.64} - \sqrt{0.64})$
= −0.41 approx
DERIVED PARAMETERS
β' (μA/V^2)
4
5
6
7
8
1 2 3 4 5 6 7 8 9 10 11 12 13
$-(V_{GS} - V_T)$

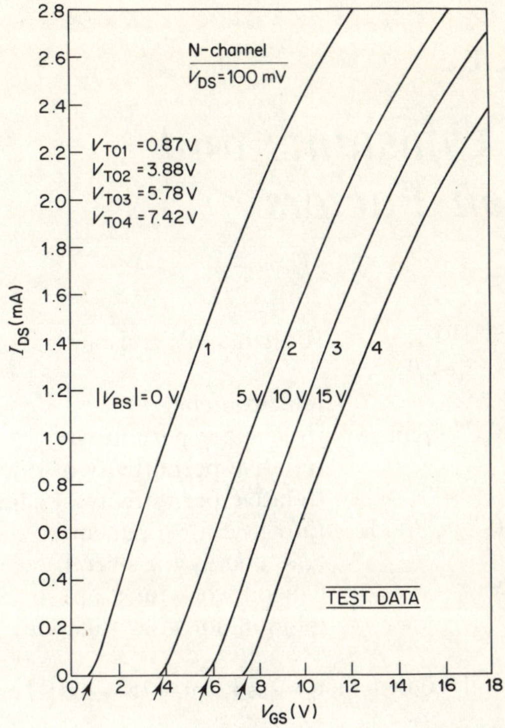
N-channel
V_{DS}=100 mV
V_{TO1} =0.87V
V_{TO2} =3.88V
V_{TO3} =5.78V
V_{TO4} =7.42V
$|V_{BS}|$=0 V
5 V
10 V
15 V
1
2
3
4
TEST DATA
I_{DS}(mA)
V_{GS}(V)

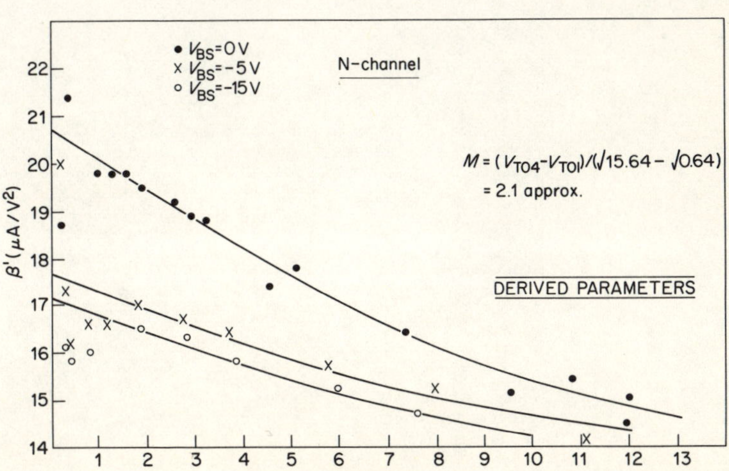
• V_{BS}=0V
x V_{BS}=−5V
o V_{BS}=−15V
N-channel
$M = (V_{TO4}-V_{TO1})/(\sqrt{15.64}-\sqrt{0.64})$
= 2.1 approx.
DERIVED PARAMETERS
$\beta'(\mu A/V^2)$
$(V_{GS}-V_T)$

APPENDIX L

Physical Constants and Conversion Factors

$k = 1.38 \times 10^{-23}$ J/K	(Boltzmann's constant)
$k/q = 8.63 \times 10^{-5}$ eV/K	
$q = 1.6 \times 10^{-19}$ C	(electron charge)
$e_0 = 8.85 \times 10^{-14}$ F/cm	(free space permittivity)
$k_{si} = 11.7$	(relative permittivity of silicon)
$k_{ox} = 3.9$	(relative permittivity of silicon dioxide)
$n_i = 1.41 \times 10^{10}$ at 300 K	(intrinsic n = p per cm^3)
$E_{go} = 1.21$	(gap energy for silicon)
$\phi_S = -4.35$ eV	(silicon work function)
$\phi_M = -3.2$ eV	(aluminium work function)

1 mil = 0.001 inch = 25.4 μm = 25.4×10^{-4} cm

1 μm = 10^{-6} m = 10^{-4} cm

1 Å = 1 angstrom unit = 10^{-10} m = 10^{-8} cm

1 $\text{mil}^2 = 6.45 \times 10^{-6}$ cm^2

APPENDIX M

Summary of Important Results

CHAPTER 1

Intrinsic concentration $n_i = n = p$ increases with temperature as:

$$n_i(T) = 3.87 \times 10^{16} T^{3/2} \cdot e^{-E_{go}/[2kT/q]} \text{ (per cm}^3\text{)}$$

with T in °K, $k = 1.38 \times 10^{-23}$, $E_{go} = 1.21$ for Si. For non-intrinsic conditions, $n \cdot p = n_i^2(T)$ still holds with $n \approx N_D$ for donor (n-type) doping and with $p \approx N_A$ for acceptor (p-type) doping.

Mobile carriers *drift* under the influence of an electric field with $J = \sigma E$ and $\sigma = q(p\mu_p + n\mu_n)$. Resistivity is $\rho = 1/\sigma$ (Ω-cm). Resistivity data (Appendix D) gives ρ vs. N_B for uniform doping.

Non-uniform carrier concentration causes *diffusion* with

$$J = q\left(D_n \frac{dn}{dx} - D_p \frac{dp}{dx}\right)$$

Analysis of a p–n junction yields the *diode equation*

$$I = I_s(e^{qV/kT} - 1)$$

valid for low to medium current levels. Contact potential

$$\psi_0 = \frac{kT}{q} \text{Ln}\left(\frac{N_A N_D}{n_i^2}\right)$$

Theoretical approximations to *depletion layer widths* are:

$$X_{mL} = \left(\frac{2k_{si}e_0(\psi_0 - V)}{qN_B}\right)^{1/2} \qquad X_{mL} = \left(\frac{12k_{si}e_0(\psi_0 - V)}{q\,|dN/dx|}\right)^{1/3}$$

Step junction — Graded junction

with associated capacitances

$$C = \frac{k_{si}e_0}{X_{mL}} \text{ (F/cm}^2\text{)}$$

Avalanche breakdown of a junction obeys

$$BV = 2.7 \times 10^{12} N_B^{-2/3} \text{ V}$$

reasonably well. N_B is bulk doping level.

CHAPTER 2

$V_{DS} > (V_{GS} - V_T)$: *Saturated operation* with $V_{CHE} = (V_{GS} - V_T)$ independent of V_{DS}.
$V_{DS} < (V_{GS} - V_T)$: *Non-saturated operation* with V_{DS} exercising some control over current.

For inverters with saturated loads, the 'high' output voltage is $(V_{DD} - V_T)$ rather than V_{DD}. Separate gate supply V_{GG} gives a non-saturated load if $V_{GG} > (V_{DD} + V_T)$. Then, v_o(high), $= V_{OH}$, goes to V_{DD}. Depletion loads give an output high level of V_{DD}.

CMOS eliminates static power dissipation.

CHAPTER 4

Diffusion profiles:

$$N(x, t) = N_s \cdot \text{erfc}\left(\frac{x}{2\sqrt{(Dt)}}\right) \qquad \text{Constant source}$$

$$N(x, t) = \frac{Q}{\sqrt{(\pi Dt)}} e^{-x^2/4Dt} \qquad \text{(Gaussian) Limited source}$$

$$Q = 2N_s(D_p t_p/\pi)^{1/2}$$

Sheet resistance $R_s = \bar{\rho}/x_j = 1/(\bar{\sigma}x_j)$, $\Omega/\square$
Irvin's Curves (Appendix A) relate N_s, N_B, $\bar{\sigma}$, x/x_j and profile type.
Lawrence and Warner Curves (Appendix B) give X_{mL} and C.

For resistivity measurement by the four-point probe method:

$$\bar{\rho} = 4.53\left(\frac{V}{I}\right)x_j \qquad (\Omega\ \text{cm})$$

CHAPTER 5

$$Q_B = [2k_{si}e_0 q N_B(2\phi_F)]^{1/2} \text{ depletion layer charge/cm}^2.$$

$$C_{ox} = k_{ox}e_0/t_{ox}\ \text{F/cm}^2, \qquad \phi_F = \frac{k_T}{q}\text{Ln}\left(\frac{N_B}{n_i}\right)\text{V}$$

Threshold calculations:

V_{TO}	p-Channel	n-Channel
Al gate	$-0.06 - \phi_F - \frac{Q_B}{C_{ox}} - \frac{Q_{SS}}{C_{ox}}$	$-0.6 + \phi_F + \frac{Q_B}{C_{ox}} - \frac{Q_{SS}}{C_{ox}}$
Si gate	$+0.55 - \phi_F - \frac{Q_B}{C_{ox}} - \frac{Q_{SS}}{C_{ox}}$	$-0.55 + \phi_F + \frac{Q_B}{C_{ox}} - \frac{Q_{SS}}{C_{ox}}$

All symbols describe positive numbers (except V_{TO}).

Theoretical V_{TO} curves are given in Appendix J for aluminum gate. Figures are altered by using silicon gate as follows:

p-Channel	add +1.15 V to V_{TO}
n-Channel	add +0.05 V to V_{TO}

Introducing *body-effect factor*

$$M = \frac{t_{ox}}{k_{ox}}\left(\frac{2k_{si}qN_B}{e_0}\right)^{1/2}$$

$$V_{TCH} = V_{TO} + M[(V_{CB} + 2\phi_F)^{1/2} - (2\phi_F)^{1/2}]$$

V_{TO} has a small negative temperature coefficient (TC).

CHAPTER 6

The current-control equations:

Non-saturated ($V_{GS} - V_{TO} > V_{DS}$)

$$I_{DS} = \beta'\left(\frac{W}{L}\right)[(V_{GS} - V_T)V_{DS} - 0.5V_{DS}^2]$$

Saturated ($V_{GS} - V_{TO} < V_{DS}$)

$$I_{DS} = \frac{\beta'}{2}\left(\frac{W}{L - L'}\right)[V_{GS} - V_T]^2$$

$$\beta' = C_{ox}\bar{\mu} = k_{ox}e_0\bar{\mu}/t_{ox} \qquad (A/V^2)$$

$$L' = \sqrt{(2k_{si}e_0[V_{DS} - (V_{GS} - V_{TCE})]/qN_B)}$$

Body-effect at channel-end is set by V_{CB}:

$$V_{CB} = (V_{GS} - V_{TCE}) + V_{SB}$$

such that

$$V_{TCE} = V_{TO} + M[(V_{GS} - V_{TCE} + V_{SB} + 2\phi_F)^{1/2} - (2\phi_F)^{1/2}]$$

Solve iteratively for V_{TCE}.

Comparison with an exact model shows that the current equations give accurate results when V_T is chosen as the mean of threshold values computed for opposite ends of the channel.

The gain factor β' depends on many factors:

(a) Transverse fields.
(b) N_B and doping type.
(c) Crystal orientation.
(d) Temperature.
(e) Gate oxide thickness.
(f) Ion implant.

CHAPTER 7

Inverter static transfer characteristics are implicitly defined by the statement $i_D = i_L$. A generalized expression for i_L in the case of enhancement loads is:

$$i_L = I_{SC}(1 - v_o/V_{OH})(1 - mv_o/V_{OH}).$$

Then:

$$\frac{\beta_D}{2}(V_i - V_T)^2 = I_{SC}\left(1 - \frac{v_o}{V_{OH}}\right)\left(1 - m\frac{v_o}{V_{OH}}\right)$$

Saturated driver $\qquad (v_i - V_T) < v_o$

$$\beta_D[(v_i - V_T)v_o - 0.5v_o^2] = I_{SC}\left(1 - \frac{v_o}{V_{OH}}\right)\left(1 - m\frac{v_o}{V_{OH}}\right)$$

Non-saturated driver $\qquad (v_i - V_T) > v_o$

with parameters defined as follows:

	Resistive load, R_L	Saturated MOS load	Non-saturated MOS load
$V_{OH} =$	V_{DD}	$V_{GG} - V_{TL}$	V_{DD}
$m =$	0.0	1.0	$\dfrac{V_{DD}}{2(V_{GG} - V_{TL}) - V_{DD}}$
$I_{SC} =$	$\dfrac{V_{DD}}{R_L}$	$\dfrac{\beta_L}{2}(V_{GG} - V_{TL})^2$	$\dfrac{\beta_L}{2m} V_{DD}^2$
$T_{SC} =$	$R_L C_L$	$\dfrac{2C_L}{\beta_L(V_{GG} - V_{TL})}$	$\dfrac{2mC_L}{\beta_L V_{DD}}$

CHAPTER 8

Rise time t_r for enhancement loads:

$$t_r = \frac{T_{SC}}{1-m} \mathrm{Ln}\left[\frac{1 - mv_{on}}{1 - v_{on}}\right], \qquad m < 1.0 \qquad \text{(Non-saturated)}$$

$$t_r = T_{SC} \cdot v_{on}/(1 - v_{on}), \qquad m = 1.0, \qquad \text{(Saturated load)}$$

with

$$v_{on} = v_o/V_{OH} \quad \text{and} \quad T_{SC} = V_{OH}C_L/I_{SC}.$$

Rise time for depletion loads:

$$\frac{t_r}{T_r}=\frac{t_s}{T_r}+\frac{t_{ns}}{T_r}=\frac{2(1-K_D)}{K_D^2}+\frac{1}{K_D}\,\mathrm{Ln}\left(\frac{2K_D-(1-v_{o2}/V_{DD})}{1-v_{o2}/V_{DD}}\right)$$

where

$$T_r=(T_D)(K_D)=(C_L/B_L V_I)(V_I/V_{DD})=(C_L/B_L V_{DD}).$$

See Fig. C.4 for graphical presentation of these results.
Fall Time calculations:

$$\frac{t_f}{T_f}=\frac{t_s}{T_f}+\frac{t_{ns}}{T_f}=2\left(\frac{v_{o1}}{v_i-V_T}-1\right)+\mathrm{Ln}\left(\frac{2(v_i-V_T)}{v_{o2}}-1\right)$$

with $T_f=C_L/\beta_D(v_i-V_T)$. See Fig. C.5.

$$\text{Transit time}=\frac{L^2}{\bar{\mu}V_{DS}}\ \text{for channel carriers.}$$

CHAPTER 10

As an electronic switch: $R_{on}=1/\beta(V_{GS}-V_T)$.

Transconductance for saturated device: $g_m=\beta(V_{GS}-V_T)=\sqrt{(2\beta\,.\,I_{DS})}$

Output resistance at drain for saturated device:

$$r_o=\frac{2L}{I_o}\left(1-\frac{L'}{L}\right)^2\frac{[V_{DS}-(V_{GS}-V_T)]^{1/2}}{[2k_{si}e_0/qN_B]^{1/2}}$$

where

$$I_o=\frac{\beta'}{2}\left(\frac{W}{L}\right)(V_{GS}-V_T)^2,$$

the saturation current.

Index